Differentiability in Banach Spaces, Differential Forms and Applications

Celso Melchiades Doria

Differentiability in Banach Spaces, Differential Forms and Applications

 Springer

Celso Melchiades Doria
Department of Mathematics, CFM
Universidade Federal de Santa Catarina
Florianópolis, Santa Catarina, Brazil

ISBN 978-3-030-77836-1   ISBN 978-3-030-77834-7   (eBook)
https://doi.org/10.1007/978-3-030-77834-7

Mathematics Subject Classification: 46-01, 47-01, 58-01

This Springer imprint is published by the registered company Springer Nature Switzerland AG
The registered company address is: Gewerbestrasse 11, 6330 Cham, Switzerland

Preface

This textbook introduces the techniques of differentiability on Banach spaces, integration of maps and some applications. In the first introduction to Calculus, the derivation and integration of the functions of one real variable are the main concern; these techniques are then extended to functions and maps of several real variables. For functions of a real variable, the main results of the theory are (a) the existence of maximum and minimum for a differentiable function defined on a compact set, and (b) the Fundamental Theorem of Calculus. When dealing with several variables, the main results are similar. Indeed, they are generalizations of cases (a) and (b), where case (b) is known as the Stokes Theorem. For functions of several variables, we should stress the importance of the Inverse Function Theorem.

The text is divided into seven chapters and three appendices. In Chap. 1, we develop the theory of differentiation of functions and maps defined on the Euclidean space $\mathbb{R}^n$. An application of the theory of differentiable maps is given to prove the Fundamental Theorem of Algebra without using properties of the field to be algebraically complete. The last section contains a description and some results related to the Jacobian Conjecture, which has been open since 1939. In Chap. 2, linear operators defined on Banach spaces are introduced, and particular classes of operators are clearly explained and illustrated. Chapter 3 is about differentiation of functions and maps defined on Banach spaces, and applications are given. The theory has several applications in several areas; we present only a few because the applications are usually very extensive and require specific knowledge of the contents. Chapter 4 contains an introduction to Vector Fields, the basic operations, the structure of the Lie algebra and its interpretation in terms of Linear Differential Operators. In Chap. 5, we review the formalism of vector calculus in which we write down the classical theorems of integration and then show their passage to the formalism of differential forms. In Chap. 6, we introduce the Exterior Algebra of Differential Forms to prove Stokes theorem; as an example of this application, we define the De Rham Cohomology groups to compute the groups of the sphere S^n and closed surfaces of genus g. In Chap. 7, some applications of the Stokes Theorem and differential forms are given in the introduction to Harmonic Functions, Maxwell's Equations, and Helmholtz's Theorem.

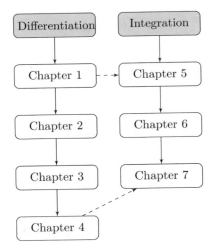

Fig.1 Chapter's flowchart

By dividing the content into two central themes (1) Differentiation and (2) Integration, we use the flowchart in Fig. 1 to show the interdependence between chapters.

There are also three appendices to make the textbook as self-contained as possible, and to support the reader in gaining an understanding of the basic terminology, the notations, the concepts and some basic theorems used in the text.

The appendices are not supposed to be read; instead they should serve as a guiding tool to recall the basics according to the needs of the readers. In Appendix A, we fix some notations, enunciate some theorems largely used throughout the other chapters, and further develop some elementary content according to our needs. In this way, Appendix A is recommended as a reference source for prerequisites and references. Appendix B introduces very basic concepts from the differentiable manifold and Lie groups. Appendix C deals with tensor algebra, which is a basic concept to have a full understanding of the content in Chap. 6.

Florianópolis, Brazil

Celso Melchiades Doria

Introduction

At the turn of the century, at the end of 1999, there were many speculations about results of greater importance that were reached in the 1st Millenium. One day, while waiting for a medical appointment, there was a magazine listing some results that were considered to be of great significance among those obtained for the development of human knowledge. To my surprise and joy, one of the noted results was the Fundamental Theorem of Calculus. I had not thought of this possibility, but I immediately agreed to its inclusion, not only because I am a mathematician but also because it is a fact that all of Classical Mechanics, Thermodynamics and Electromagnetism were developed using calculus. Consequently, technological advancements achieved in the exact sciences and social sciences depended largely on the development of calculus.

In general, when we refer to calculus, we are discussing the techniques of differentiation and integration. In most textbooks, the study of the derivative precedes that of the integral, but not historically. The method of exhaustion developed by Archimedes was an infinite sum process, analogous to that used nowadays to define the integral of a function.

The concept of derivative of a function appeared in the sixteenth century after the emergence of analytical geometry to calculate the relative rate of change of a quantity. In the period noted, many ideas in physics were evolving rapidly due to the scientific method. In this period, the development of mechanics was latent. The Pioneers of Calculus as we know it today were Sir Isaac Newton (1642–1727), who developed the Method of Flows, and Gottfried Wilhelm von Leibniz (1646–1716) who developed Calculus, as he named it, and also gave a good part of the notation used to this day. Newton discovered the basic Laws of Classical Mechanics, then applied them together with the Method of Flows to demonstrate Kepler's Laws.

Newton's 2nd Law states that a Force on a body of mass m generates a relative rate of change of velocity in relation to time. More precisely, in our current mathematical language, the second Law states that the force $\overrightarrow{F}$ acting on a body of mass m is given by $\overrightarrow{F} = m\frac{d\overrightarrow{v}}{dt}$, where $\overrightarrow{v}$ is the velocity vector of the body. Thus, the concept of the derivative is essential for formulating Newtons' Law. To understand the dynamic behavior of a variable from its relative variation proved to be very efficient, in that the most important information follows from the local data of the variable studied.

The idea of studying a phenomenon from knowing how it changes with respect to a parameter requires answering the following question: suppose that the variation of a quantity that has been studied is known, can we determine the quantity? This question is partially answered in the Fundamental Theorem of Calculus which reveals the connection between the concepts of derivative and integral. While the integral gives global information, the derivative gives local information. The local nature of differentiation and integration are complementary behaviors that unite in various applications.

In the nineteenth century, classical mechanics and calculus were maturing, except for the questions of the fundamentals of calculus such as the understanding of real numbers, the convergence of series and limits of functions, all of which were later treated in Mathematical Analysis. The basic laws of Electromagnetism were formulated regarding mathematically expressed laws at the outset of the nineteenth century and culminated with Maxwell's Equations published in 1861. Electromagnetism was one of the main sources of motivation for the development of Vector Calculus, as well as the development of Fluid Mechanics, the Lagrangian formalism for Lagrangian Mechanics and the Quaternions discovered by Sir William Rowan Hamilton.

Chronologically, the Electromagnetism equations were written using local coordinates, which then evolved to vector notation, and finally to the formulation using differential forms. Electromagnetism was important because of its applicability, and this boosted experimental and theoretical knowledge of the theory. Due to the electromagnetic theory, the industrial revolution went from steam engines to electric motors that led to an unprecedented development in manufacturing and also to communications, where the telegraph has become essential.

It was due to Maxwell's Equations that the wave behavior of the electrical and magnetic fields had been discovered as well as the fact that both travel at the speed of light. The mathematical richness of electromagnetism revealed several structures that contributed to the development of new ideas and new methods, for example, for the current format of integration theorems that we find in calculus textbooks.

Differential forms appeared later and revealed a more precise and succinct language. The original formulation of Maxwell's Equations, as published by James Clerk Maxwell (1831–1879) in 1861, consisted of 20 equations; years later, Olivier Heaviside (1850–1925) introduced the vector operators $\nabla \times$ (Curl) and ∇ (Divergent) and reduced the equations to four equations. Later, with the use of differential forms, the equations were reduced to only two equations.

Calculus has its technical limits, and these limits are the scope of the issues addressed in Mathematical Analysis. Essentially, the fundamentals of Calculus depend on the concept of limit, which was formalized only in the nineteenth century. These fundamentals and their consequences are fundamental for, for example, to correctly extend the concepts and methods of Calculus in Euclidean spaces to the Calculus of Variations and to study functions and maps in Banach Spaces.

The development of Calculus motivated the development of several other areas of mathematics. Today we can claim there is enough apparatus of techniques and tools sufficient to solve various theoretical problems in Linear Algebra, which is not true when the problem belongs to the non-linear world. Non-linear questions can be stratified among those where linear techniques and tools are efficient, for example, in non-linear phenomena whose linear approximation is useful to the study and in those non-linear phenomena whose approximation is hopeless. In the latter case, non-linear problems are much more difficult and rarely are embedded in a global theory, that is, each question is a problem in itself.

With the development of Quantum Mechanics and the evolution of optimization problems arising in several areas, it became essential to develop the calculus for the spaces of functions, which in this text will be the Banach spaces, much more general and more abstract than the Euclidean space $\mathbb{R}^n$. In many cases, these are the Hilbert spaces, particular cases of Banach spaces.

As knowledge evolved, questions arose inducing such new areas as Algebraic Topology, Differential Topology and Geometric Topology. It has become evident in various models and theories that the topological spaces and geometric properties would be fundamental to our understanding. The first proof by Gauss of the Fundamental Theorem of Algebra used the index of a curve showing that topology plays an important role; the same thing happens later with the development of Calculus of a function of one complex variable, and the Theory of Dynamical Systems. Concepts and techniques evolved very efficiently. We introduce De Rham Cohomology Groups, which contain information on the topology space and make extensive use of the local and global nature of the derivative and integral operators.

Mathematics is a language to quantify, as such its domain has the function of improving understanding and efficiency to solve a problem of a quantifying nature. However, only the domain of language does not reveal the ways to understand the inherent phenomena of a model and its applications. Something similar occurs in other theories and areas of human knowledge where mathematics is present. Of course, math benefits from this interaction. As the saying goes "one hand washes the other."

Contents

Chapter 1
Differentiation in $\mathbb{R}^n$

The analysis of the behavior of a function is efficiently carried out when we study the way in which the function varies. In this chapter, techniques used in studying functions of one real variable $f : I \to \mathbb{R}$, defined in an open interval $I \subset \mathbb{R}$, are extended for functions of several real variables $f : U \to \mathbb{R}^m$ defined over an open subset $U \subset \mathbb{R}^n$. Several real variables is understood to mean a finite number of variables $(x_1, \ldots, x_n) \in \mathbb{R}^n$. The simple topological nature of $\mathbb{R}^n$ allows the theory to be more easily understood as all the concepts and techniques. The same concepts and techniques will be studied in the chapters ahead within the framework of Banach spaces.

1 Differentiability of Functions $f : \mathbb{R}^n \to \mathbb{R}$

When we study the continuity of a function of one real variable in the neighborhood of a given point $x_0 \in \mathbb{R}$ for $\epsilon > 0$, we consider the interval $I_\epsilon(x_0) = (x_0^-, x_0^+)$, $x_0^{\pm} = x_0 \pm \epsilon$, to analyze the values $f(x_0^-)$ and $f(x_0^+)$ taken by f in the lateral neighborhoods (x_0^-, x_0) (left-hand side) and (x_0, x_0^+) (right-hand side). In this case, the continuity of f at x_0 follows by proving that the lateral limits are equal to the value of the function at the point, i.e.,

$$\lim_{\epsilon \to 0} f(x_0^+) = \lim_{\epsilon \to 0} f(x_0^-) = f(x_0).$$

In the case of functions of several variables, there are too many directions to be analyzed, as shown in the next examples;

© Springer Nature Switzerland AG 2021

C. M. Doria, *Differentiability in Banach Spaces, Differential Forms and Applications*,

https://doi.org/10.1007/978-3-030-77834-7_1

Examples
(1) Let $f : \mathbb{R}^2 \to \mathbb{R}$ be given by

$$f(x, y) = \begin{cases} \frac{xy}{x^2+y^2}, & (x, y) \neq (0, 0), \\ 0, & (x, y) = (0, 0). \end{cases} \tag{1}$$

To verify the continuity of $f(x, y)$ at the origin, we analyze the values of $f(x, y)$ when (x, y) is getting close to $(0, 0)$ along straight lines. For $\alpha \in \mathbb{R}$, considering $\gamma(t) = (t, \alpha t)$, we take the limit

$$\lim_{t \to 0} f(\gamma(t)) = \lim_{t \to 0} \frac{\alpha}{1 + \alpha^2} = \frac{\alpha}{1 + \alpha^2}.$$

Since the limit depends on the slope of the line, it follows that $f(x, y)$ is not continuous at the origin, as shown in Fig. 1.
(2) The behavior of the values of a function in the neighborhood is quite subtle, as the example below shows:

$$f(x, y) = \begin{cases} \frac{x^2 y}{x^4+y^2}, & (x, y) \neq (0, 0), \\ 0, & (x, y) = (0, 0). \end{cases}$$

When we approximate (x, y) of $(0, 0)$ on the straight lines $\gamma(t) = (t, \alpha t)$, $\alpha \in \mathbb{R}$, we get $\lim_{t \to 0} f(\gamma(t)) = 0$. This suggests that f is continuous at the origin. However, in studying the parabolic approximation $\gamma(t) = (t, \beta t^2)$, we get the limit to be $\lim_{t \to 0} f(\gamma(t)) = \frac{\beta}{1+\beta^2}$. That is, when we approximate (x, y) of $(0, 0)$ by moving on to the parabolas, we can approach any value $r \in [0, 1/2)$, as shown in Fig. 2. So, the function $f(x, y)$ is not continuous at $(0, 0)$.
   The examples above show how complicated it might be to study the behavior of a function in the neighborhood of a point. It becomes clear that the way we approximate to the point requires learning about continuity.

1.1 Directional Derivatives

Let $U \subset \mathbb{R}^n$ be an open subset with coordinates $(x_1, \ldots, x_n)$; let $f : U \to \mathbb{R}$ be a function and $\gamma : (-\epsilon, \epsilon) \to U$ a curve such that $\gamma(0) = p \in U$ and $\gamma'(0) = \vec{v}$ is a vector in $\mathbb{R}^n$.

Definition 1 The directional derivative of f at $p \in U$ and in the direction of the vector $\vec{v}$ is

$$\frac{\partial f}{\partial v}(p) = \lim_{t \to 0} \frac{f(\gamma(t)) - f(p)}{t}. \tag{2}$$

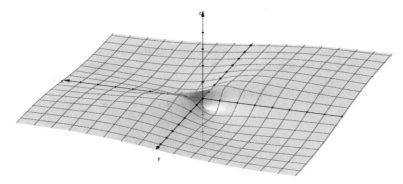

Fig. 1 Example 1

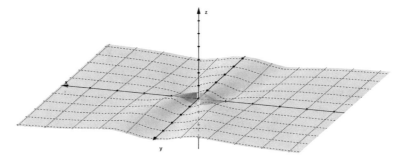

Fig. 2 Example 2

According to the definition, the directional derivative $\frac{\partial f}{\partial v}(p)$ depends on the curve γ. As discussed in the previous examples, dependence on γ is rather complicated and not easily understood. In what follows, we will examine the question of determining the class of functions such that $\frac{\partial f}{\partial v}(p)$ depends only on the point p and on the vector $\vec{v}$. Initially, we use the curve $\gamma(t) = p + t\vec{v}$ to compute the directional derivative $\frac{\partial f}{\partial v}(p)$.

Definition 2 Consider $\beta = \{e_1, \ldots, e_n\}$ the canonical basis of $\mathbb{R}^n$ and $p = (p_1, \ldots, p_n) \in U$. The partial derivative of f with respect to the variable x_i at p is

$$\frac{\partial f}{\partial x_i}(p) = \lim_{t \to 0} \frac{f(p + te_i) - f(p)}{t} = \lim_{t \to 0} \frac{f(p_1, \ldots, p_i + t, \ldots, p_n) - f(p_1, \ldots, p_n)}{t}.$$

In developing the concept of a differentiable function of several variables, there are two relevant aspects to be addressed:

(1) The independence of $\frac{\partial f}{\partial x_i}(p)$ on γ.

(2) The continuity of f at p. The simple existence of partial derivatives at a point p does not imply continuity of the function at p. The function f defined in Eq. (1) is not continuous at $(0, 0)$, although the partial derivatives (limits) of f exist at $(0, 0)$:

$$\frac{\partial f}{\partial x}(0, 0) = \lim_{t \to 0} \frac{f(t, 0) - f(0, 0)}{t} = 0, \quad \frac{\partial f}{\partial y}(0, 0) = \lim_{t \to 0} \frac{f(0, t) - f(0, 0)}{t} = 0.$$

Assuming the existence of the partial derivatives of the function $f : U \to \mathbb{R}$ at each point of U, we associate f with the n functions $\frac{\partial f}{\partial x_1}(x), \ldots, \frac{\partial f}{\partial x_n}(x) : U \to \mathbb{R}$.

Definition 3 A function $f : U \to \mathbb{R}$ is differentiable of class C^1 in U if all partial derivatives of the functions $\frac{\partial f}{\partial x_i}(x)$, $i = 1, \ldots, n$, are continuous for every point $x \in U$. Consider $C^1(U)$ to be the set of all differentiable functions in U.

Ignoring dependence on γ and taking $\gamma(t) = p + t\vec{v}$, we work out the following case: $U \subset \mathbb{R}^2$, $f : U \to \mathbb{R}^2$ and $f \in C^1(U)$. Let $p = (x_0, y_0) \in U$ and $\vec{v} = v_1 e_1 + v_2 e_2 \in \mathbb{R}^n$:

$$\frac{\partial f}{\partial v}(p) = \lim_{t \to 0} \frac{f(x_0 + tv_1, y_0 + tv_2) - f(x_0, y_0)}{t}$$

$$= \lim_{t \to 0} \frac{1}{t} \left\{ \left[f(x_0 + tv_1, y_0 + tv_2) - f(x_0, y_0 + tv_2) \right] + \left[f(x_0, y_0 + tv_2) - f(x_0, y_0) \right] \right\}$$

$$= \lim_{t \to 0} \frac{1}{t} \left\{ \frac{\partial f}{\partial x}(c_1, y_0 + tv_2).(tv_1) + \frac{\partial f}{\partial y}(x_0, c_2).(tv_2) \right\},$$

and now the last equality follows from the Mean Value Theorem, which guarantees the existence of c_1, c_2, such that $x_0 < c_1 < x_0 + tv_1$ and $y_0 < c_2 < y_0 + tv_2$. When we pass to the limit $t \to 0$, we get

$$\frac{\partial f}{\partial v}(p) = \frac{\partial f}{\partial x}(p).v_1 + \frac{\partial f}{\partial y}(p).v_2. \tag{3}$$

The right-hand side of Eq. (3) depends only on the values of p and $\vec{v}$. Using the internal product $< ., . >: \mathbb{R}^2 \times \mathbb{R}^2 \to \mathbb{R}$, Eq. (3) is written as

$$\frac{\partial f}{\partial v}(p) =< \left(\frac{\partial f}{\partial x}(p), \frac{\partial f}{\partial y}(p) \right), (v_1, v_2) > .$$

Similarly for the general case $f = f(x_1, \ldots, x_n)$, we have the identity

$$\frac{\partial f}{\partial v}(p) =< \left(\frac{\partial f}{\partial x_1}(p), \ldots, \frac{\partial f}{\partial x_n}(p) \right), (v_1, \ldots, v_n) > .$$

Motivated by the above, we consider the following situation: $U \subset \mathbb{R}^3$, $f : U \to \mathbb{R}$ of class C^1, $\gamma : (-\epsilon, \epsilon) \to \mathbb{R}$, $\gamma(t) = (\gamma_1,(t), \gamma_2(t), \gamma_3(t))$, a C^1-curve such that $\gamma(0) = p = (p_1, p_2, p_3)$ and $\gamma'(0) = \vec{v} = (v_1, v_2, v_3)$:

$$\frac{\partial f}{\partial v}(p) = \lim_{t \to 0} \frac{f(\gamma(t)) - f(p)}{t} = \lim_{t \to 0} \frac{f((\gamma_1,(t), \gamma_2(t), \gamma_3(t))) - f(p_1, p_2, p_3)}{t}$$

$$= \lim_{t \to 0} \frac{1}{t} \left\{ f((\gamma_1,(t), \gamma_2(t), \gamma_3(t))) - f(p_1, \gamma_2(t), \gamma_3(t)) \right\}$$

$$+ \lim_{t \to 0} \frac{1}{t} \left\{ f(p_1, \gamma_2(t), \gamma_3(t)) - f(p_1, p_2, \gamma_3(t)) \right\} + + \lim_{t \to 0} \frac{1}{t} \left\{ f(p_1, p_2, \gamma_3(t)) - f(p_1, p_2, p_3) \right\}$$

$$= \frac{\partial f}{\partial x_1}(c_1, \gamma_2(t), \gamma_3(t)) . \frac{\gamma_1(t) - p_1}{t} + \frac{\partial f}{\partial x_2}(p_1, c_2(t), \gamma_3(t)) . \frac{\gamma_2(t) - p_2}{t}$$

$$+ \frac{\partial f}{\partial x_3}(p_1, p_2(t), c_3(t)) . \frac{\gamma_3(t) - p_3}{t},$$

and now the Mean Value Theorem guarantees that c_1, c_2, c_3 exists such that $c_1 \in (p_1, \gamma_1(t))$, $c_2 \in (p_2, \gamma_2(t))$ and $c_3 \in (p_3, \gamma_3(t))$. Once $\gamma \in C^1$, we have d_1, d_2 and d_3 in $(0, t)$ such that

$$\frac{\partial f}{\partial v}(p) = \lim_{t \to 0} \left[\frac{\partial f}{\partial x_1}(c_1, \gamma_2(t), \gamma_3(t)) . \frac{\gamma_1'(d_1).t}{t} + \frac{\partial f}{\partial x_2}(p_1, c_2(t), \gamma_3(t)) . \frac{\gamma_2'(d_2).t}{t} \right]$$

$$+ \lim_{t \to 0} \left[\frac{\partial f}{\partial x_3}(p_1, p_2(t), c_3(t)) . \frac{\gamma_3'(d_3).t}{t} \right] = \frac{\partial f}{\partial x_1}(p).v_1 + \frac{\partial f}{\partial x_2}(p).v_2 + \frac{\partial f}{\partial x_3}(p).v_3$$

$$= < \left(\frac{\partial f}{\partial x_1}(p), \frac{\partial f}{\partial x_2}(p), \frac{\partial f}{\partial x_3}(p) \right), (v_1, v_2, v_3) >,$$

that is, under the hypothesis above, the directional derivative $\frac{\partial f}{\partial v}(p)$ depends only on $p \in U$ and also on $\vec{v}$. In the case of the function $f : \mathbb{R}^n \to \mathbb{R}$ and $v = \sum_i v_i e_i$, we get

$$\frac{\partial f}{\partial v}(p) = \sum_{i=1}^{n} \frac{\partial f}{\partial x_2}(p).v_i. \tag{4}$$

1.2 Differentiable Functions

The tangent plane at $p \in U$ is the vector space

$$T_p U = \{ \vec{v} \in \mathbb{R}^n \mid \exists \gamma : (-\epsilon, \epsilon) \to U, \ \gamma(0) = p, \ \gamma'(0) = \vec{v} \}.$$

Definition 4 A function $f : U \to \mathbb{R}$ is differentiable at the point $p \in U$ if there is a linear functional $df_p : T_p U \to \mathbb{R}$, such that

$$f(p + \vec{v}) - f(p) = df_p.\vec{v} + r(\vec{v}) \tag{5}$$

and $\lim_{\vec{v}\to 0} \frac{r(\vec{v})}{|\vec{v}|} = 0$ for all $\vec{v} \in T_p U$. The linear operator df_x is the differential (or the derivative) of $f(x)$ at x.

The uniqueness of df_p is achieved directly from the definition.

Theorem 1 *If $f : U \to \mathbb{R}$ is differentiable at the point $p \in U$, then f is continuous at p.*

Proof It follows from the definition that $\lim_{\vec{v}\to 0} r(\vec{v}) = \lim_{\vec{v}\to 0} \frac{r(\vec{v})}{|\vec{v}|} . \mid \vec{v} \mid = 0$. Hence

$$\lim_{\vec{v}\to 0}\left[f(p + \vec{v}) - f(p) \right] = \lim_{\vec{v}\to 0} df_p . \vec{v} + \lim_{\vec{v}\to 0} r(\vec{v}) = 0.$$

$\square$

In what follows, vectors are designated as v instead of $\vec{v}$, unless the context requires otherwise. Considering the unit vector $\hat{v} = \frac{v}{|v|}$, the identity below shows the relation between the differential $df_p . \hat{v} = \frac{\partial f}{\partial \hat{v}}$ and the directional derivatives,

$$\frac{r(v)}{\mid v \mid} = \frac{f(p+ \mid v \mid \hat{v}) - f(p)}{\mid v \mid} - df_p . \frac{v}{\mid v \mid}.$$

Theorem 2 *If $f \in C^1(U)$, then f is differentiable.*

Proof Let $\beta = \{e_1, \ldots, e_n\}$ be the canonical basis of $\mathbb{R}^n$. Taking $x = (x_1, \ldots, x_n) \in U$ and $v = \sum_i v_i e_i \in T_x U$, define

$$r(v) = r(v_1, \ldots, v_n) = [f(x_1 + v_1, \ldots, x_n + v_n) - f(x_1, \ldots, x_n)] - \sum_{i=1}^{n} \frac{\partial f}{\partial x_i}(x) . v_i .$$

So the Eq. (4) implies

$$\lim_{v\to 0} \frac{r(v)}{\mid v \mid} = \lim_{v\to 0} \frac{f(x_1 + v_1, \ldots, x_n + v_n) - f(x_1, \ldots, x_n)}{\mid v \mid} - \sum_{i=1}^{n} \frac{\partial f}{\partial x_i}(x) . \frac{v_i}{\mid v \mid}$$

$$= \frac{\partial f}{\partial \hat{v}} - \sum_{i=1}^{n} \frac{\partial f}{\partial x_i}(x) . \frac{v_i}{\mid v \mid} = 0.$$

Considering $df_x . \hat{v} = \frac{\partial f}{\partial \hat{v}} = \sum_{i=1}^{n} \frac{\partial f}{\partial x_i}(x) . \frac{v_i}{|v|}$, which is linear on v, the linear functional $df_x : T_x U \to \mathbb{R}$ is defined as

$$df_x . v = \sum_{i=1}^{n} \frac{\partial f}{\partial x_i}(x) . v_i .$$

Therefore $f(p + v) - f(p) = df_p . v + r(v)$ and $\lim_{v\to 0} \frac{r(v)}{|v|} = 0$. Hence f is differentiable.

$\square$

Using the inner product defined on $\mathbb{R}^n$, the differential $df_x : T_x U \to \mathbb{R}$ is given by

$$df_x.v =< \left(\frac{\partial f}{\partial x_1}(x), \ldots, \frac{\partial f}{\partial x_n}(x) \right), (v_1, \ldots, v_n) >, \ \forall v \in T_x U. \tag{6}$$

Definition 5 Let $f \in C^1(U)$ and $\beta = \{e_1, \ldots, e_n\}$ be the canonical basis of $\mathbb{R}^n$. The gradient vector of f at $p \in U$ is

$$\nabla f(p) = \sum_{i=1}^{n} \frac{\partial f}{\partial x_i}(p).e_i = \left(\frac{\partial f}{\partial x_1}, \ldots, \frac{\partial f}{\partial x_n} \right)(p).$$

Therefore $df_p.v =< \nabla f(p), v >$.

Taking an orthonormal basis $\beta = \{e_1, \ldots, e_n\}$, the inner product induces the isomorphism $\mathcal{P}^{-1} : T_p^* U \to T_p U$, $\mathcal{P}^{-1}(\phi) = (\phi(e_1), \ldots, \phi(e_n))$. Since $df_p(e_i) = \frac{\partial f}{\partial x_i}(p)$, we have $\mathcal{P}^{-1}(df_p) = \nabla f(p)$. The dependence of the gradient vector on the inner product is often ignored, since in R^n the canonical inner product is always the one being used. The gradient operator $\nabla : C^1(U) \to C^0(U)$, $f \to \nabla f$, satisfies the following identities: let $a, b \in \mathbb{R}$ and $f, g \in C^1(U)$:

(1) $\nabla(af + bg) = a\nabla f + b\nabla g$ (linearity).
(2) $\nabla(f.g) = \nabla(f).g + f.\nabla(g)$ (Leibniz's rule).

The differential of a function $f \in C^1(U)$ induces the application $x \to df_x$. Let $\mathcal{L}(\mathbb{R}^n; \mathbb{R})$ be spaces of linear functionals defined on $\mathbb{R}^n$. The differentiability of f requires the continuity of the application $df : U \to L(\mathbb{R}^n; \mathbb{R})$, $x \to df_x$. The topological properties of the space of linear operators is studied in Chap. 3.

To study the variation of a function, we start by choosing a curve in the domain, as illustrated in the following: consider $p \in U$ and $\gamma : (-\epsilon, \epsilon) \to U$ to be a C^1-curve, $\gamma(t) = (\gamma_1(t), \ldots, \gamma_n(t))$ and such that $\gamma(0) = p$ and $\gamma'(0) = v$. The relative rate of change in the variable t for the function $h : (-\epsilon, \epsilon) \to \mathbb{R}$, $h(t) = f(\gamma(t))$, is given at each instant t by the derivative $h'(t)$ of the function obtained as follows:

$$h'(t) = \frac{d[f(\gamma(t))]}{dt} = df_{\gamma(t)}.\gamma'(t) =< \nabla f(\gamma(t)), \gamma'(t) >= \sum_{i=1}^{n} \frac{\partial f}{\partial x_i}(\gamma(t)).\frac{d[\gamma_i(t)]}{dt}. \tag{7}$$

If $f \in C^1(U)$ is constant, then $\nabla f(x) = 0$, $\forall x \in U$. To verify the reverse claim, it is necessary to assume that U is a connected set and $f \in C^1(U)$, that is for all pairs of points $x, y \in U$, we have a C^0-curve $\gamma : [0, 1] \to U$ such that $\gamma(0) = x, \gamma(1) = y$ and $\gamma([0, 1]) \subset U$. Initially, we assume that U is convex and $\gamma(t) = x + t(y - x)$. In this way, the Mean Value Theorem guarantees the existence of $c \in (0, 1)$ such that

$$h(1) - h(0) = h'(c) =< \nabla f(\gamma(c)), \gamma'(c) >=< \nabla f(\gamma(c)), y - x > .$$

Since the partial derivatives of f are continuous, $M = \max_{t \in [0,1]} \mid \nabla f(\gamma(t)) \mid$. Since U is convex for every pair $x, y \in U$, we get

$$\mid f(y) - f(x) \mid \leq M \mid y - x \mid .\qquad(8)$$

Proposition 1 *Let $U \subset \mathbb{R}^n$ be a connected open set and $f : U \to \mathbb{R}$ a C^1-function. If $\nabla f(x) = 0$ for all $x \in U$, then f is constant.*

Proof We first assume U is convex; it follows from the Inequality (8) that

$$\mid f(y) - f(x) \mid \leq \max_{x \in U} \mid \nabla f(x) \mid . \mid y - x \mid = 0$$

and $f(y) = f(x)$, for all $x, y \in U$. Assuming U is connected, then the conclusion follows from the local convexity. First, we fix the sets $\{p_\lambda \in U \mid \lambda \in \Lambda\}$ and $\{\epsilon_\lambda \in (0, \infty)\}$ in such a way that the family of open balls $\{B_\lambda = B_{\epsilon_\lambda}(p_\lambda) \mid \lambda \in \Lambda\}$ is an open cover of U. Choose $\lambda_0 \in \Lambda$ so that $B_{\lambda_0} \subset U$. Since B_{λ_0} is convex, we have $c \in \mathbb{R}$ such that $f(x) = c$, for all $x \in B_{\lambda_0}$. The set $U_c = \{x \in U \mid f(x) = c\}$ is open and closed as a subset in U, and therefore $U_c = U$.   □

Our goal is to study differentiable functions, so we will always assume that $f \in C^1(U)$. To understand the behavior of a function, it is important to know the properties of the subsets of U in which the function is constant. Given a value $c \in \mathrm{Im}(f)$, we define the c-level set as $f^{-1}(c) = \{x \in U \mid f(x) = c\}$. Indeed, Sard's theorem [21] affirms that the level sets $f^{-1}(c)$ are hypersurfaces $\mathbb{R}^n$ for a dense set of values $c \in \mathrm{Im}(f)$; in case $n = 2$, the set $f^{-1}(c)$ is a level curve and for $n = 3$, it is a level surface. The characteristics of the set of levels of a function are important to address optimization questions, for example to find local maximums and local minimums of f.

The following remarks derive from Eq. (7);

(1) Let $\theta(t)$ be the angle formed by the vectors $\nabla f(\gamma(t))$ and $\gamma'(t)$, so
$$h'(t) = \frac{d[f(\gamma(t))]}{dt} = \; <\nabla f(\gamma(t)), \gamma'(t)> \; = \mid \nabla f(\gamma(t)) \mid . \mid \gamma'(t) \mid \cos(\theta(t)),$$

(2) The function has the highest growth rate in the direction of the vector $\nabla(\gamma(t))$. If $\gamma'(t) = \nabla f(\gamma(t))$, then the derivative of the function $h(t) = f(\gamma(t))$ is given by
$$h'(t) = df_{\gamma(t)}.\gamma'(t) = \mid \nabla f(\gamma(t)) \mid^2 > 0 \; \Rightarrow \; h(t) \text{ increases.}$$

The growth of f is more accentuated in the direction of the gradient vector because $\cos(\theta(t)) = 1$. Of course, f decreases sharply in the opposite direction of the gradient.

(3) If $\gamma(t) \subset f^{-1}(c)$, then $\nabla f(\gamma(t))$ is orthogonal to the level set $f^{-1}(c)$ for all $t \in (-\epsilon, \epsilon)$, since
$$0 = \frac{d[f(\gamma(t))]}{dt} = \; <\nabla f(\gamma(t)), \gamma'(t)> \; .$$

Exercises

(1) Sketch the level sets and graphs of the following functions;

(i) $f(x, y) = x^2 + y^2$,

(ii) $f(x, y) = x^2 - y^2$,

(iii) $f(x, y) = -x^2 - y^2$

(iv) $f(x, y) = x^2(x - 1)^2 + y^2$,

(v) $f(x, y) = (x^2 + y - 11)^2 + (x + y^2 - 7)^2$.

(2) Consider $0 \leq k \leq \infty$ and $I_k = (i_1, \ldots, i_k) \in \{1, \ldots, n\}^k$. Let $C^k(U)$ be the set of functions $f : U \to \mathbb{R}$ with partial derivatives $\frac{\partial^k f}{\partial x_{i_1} \ldots \partial x_{i_k}}$ that are continuous for all multi-index $I_k \in \{1, \ldots, n\}^k$, and let $C^\infty(U)$ be the set of $f \in C^k(U)$ for every $k \in \mathbb{N}$. Prove that $C^k(U)$ is a ring for all $k \in \mathbb{N}$.

(3) A homogeneous function of degree k is a function $f : \mathbb{R}^n \to \mathbb{R}$ that satisfies the condition $f(tx) = t^k f(x)$ for any $t \in \mathbb{R}$ and $x \in \mathbb{R}^n$. Show that

$$k.f(x) = \sum_{i=1}^{n} x_i \frac{\partial f}{\partial x_i}.$$

(4) Let $f : \mathbb{R}^n \to \mathbb{R}$ be a differentiable function at the origin. If $f(tx) = tf(x)$ for all $t > 0$ and all $x \in \mathbb{R}^n$, prove f is linear. Use this result to prove that the function

$$f(x, y) = \begin{cases} \frac{x^3}{x^2+y^2}, & (x, y) \neq (0, 0), \\ 0, & (x, y) = (0, 0) \end{cases}$$

is not differentiable at the origin.

(5) Let $f(x, y, z) = e^{x^2 + \sqrt{yx} + \ln(z)}$. Find the partial derivatives $\frac{\partial f}{\partial x}$, $\frac{\partial f}{\partial y}$ and $\frac{\partial f}{\partial z}$. Define $h(t) = f(t, t^2, t^3)$ and compute $\frac{dh}{dt}$.

(6) Consider the function $f(x, y) = x.y$. Find the total derivative df and the increment $\Delta f = f(x + \Delta x, y + \Delta y) - f(x, y)$. Next, compare the results obtained using the fact that $dx \sim \Delta x$, $dy \sim \Delta y$.

(7) Find the approximate value of the variation of the hypotenuse of a triangle with sides that measure 3 cm and 4 cm; the smaller side is decreased by 0.5 cm and the largest side is raised by 0.75 cm.

(8) Let $\mathcal{L}(\mathbb{R}^n; \mathbb{R})$ be the space of linear functionals. Prove that $f \in C^1(U)$ if and only if the map $df : U \to \mathcal{L}(\mathbb{R}^n; \mathbb{R})$, $x \to df_x$ is continuous (hint: $\mathcal{L}(\mathbb{R}^n; \mathbb{R})$ can be identified with the space $M(n, 1)$ of real matrices $n \times 1$).

(9) The Laplacian operator $\triangle : C^2(U) \to C^0(U)$ defined on an open subset $U \subset \mathbb{R}^n$ is defined as $\triangle f = \sum_{i=1}^{n} \frac{\partial^2 f}{\partial x_i^2}$. A function f is harmonic if it satisfies the partial differential equation (PDE) $\triangle f = 0$ in U. Consider $U = \mathbb{R}^2$, and prove that $u(x, y) = e^x \sin(y)$ is harmonic in $\mathbb{R}^2$.

(10) The *PDE* governing the evolution of heat in space is $\Delta v - \frac{\partial v}{\partial t} = 0$. Prove that the function $f(x, t) = \frac{1}{8\pi^{3/2}(t_0-t)^{3/2}} e^{-\frac{r^2}{4(t_0-t)}}$, with $r = \sqrt{(x - x_0)^2 + (y - y_0)^2 + (z - z_0)^2}$ satisfies the heat equation.

(11) Consider $U = \mathbb{R} \times \mathbb{R}$. The uni-dimensional wave equation is the *PDE*

$$\frac{\partial^2 f}{\partial x^2} - \frac{\partial^2 f}{\partial t^2} = 0.$$

Suppose $f \in C^2(U)$ is a solution of the wave equation. Show that f satisfies the equation $\frac{\partial^2 f}{\partial x' \partial t'} = 0$ after changing the coordinates to $x' = x + t$, $t' = x - t$. Conclude that there are functions $\phi, \psi : U \times I \to \mathbb{R} \in C^2(U \times I)$, defined on an interval I, such that $f(x, t) = \phi(x + t) + \psi(x - t)$.

(12) Consider $C \subset U$ to be a closed subset and let $f \in C^1(U)$. Extend the concept of differentiability for the function $\bar{f} : C \to \mathbb{R}$.

(13) If $\gamma : \mathbb{R} \to U$ is a non-constant curve such that $\gamma'(t) = \nabla f(\gamma(t))$, prove that $\gamma(t)$ cannot be periodic.

(14) (Maximum descent method) Given $A \in M_n(\mathbb{R})$ and $y_0 \in \mathbb{R}^n$, consider the linear equation $A.x = y_0$. For a solution, we consider the problem of finding the minimum of the function $f(x) = \frac{1}{2} \mid A.x - y_0 \mid^2$;

(a) Prove that $\nabla f(x) = A^t(A.x - y_0)$.

(b) Take $x_0 \in U$ and consider the straight line $r_0(t) = x_0 - t\nabla f(x_0)$. Conclude that the function $f_0(t) = \frac{1}{2} \mid A.(x_0 - t\nabla f(x_0)) - y_0 \mid^2$ attains its minimum at

$$t_0 = \frac{\mid A^t(A.x_0 - y_0) \mid}{\mid A^t A(A.x_0 - y_0) \mid}.$$

(c) Define $x_1 = x_0 - t_0\nabla f(x_0)$ and prove that $f(x_1) \leq f(x_0)$.

(d) Applying the minimizing procedure of the previous item, define the sequence

$$x_{n+1} = x_n - \frac{\mid A^t(A.x_n - y_0) \mid}{\mid A^t A(A.x_n - y_0) \mid} A^t(A.x_n - y_0), \quad n \in \mathbb{N},$$

and prove that the sequence $\{x_n\}_{n\in\mathbb{N}}$ converges to a solution of the linear equation $A.x = y_0$.

1.3 Differentials

Taking in $U \subset \mathbb{R}^n$ an orthogonal coordinate system $(x_1, \ldots, x_n)$ for $1 \leq i \leq n$, we get the linear functionals

$$dx_i : T_pU \to \mathbb{R},$$
$$v = \sum_i v_i e_i \to dx_i(v) = v_i. \tag{9}$$

In this coordinate system, the differential of f is

$$df = \sum_{i=1}^{n} \frac{\partial f}{\partial x_i} . dx_i . \tag{10}$$

Shifting the point (x, y) to the point $(x + \triangle x, y + \triangle y)$ yields a variation of f:

$$\triangle_{(x,y)} f = f(x + \triangle x, y + \triangle y) - f(x, y).$$

The Mean Value Theorem assures us that there are $\alpha \in (x, x + \triangle x)$ and $\beta \in (y, y + \triangle y)$ such that

$$\triangle_{(x,y)} f = \frac{\partial f}{\partial x}(\alpha, y + \triangle y).\triangle x + \frac{\partial f}{\partial y}(x, \beta).\triangle y.$$

That is, if $\triangle x \sim 0$ and $\triangle y \sim 0$, then $\triangle_{(x,y)} f \sim d f_{(x,y)}$ ($\sim$ "is approximately equal to")

$$\triangle(f(\gamma(t))) \sim \sum_{i=1}^{n} \frac{\partial f}{\partial x_i}(\gamma(t)).\gamma'(t)\triangle t.$$

The total derivative and the partial derivative are often used in the language of physicists. The partial derivative of a function f, relative to the variable x, is simply $\frac{\partial f}{\partial x}$; the total derivative is the differential of f. The difference is clear, considering the following example: the temperature on a two-dimensional plate defines a function $T = T(x, y, t)$ depending on the coordinates (x, y) of the point where it is being measured and the instant t of the measurement. The partial derivative $\frac{\partial T}{\partial t}$ measures the rate of change of the temperature relative to t at a fixed point (x, y). The total derivative $\frac{dT}{dt}$ measures the rate of change of the temperature along a trajectory $\gamma(t) = (x(t), y(t))$; $\frac{dT}{dt} = \frac{\partial T}{\partial x}\frac{dx}{dt} + \frac{\partial T}{\partial x}\frac{dy}{dt} + \frac{\partial T}{\partial t}$. The differential allows us to calculate any variation of the function.

1.4 Multiple Derivatives

In the same way that the functions $\frac{\partial f}{\partial x_i}$ were generated from the function $f \in \mathrm{C}^1(U)$, many others can be generated by taking more partial derivatives $\frac{\partial}{\partial x_j}\left(\frac{\partial f}{\partial x_i}\right)$. Assuming $\frac{\partial f}{\partial x_i} \in \mathrm{C}^1(U), \forall \, i = 0, \dots, n$, n^2 functions can be defined by

$$\frac{\partial^2 f}{\partial x_j \partial x_i} = \frac{\partial}{\partial x_j}\left(\frac{\partial f}{\partial x_i}\right) \in C^0(U), \ (i, j) \in \{1, \dots, n\}^2.$$

Consider $1 \leq k \leq n$; for each multi-index $(i_1, \ldots, i_k) \in \{1, \ldots, n\}^k$, the following functions are defined inductively by

$$\frac{\partial^k f}{\partial x_{i_k} \ldots \partial x_{i_1}} = \frac{\partial}{\partial x_{i_k}} \left(\frac{\partial^{k-1} f}{\partial x_{i_{k-1}} \ldots \partial x_{i_1}} \right).$$

A function $f : U \to \mathbb{R}$ has class $C^2(U)$ if the following items are satisfied, for all $(i_1, i_2) \in \{1, \ldots, n\}^2$;

(i) $\frac{\partial f}{\partial x_i} \in C^1(U)$, $\forall\, i \in \{1, \ldots, n\}$ and (ii) $\frac{\partial^2 f}{\partial x_{i_1} \partial x_{i_2}} \in C^0(U)$.

Sucessively, $f \in C^k(U)$. If $f \in C^k(U)$ for all $k \in \mathbb{N}$, we say f is smooth and $f \in C^\infty(U)$.

The following result, known as Schwarz's Theorem, gives a sufficient condition for the partial derivatives to commute.

Theorem 3 (Schwarz). *If $f \in C^2(U)$, then*

$$\frac{\partial}{\partial x_j} \left(\frac{\partial f}{\partial x_i} \right) = \frac{\partial}{\partial x_i} \left(\frac{\partial f}{\partial x_j} \right)$$

for all $i, j \in \{1, \ldots, n\}$ and $x \in U$.

The next lemma goes half-way towards proving Schwarz's theorem.

Lemma 1 *Let $U \subset \mathbb{R}^n$ be an open subset and let $f \in C^1(U \times [a, b])$. For every $i = 1, \ldots, n$, the partial derivative $\frac{\partial \phi}{\partial x_i}$ of the function $\phi : U \to \mathbb{R}$, $\phi(x) = \int_a^b f(x, t)dt$ at $x \in U$ is*

$$\frac{\partial \phi}{\partial x_i}(x) = \int_a^b \frac{\partial f}{\partial x_i}(x, t)dt. \tag{11}$$

Proof Let e_i be an element of the canonical basis $\mathbb{R}^n$ and assume that $[x, x + se_i] \subset U$, so we have

$$\frac{\phi(x + se_i) - \phi(x)}{s} - \int_a^b \frac{\partial f}{\partial x_i}(x, t)dt = \int_a^b \left[\frac{f(x + se_i, t) - f(x, t)}{s} - \frac{\partial f}{\partial x_i}(x, t) \right] dt.$$

By the Mean Value Theorem, we have $r_0 \in (0, 1)$ such that

$$\frac{f(x + se_i, t) - f(x, t)}{s} = \frac{\partial f}{\partial x_i}(x + r_0 se_i, t).$$

Therefore

$$\frac{\phi(x + se_i) - \phi(x)}{s} - \int_a^b \frac{\partial f}{\partial x_i}(x, t)dt = \int_a^b \left[\frac{\partial f}{\partial x_i}(x + r_0 se_i, t) - \frac{\partial f}{\partial x_i}(x, t) \right] dt. \tag{12}$$

We have to control the integral term at the right-hand side of Eq. (12). To do so, we use the compactness of $[a, b]$ and the continuity of the partial derivatives, so the functions $\frac{\partial f}{\partial x_i}$, $1 \leq i \leq n$, are uniformly continuous with respect to the variable t (Appendix A, Theorem 4). Taking $\epsilon > 0$, we get $\delta > 0$ to be not dependent on t, such that

$$\left| \frac{\partial f}{\partial x_i}(x + r_0 s e_i, t) - \frac{\partial f}{\partial x_i}(x, t) \right| < \epsilon.$$

Therefore taking the limit $s \to 0$ in the identity (12), we get the identity (11). $\qquad \square$

Proof (Schwarz's Theorem) By the Fundamental Theorem of Calculus,

$$f(x, y) = f(x, b) + \int_b^y \frac{\partial f}{\partial y}(x, t) dt.$$

Deriving the equation above, we obtain

$$\frac{\partial f}{\partial x}(x, y) = \frac{\partial f}{\partial x}(x, b) + \frac{\partial}{\partial x}\left[\int_b^y \frac{\partial f}{\partial y}(x, t) dt \right]$$
$$= \frac{\partial f}{\partial x}(x, b) + \left[\int_b^y \frac{\partial}{\partial x}\frac{\partial f}{\partial y}(x, t) dt \right].$$

In deriving the above equation with respect to y, it follows that $\frac{\partial}{\partial y}\frac{\partial f}{\partial x} = \frac{\partial}{\partial x}\frac{\partial f}{\partial y}$. $\square$

As the following function shows, the Schwarz theorem is false in general:

$$f(x, y) = \begin{cases} \frac{xy(x^2 - y^2)}{x^2 + y^2}, & (x, y) \neq (0, 0), \\ 0, & (x, y) = (0, 0). \end{cases} \tag{13}$$

In this case, the partial derivatives are $\frac{\partial f}{\partial x}(0, y) = -y$ and $\frac{\partial f}{\partial x}(x, 0) = x$, while $\frac{\partial^2 f}{\partial y \partial x}(0, 0) = -1$ and $\frac{\partial^2 f}{\partial x \partial y}(0, 0) = 1$.

Exercises

(1) State and prove Schwarz's theorem for the general case. Considering a bijection $\tau : \{1, \ldots, k\} \to \{1, \ldots, k\}$, then we have

$$\frac{\partial^k f}{\partial x_{i_{\tau(1)}} \ldots \partial x_{i_{\tau(k)}}} = \frac{\partial^k f}{\partial x_{i_1} \ldots \partial x_{i_k}}.$$

(2) Prove that the function defined in (13) is not C^1.

1.5   Higher Order Differentials

The 2nd-order differential of f is

$$d^2 f = d\left(\sum_{j=1}^{n} \frac{\partial f}{\partial x_i} dx_i\right) = \sum_{j=1}^{n} d\left(\frac{\partial f}{\partial x_i}\right) dx_i = \sum_{i,j=1}^{n} \frac{\partial^2 f}{\partial x_i \partial x_j} dx_i dx_j.$$

Inductively, we define the differential of order k by $d^k f = d(d^{k-1} f)$, and $I_k = (i_1, \ldots, i_k) \in \{1, \ldots, n\}^k$. Therefore we have

$$d^k f = \sum_{I_k} \frac{\partial^k f}{\partial x_{i_1} \ldots \partial x_{i_k}} dx_{i_1} \ldots dx_{i_k}. \tag{14}$$

Of course, $dx_i dx_j = dx_j dx_i$ for all pairs i, j. For $p \in U$, the differential $d^2 f$ induces a bilinear form $d^2 f_p : T_p U \times T_p U \to \mathbb{R}$; for any vectors $u = \sum_i u_i e_i$ and $v = \sum_j v_j e_j \in T_p U$,

$$(d^2 f_p).(u, v) = \sum_{i,j=1}^{n} \frac{\partial^2 f}{\partial x_i \partial x_j} dx_i dx_j (u, v) = \sum_{i,j=1}^{n} \frac{\partial^2 f}{\partial x_i \partial x_j} u_i v_j = u.H(p).v^t,$$

The 2nd-order differential of f defines the symmetric matrix $H(p) = \left(\frac{\partial^2 f}{\partial x_i \partial x_j}(p)\right) \in S_n$ called the Hessian matrix of f at p. The condition $f \in C^2(U)$ is equivalent to the continuity of the map $d^2 f : U \to S_n$, $p \to H(p)$.

2   Taylor's Formula

The 1st-order differential df_p gives the linear approximation of a function $f \in C^1(U)$ in a small neighborhood of $p \in U$,

$$f(p + v) \sim f(p) + \sum_{i=1}^{n} \frac{\partial f}{\partial x_i}(p) \triangle x_i.$$

According to the definition, $f \in C^1$ can be written as

$$f(p + v) = f(p) + df_p.v + r(v),$$

such that $\lim_{v \to 0} \frac{r(v)}{|v|} = 0$. Next, a quadratic approximation of $f(x)$ is described in terms of the 2nd partial derivatives.

Theorem 4 (Taylor's Formula). *Let $f \in C^2(U)$, $p \in U$ and B_δ be a ball centered at the origin with radius $\delta > 0$ such that $p + tv \in U$ for all $v \in B_\delta$. So we have a function $r : B_\delta \to \mathbb{R}$ such that*

$$f(p + v) - f(p) = df_p.v + \frac{1}{2}(d^2 f_p).(v, v) + r(v), \tag{15}$$

and $\lim_{v \to 0} \frac{r(v)}{|v|^2} = 0$.

Proof Let $p \in U$ and define

$$r(v) = f(p + v) - f(p) - df_p.v - \frac{1}{2}d^2 f_p(v, v).$$

The function $r = r(v)$ is C^2-differentiable. Consider $v = (x_1, \ldots, x_n)$, so $r(0) = 0$ and

$$\frac{\partial r}{\partial x_i}(v) = \frac{\partial f}{\partial x_i}(p + v) - \frac{\partial f}{\partial x_i}(p) - \sum_{i=1}^{n} \frac{\partial^2 f}{\partial x_j \partial x_i}(p)x_j \Rightarrow \frac{\partial r}{\partial x_i}(0) = 0.$$

Deriving the equation above once more, we get

$$\frac{\partial^2 r}{\partial x_j \partial x_i}(v) = \frac{\partial^2 f}{\partial x_j \partial x_i}(p + v) - \frac{\partial^2 f}{\partial x_j \partial x_i}(p) \Rightarrow \frac{\partial^2 r}{\partial x_j \partial x_i}(0) = 0. \tag{16}$$

We check the condition $\lim_{v \to 0} \frac{r(v)}{|v|^2} = 0$. Since $r \in C^2(B_\delta)$, we have $R \in C^0(B_\delta)$ such that

$$\frac{\partial r}{\partial x_i}(v) - \frac{\partial r}{\partial x_i}(0) = d\left(\frac{\partial r}{\partial x_i}\right)_0.v + R(v), \quad \lim_{v \to 0} \frac{R(v)}{|v|} = 0.$$

Therefore $r(0) = \frac{\partial r}{\partial x_i}(0) = \frac{\partial^2 r}{\partial x_j \partial x_i}(0) = 0$ implies $\frac{\partial r}{\partial x_i}(v) = R(v)$ and $\lim_{v \to 0} \frac{\frac{\partial r}{\partial x_i}(v)}{|v|} = \lim_{v \to 0} \frac{R(v)}{|v|} = 0$. Since δ is sufficiently small, the solution to r can be approximated by its differential. Consider the curve $\gamma(t) = tv$, satisfying $\gamma(0) = 0$ and $\gamma(1) = v$, and the function $h(t) = r(\gamma(t)) = r(tv)$. By the Mean Value Theorem there is $c \in (0, 1)$ such that $h(1) - h(0) = h'(c)$, that is, $r(v) = \sum_{i=1}^{n} \frac{\partial r}{\partial x_i}(cv)x_i$. Therefore

$$\frac{r(v)}{|v|^2} = \sum_{i=1}^{n} \frac{\frac{\partial r}{\partial x_i}(cv)}{|cv|} \cdot \frac{cx_i}{|v|}.$$

Since $\lim_{v \to 0} \frac{\frac{\partial r}{\partial x_i}(cv)}{|cv|} = 0$ and $0 \leq |c.\frac{v}{|v|}| < |c| < 1$, it follows that $\lim_{v \to 0} \frac{r(v)}{|v|^2} = 0$. $\square$

Equation 15 represents the 2nd order for Taylor's formula of f centered at p. Taylor's formula of order k, centered at $p \in U$, of $f \in C^k(U)$ is

$$T_k(f; p) = f(p) + \sum_{i=1}^{k} \frac{1}{i!} (d^i f_p).(v, .\overset{i}{.}., v).$$

The general formula $T_k(f, p)$ is left as an exercise at the end of the section. The Taylor Series can be defined by taking $k = \infty$ in Taylor's formula. A function f is analytic at $x \in U$ if

$$f(x) = \lim_{k \to \infty} T_k(f; x),$$

and is analytic in U if the limit above is true for every $x \in U$. The set of analytic functions defined in U is denoted $C^\omega(U)$.

Exercises

(1) Let $p = (p_1, \ldots, p_n)$ and let $B_\delta(p)$ be a ball of radius $\delta > 0$ centered at p. Let $f : B_\delta \to \mathbb{R}$ be a C^k-function and let $F : (0, \delta) \to \mathbb{R}$ be given by $F(t) = f(p + t(x - p))$. Apply Taylor's formula for a one-variable function to

$$F(t) = \sum_{i=0}^{k} \frac{F^i(0)}{i!} t^i + r_k(t),$$

$r_k(t) = \frac{1}{(k+1)!} F^{k+1}(ct), 0 < c < 1$, to obtain the Taylor series of f.

(2) Generalize the Taylor formula to

$$f(p + v) - f(p) = \sum_{i=1}^{k} \frac{1}{i!} (d^i f_p).(v, .\overset{i}{.}., v) + r(v),$$

such that $\lim_{v \to 0} \frac{r(v)}{|v|^k} = 0$.

(3) Consider $D_i = \frac{\partial}{\partial x_i} : C^k(U) \to C^{k-1}(U), 1 \leq i \leq n$. Define V as the real vector space generated by the formal sums $\sum_{i=1}^{n} c_i D_i$, such that $c_i \in \mathbb{R}, i \in \{1, \ldots, n\}$. Define the product of an operation on V as follows:

$$D_i D_j = \frac{\partial}{\partial x_i} \frac{\partial}{\partial x_j} = \frac{\partial^2}{\partial x_i \partial x_j} = D_{ij}, \quad \ldots \quad, D_{i_1 \ldots i_k} = \frac{\partial^k}{\partial x_{i_1} \ldots \partial x_{i_k}}.$$

Prove the identity below for functions in $C^k(U)$: consider $I_k = (i_1, \ldots, i_k)$ such that $i_1 + \ldots + i_k = k$:

$$\left(c_1 D_1 + \ldots + c_n D_n\right)^k = \sum_{I_k} \frac{k!}{i_1! i_2! \ldots i_k!} D_1^{i_1} D_2^{i_2} \ldots D_k^{i_k}.$$

(a) In item (3), consider $v = (v_1, \ldots, v_n) \in \mathbb{R}^n$ and define the product

$$H.\nabla = (v_1, \ldots, v_n).(D_1, \ldots, D_n) = v_1 D_1 + \ldots + v_n D_n.$$

Prove that the derivative of the function of $g(t) = f(p + tv)$ is

$$\frac{dg(t)}{dt} =< \nabla f(p + tv), v >= (v.\nabla).f(p + tv).$$

(b) Consider $r \in \mathbb{N}$ and $f \in C^r(U)$. If $p \in U$ and $v \in T_p U$, prove that

$$\frac{d^r}{dt}(f(p + tv)) = (v.\nabla)^r f(p + tv).$$

(c) Find an estimate for the term $R(v) = \frac{1}{k!}(v.\nabla)^i f(p + \alpha v)$.

3 Critical Points and Local Extremes

Let $U \subset \mathbb{R}^n$ be an open set with coordinates $(x_1, \ldots, x_n)$ and $f \in C^k(U)$, $1 \leq k \leq \infty$. Whenever the differential df_p is non-null at a point $p \in U$, the local properties of f near p can be extracted from its linear term in Taylor's formula. However, if $df_p = 0$, then the local properties depend upon the Hessian matrix $H(p)$.

Definition 6 The point $p \in U$ is a critical point of f if the linear functional $df_p : T_p U \to \mathbb{R}$ is not surjective. Consider $\mathrm{Cr}(f)$ the set of critical points of f.

It is straightforward from the definition that if $p \in \mathrm{Cr}(f)$, then $df_p = 0$ and, consequently, $\nabla f(p) = 0$. Depending on the context, p also is called a singular point. Taking $p \in \mathrm{Cr}(f)$, the 2nd-order of Taylor's formula of f centered at p is

$$f(p + v) - f(p) = \frac{1}{2} < v, H(p).v > +r(v).$$

Taking the canonical basis of $\mathbb{R}^n$, the operator $H(p) : T_p U \to T_p U$ is represented by the Hessian matrix $H(p) = \left(\frac{\partial^2 f}{\partial x_i \partial x_j}(p)\right)$ of f at p.

Definition 7 Let $p \in U$, and for any small $\delta > 0$, consider the ball $B_\delta(p) \subset U$:
(i) p is a local maximum of f if there is $\delta > 0$ such that $f(x) \leq f(p)$ for all $x \in B_\delta(p)$.

minimum

Fig. 3 $f(x, y) = x^2 + y^2$

(ii) p is a local minimum of f if there is $\delta > 0$ such that $f(x) \geq f(p)$ for all $x \in B_\delta(p)$.

We consider the case p to be a local maximum, so $f(p + t\hat{v}) - f(p) \leq 0$ for all $|t| < \delta$ and $\hat{v} \in T_p U$ is a unitary vector. By the Mean Value Theorem, we have $c \in (0, 1)$ such that

$$f(p + t\hat{v}) - f(p) = \frac{\partial f}{\partial \hat{v}}(p + c\hat{v}).t \leq 0.$$

Therefore we consider the following cases: (i) if $t > 0$, then $\frac{\partial f}{\partial \hat{v}}(p + c\hat{v}) \leq 0$ and (ii) if $t < 0$, then $\frac{\partial f}{\partial \hat{v}}(p + c\hat{v}) \geq 0$. So taking the limit $t \to 0$ yields $\frac{\partial f}{\partial \hat{v}}(p) = 0$ for any $\hat{v}$. Hence $\nabla f(p) = 0$ and $df_p = 0$, so p is a critical point. The same is true if we assume that p is a local minimum. However p being a critical point does not mean it is a local maximum or a local minimum, as shown in the next examples.

Example 1 From the classification[1] of quadratic forms in $\mathbb{R}^n$, the examples below are canonical models of critical points for a function $f : \mathbb{R}^2 \to \mathbb{R}$, in which the Hessian is a non-singular matrix. Each case is illustrated in Figs. 3, 4 and 5.

(1) $f(x, y) = x^2 + y^2$;
  Since $\nabla f(x, y) = 2(x, y)$, $p = (0, 0)$ is the only critical point of f. Clearly, $p = 0$ is a local minimum and the Hessian $H(0)$ is

$$H(0, 0) = \begin{pmatrix} 2 & 0 \\ 0 & 2 \end{pmatrix}.$$

---

[1]Theorem of Sylvester, Appendix A.

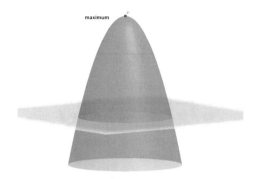

Fig. 4 $f(x, y) = -x^2 - y^2$

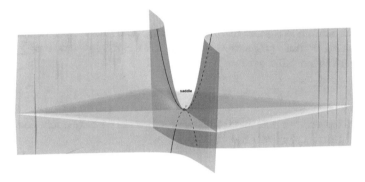

Fig. 5 $f(x, y) = x^2 - y^2$

(2) $f(x, y) = -x^2 - y^2$;

In this case, $\nabla f(x, y) = -2(x, y)$ and $p = 0$ is the only critical point and a local maximum. So,

$$H(0, 0) = \begin{pmatrix} -2 & 0 \\ 0 & -2 \end{pmatrix}.$$

(3) $f(x, y) = x^2 - y^2$;

Analogously, the identity $\nabla f(x, y) = 2(x, -y)$ yields that $p = 0$ is a critical point; however, it is neither a local maximum nor a local minimum of f. In this case, $p = 0$ is called a saddle point of f. The Hessian matrix at $p = 0$ is

$$H(0, 0) = \begin{pmatrix} 2 & 0 \\ 0 & -2 \end{pmatrix}.$$

Since the Hessian matrix $H(p)$ is symmetric, it is diagonalizable, and all eigenvalues are real numbers; that is, there is an orthonormal basis of $T_p U$ in which the elements are eigenvectors of $H(p)$. The spectrum of $H(p)$ is the set $\sigma(H(p)) = \{\lambda \in \mathbb{R} \mid \exists v \neq 0 \in \mathbb{R}^n, \ H(p).v = \lambda v\}$ of the eigenvalues of $H(p)$.

Definition 8 Let $T : \mathbb{R}^n \to \mathbb{R}^n$ be a symmetric operator ($\sigma(T) \subset \mathbb{R}$);

(1) T is non-negative if $\sigma(T) \subset [0, \infty)$, and is positive if $\sigma(T) \subset (0, \infty)$.
(2) T is non-positive if $\sigma(T) \subset (-\infty, 0]$, and is negative if $\sigma(T) \subset (-\infty, 0)$.

Theorem 5 *Let $p \in U$ be a critical point of $f \in C^2(U)$;*
(i) if $H(p)$ is positive, then p is a local minimum of f.
(ii) if $H(p)$ is negative, then p is a local maximum of f.

Proof Let $\beta = \{e_1, \ldots, e_n\}$ be a basis of eigenvectors of $H(p)$, and therefore $H(e_i) = \lambda_i e_i$. Consider that the function

$$g : S^{n-1} \to \mathbb{R},$$
$$g(\hat{v}) = < \hat{v}, H(\hat{v}) >$$

attains its maximum at $\hat{v}_M$ and its minimum at $\hat{v}_m$. Define the extreme values of the spectrum as $\lambda_m = \min \sigma(H)$ and $\lambda_M = \max \sigma(H)$. For any vector $v = \sum_i v_i e_i$, the function g takes the value $g(v) = \sum_i \lambda_i v_i^2$, so

$$g(\hat{v}_m) = \lambda_m \leq g(\hat{v}) \leq \lambda_M = g(\hat{v}_M).$$

To prove item (i), we note that from Taylor's formula we have a function $\rho : B_\delta \to \mathbb{R}$ such that $\lim_{v \to 0} \rho(v) = 0$ and

$$f(p + v) - f(p) = \frac{1}{2} < v, H(p)v > + \rho(v) \mid v \mid^2$$

$$= \mid v \mid^2 \sum_{i=1}^n \lambda_i \hat{v}_i^2 + \mid v \mid^2 \rho(v) \geq \mid v \mid^2 [\lambda_m + \rho(v)].$$

Since $\lambda_m > 0$, by taking δ small enough in the identity above, we get $f(p + v) \geq f(p)$ for all $v \in B_\delta$. Item (ii) follows by the same argument. $\square$

The theorem does not hold if $H(p)$ is an indefinite quadratic form, that is, when p is neither a local maximum nor a local minimum of f. If the spectrum of $H(p)$ has negative and positive eigenvalues, and $0 \notin \sigma(H(p))$, we say that p is a saddle point of f. When $0 \in \sigma(H(p))$, the classification of quadratic forms is rather extended; moreover, the critical point is no longer isolated as shown in the following examples:

(1) $f(x, y) = x^2$,

$$H(x, y) = \begin{pmatrix} 2 & 0 \\ 0 & 0 \end{pmatrix} \Rightarrow \begin{matrix} \nabla f(p) = 0 \Rightarrow p = (0, y) \\ H(0) \text{ is indefinite} \end{matrix}, \quad p \text{ is minimum.}$$

(2) $f(x, y) = -x^2$,

$$H(x, y) = \begin{pmatrix} -2 & 0 \\ 0 & 0 \end{pmatrix} \Rightarrow \begin{matrix} \nabla f(p) = 0 \Rightarrow p = (0, y) \\ H(0) \text{ is indefinite} \end{matrix}, \quad p \text{ is maximum.}$$

(3) $f(x, y, z) = x^2 - y^2$,

$$H(x, y) = \begin{pmatrix} 2 & 0 & 0 \\ 0 & -2 & 0 \\ 0 & 0 & 0 \end{pmatrix} \Rightarrow \begin{matrix} \nabla f(p) = 0 \Rightarrow p = (0, 0, z) \\ H(0) \text{ is indefinite} \end{matrix}, \quad p \text{ is saddle.}$$

Exercises

(1) Consider the function $f(x, y) = x^3 - 3xy^2$ (monkey's saddle). Is $p = (0, 0, 0)$ a local minimum, a local maximum or a saddle point of f?

(2) For $f : \mathbb{R}^2 \to \mathbb{R}$ defined below, find the critical points and classify them.

(a) $f(x, y) = x^2(x - 1)^2 + y^2$.
(b) $f(x, y) = (x^2 + y - 11)^2 + (x + y - 7)^2$.
(c) $f(x, y) = x^3 - 3xy^2$.

(3) Let $f \in C^2(\mathbb{R}^2)$ and $p \in \text{Cr}(f)$. Show that:

(a) p is a local minimum of f if $\det(H(p)) > 0$ and $\frac{\partial^2 f}{\partial x^2}(p) > 0$.
(b) p is a local maximum f if $\det(H(p)) > 0$ and $\frac{\partial^2 f}{\partial x^2}(p) < 0$.
(c) p is a saddle point if $\det(H(p)) < 0$.
(d) if $\det(H(p)) = 0$, then it is not possible to decide about its nature.
(e) Suppose p is a strict minimum, that is, if $f(p) < f(x)$ for all $x \in B_\delta$, then $\lambda_m > 0$.
(f) Give a sufficient condition in order for p to be a strict maximum ($f(p) > f(x)$, for all $x \in U$).
(g) Give an example such that $f(p) \leq f(x), \forall x \in B_\delta$, and it is false that $\lambda_m > 0$.

3.1 Morse Functions

As before, let $U \subset \mathbb{R}^n$ be an open set and $f \in C^2(U)$. The criteria in the previous section are accurate when the Hessian matrix $H(p)$ is non-degenerate, which by definition means that $\text{Ker}(H(p)) = \{0\}$.

Definition 9 $f \in C^2(U)$ is a Morse function if the Hessian $H(p)$ is non-degenerate for all critical points $p \in \text{Cr}(f)$.

A function f is a Morse function if $0 \notin \sigma(H(p))$. In the finite dimension, this is equivalent to the condition $\det(H(p)) \neq 0$. Defining the sets $\sigma^+(H(p)) = \{\lambda \in \sigma(H(p)) \mid \lambda > 0\}$ and $\sigma^-(H(p)) = \{\lambda \in \sigma(H(p)) \mid \lambda < 0\}$, we get $\sigma(H(p)) = \sigma^+(H(p)) \cup \sigma^-(H(p))$. Let V_λ be the eigenspace associated to the eigenvalue $\lambda \in \sigma(H(p))$, and define the subspaces

$$V^+(p) = \bigoplus_{\lambda > 0} V_\lambda, \quad V^-(p) = \bigoplus_{\lambda < 0} V_\lambda.$$

The non-degeneracy condition on $H(p)$ yields the decomposition $\mathbb{R}^n = V^+(p) \oplus V^-(p)$.

Definition 10 Let $f \in C^2(U)$ be a Morse function and $p \in \mathrm{Cr}(f)$. The Morse index of f at p is $\mu(p) = \dim(V^-(p))$.

Therefore if p is a local minimum, then we have $\mu(p) = 0$, while if p is a local maximum, then $\mu(p) = n$. When p is a saddle point, the Morse index satisfies $0 < \mu(p) < n$. The local behavior of a Morse function in a neighborhood of a critical point is completely described by the spectrum of the Hessian matrix, as shown in the next result.

Lemma 2 (Morse). *Let $f \in C^2(U)$ be a Morse function and let $p \in \mathrm{Cr}(f)$ be a non-degenerate critical point. So we have an open set $W \subset U$ containing p, a neighborhood $V \ni 0$ and a diffeomorphism $\xi : V \to W$ such that (i) $\xi(0) = p$ (ii) $H : V \times V \to \mathbb{R}$ is a non-degenerate symmetric bilinear form such that*

$$f(\xi(y)) - f(p) = < y, H(y) > .$$

Applications of the Morse lemma are beyond the scope of this text. For reading the proof we recommend Lima [30]. The condition imposed on a Morse function is not as restrictive as it seems because Sard's Theorem, as in Guillemin-Pollack [21], states that the set of Morse functions are dense in $C^2(U)$. This result can be understood by observing that a slight perturbation in the function can turn a Hessian's null eigenvalue into a non-zero eigenvalue.

Exercises

(1) Let $A \in M_n(\mathbb{R})$ be a self-adjoint matrix and let $f_A : \mathbb{R}^n \to \mathbb{R}$ be the function

$$f_A(x) = \frac{< x, Ax >}{|x|^2}.$$

Therefore f_A induces a function $\hat{f}_A : S^{n-1} \to \mathbb{R}$.

(a) Find the critical points of $\hat{f}_A$.

(b) Find sufficient conditions to guarantee that $\hat{f}_A$ is a Morse function.

(c) Assume that $\hat{f}_A$ is a Morse function and find the Morse index of each critical point.

(2) The real projective plane of dimension 2 is the quotient space $\mathbb{R}P^2 = (\mathbb{R}^3\backslash\{0\})/\sim$. The equivalent relation is defined as follows: $(x', y', z') \sim (x, y, z)$ if $(x', y', z') = t.(x, y, z)$ and $t \in \mathbb{R}\backslash\{0\}$. Indeed, $\mathbb{R}P^2$ is the space of lines in $\mathbb{R}^3$ passing through the origin.

(a) Define the quotient topology in $\mathbb{R}P^2$.

(b) Prove that $\mathbb{R}P^2 = (S^2/\sim)$ given that $(x', y', z') \sim (x, y, z)$ if $(x', y', z') = (x, y, z)$ or $(x', y', z') = -(x, y, z)$. So, $\mathbb{R}P^2 = S^2/\mathbb{Z}_2$.

(c) Let $A \in S_3(\mathbb{R})$ be a symmetric matrix. Prove that $\hat{f}_A$ induces a differentiable function $\tilde{f}_A : \mathbb{R}P^2 \to \mathbb{R}$. Use the results obtained for the function f_A to find the Morse indices of the critical points of $\hat{f}_A$.

4 The Implicit Function Theorem and Applications

At the origin of many mathematical ideas is the need to solve equations. Given an equation, the set of solutions can be empty, finite, or infinite. First, let's look at the example of an equation with one variable. In order to solve the equation $e^{-x^2} - \ln(x) = 0$, $x > 0$, the best strategy is to consider the function $f(x) = e^{-x^2} - \ln(x)$ and to study its properties. The graph of f is illustrated in Fig. 6. Since f is continuous and $f(1) = e^{-1} > 0$ and $f(2) < 0$, there is a point $a \in (1, 2)$ such that $f(a) = 0$. In addition, the derivative of the function satisfies $f'(x) < 0$ for all $x \in (0, \infty)$; consequently there is a unique solution to the equation.

The example above shows how studying functions can be useful in leading to the possibility of solutions for an equation.

The solution set of an equation of one variable in which the associated function $f(x)$ is differentiable and is generically discrete because it is geometrically described as the intercept of the graph of $f(x)$ with the x-axis. In most cases a solution to an equation is not explicitly obtained through analytical methods, but once the existence of a solution has been demonstrated, numerical methods are employed to obtain an approximation as close as possible, which demands computational methods.

In the case of an equation with two variables, we must understand what it means to find a a solution. To find the solution set S for the equation $x + y = 1$ is equivalent to finding the zero set $f^{-1}(0)$ of the function $f(x, y) = x + y - 1$. In both cases, by introducing the function $\phi(x) = 1 - x$, we get $S = f^{-1}(0) = \{(x, y) \in \mathbb{R}^2 \mid x \in \mathbb{R}, y = \phi(x)\}$. The case of three variables is illustrated in the equation $x^2 + y^2 + z^2 = 1$. Letting $\phi(x, y) = \sqrt{1 - x^2 - y^2}$, the solution set can be defined as $S = \{(x, y, z) \in \mathbb{R}^3 \mid z = \phi(x, y)\}$.

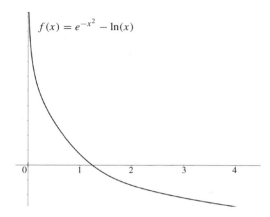

Fig. 6 $f(x) = e^{-x^2} - \ln(x)$

The last examples allow us to express one of the variables as a function of the others. This procedure is not always possible, e.g., in the equation $e^{-(x+y)^2} - \ln(x + y) = 0$ neither can y be written as a function of x nor can x be written as a function of y. Solution sets can be complex; examples with two variables define a curve and with three variables define a surface. However, once the case with one equation is well understood, the case with n-equations and m-variables can be dealt with. Before going forward, a short step back to get an understanding of linear systems is necessary.

Given the linear system

$$a_{11}x_1 + a_{12}x_2 + \ldots + a_{1n}x_n = b_1$$

$$\vdots \tag{17}$$

$$a_{m1}x_1 + a_{12}x_2 + \ldots + a_{mn}x_n = b_m,$$

assume the vectors $n_i = (a_{i1}, \ldots, a_{in})$, $1 \leq i \leq m$, are linearly independent, and the solution set is a space with dimension $(n - m)$. The linear case is the simplest case to understand. Each equation allows us to write one variable as a function of the others and replace it in the next equation. After performing this algorithm in every equation, we obtain a single equation with $(n - m)$ variables. Geometrically, this system corresponds to a set of m-distinct and non-parallel hyperplanes of $\mathbb{R}^n$, and the solution set is the intersection of the hyperplanes. If the intersection is not empty, then generically it defines an $(n - m)$ affine space. Although the systems of nonlinear equations are far more complex, the linear methods are useful to give us some insight or even to find a solution in many examples.

We go through more examples of equations defined by functions $f \in C^\infty(U)$, $U \subset \mathbb{R}^2$ an open subset;

Example 2 (1) $f : \mathbb{R}^2 \to \mathbb{R}$, $f(x, y) = x^2 + y^2 - 1$. The set $f^{-1}(0)$ is the circumference of radius 1. The functions $g_\pm(x) = \pm\sqrt{1 - x^2}$ define the sets $U^\pm = \{(x, g^\pm(x)) \mid x \in (-1, 1)\}$, respectively, such that $f^{-1}(0) = U^+ \cup U^- \cup \{-1, 1\}$. Restricting to looking at each set $U^\pm$, the solution set $f^{-1}(0)$ is the graph of $g^\pm$; however, it is not possible to extend the functions $g_\pm$ on a larger open interval $(-1 - \epsilon, 1 + \epsilon)$. The reason for not being able to extend results comes from the fact that in the neighborhood of points $(-1, 0)$ and $(1, 0)$, the circumference cannot be described as the graph of a function depending on x. However, in the neighborhood of each point $Q_\pm = (\pm 1, 0)$, we can express $x = \pm\sqrt{1 - y^2}$, that is, we have $(h_\pm(y), y)$. This example is the most elementary; it contains the basic elements to understand the sufficient conditions needed to solve an equation.

(2) $f : \mathbb{R}^2 \to \mathbb{R}$, $f(x, y) = x^2 - xy + y^2 - 1$. In this case, neither x nor y can be expressed explicitly as a function of each other by taking $f(x, y) = 0$. The solution set defines the curve shown in Fig. 7. We notice that the curve is not a graph as a whole, though locally for some intervals, it stands for a graph.

(3) $f : \mathbb{R}^2 \to \mathbb{R}$, $f(x, y) = y^5 + 2y - x^3 + x$. Figure 8 shows that the curve defined by the solution set of the equation $f(x, y) = 0$ is a graph. This is easily checked since $f(., y)$ is injective and increasing as a consequence of

$$\frac{\partial f}{\partial y}(x_0, y) = 5y^4 + 2 > 0, \text{ for all } y.$$

So for each $x = x_0$, there is a unique y_0 such that $f(x_0, y_0) = 0$, and $\lim_{y \to -\infty} f(x_0, y) = -\infty$ and $\lim_{y \to \infty} f(x_0, y) = \infty$. Hence $f(x_0, y)$ is a strictly increasing continuous function with $\mathbb{R}$ as the image. However, there is no function g such that $y = g(x)$.

Next, the Implicit Function Theorem gives a sufficient condition for solving the equation $f(x_1, \ldots, x_{n+1}) = c$ when $f \in C^k(U)$, $k \geq 1$. Consider $\mathbb{R}^{n+1} = \mathbb{R}^n \times \mathbb{R} = \{(x, y) \mid x = (x_1, \ldots, x_n) \in \mathbb{R}^n, \ y \in \mathbb{R}\}$ and $U \subset \mathbb{R}^{n+1}$ as an open set.

Theorem 6 (ImFT). *Let* $f \in C^k(U)$, $k \geq 1$. *Let* $(x_0, y_0) \in U$ *be such that* $f(x_0, y_0) = c$ *and assume that* $\frac{\partial f}{\partial y}(x_0, y_0) \neq 0$. *So we have a ball* $B = B_\delta(x_0) \subset \mathbb{R}^n$ *and an interval* $J = (y_0 - \epsilon, y_0 + \epsilon)$ *with the following properties:*

(1) $B \times \overline{J} \subset U$ *and* $\frac{\partial f}{\partial y}(x, y) \neq 0$ *for all* $(x, y) \in B \times \overline{J}$.
(2) *There is a* C^k-*function* $\xi : B \to J$ *such that* $f(x, y) = f(x, \xi(x)) = c$.
(3) *The partial derivatives of* ξ *are*

$$\frac{\partial \xi}{\partial x_i}(x) = -\frac{\frac{\partial f}{\partial x_i}(x, \xi(x))}{\frac{\partial f}{\partial y}(x, \xi(x))}. \tag{18}$$

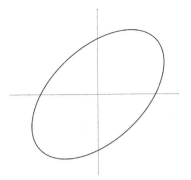

Fig. 7 $x^2 - xy + y^2 = 1$

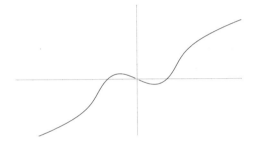

Fig. 8 $y^5 + 2y - x^3 + x = 0$

Proof Assume $\frac{\partial f}{\partial y}(x_0, y_0) > 0$. Since $\frac{\partial f}{\partial y}$ is continuous, we have $\delta > 0$ and $\epsilon > 0$ such that $\frac{\partial f}{\partial y}(x, y) > 0$ for all $(x, y) \in B_\delta(x) \times (y_0 - \epsilon, y_0 + \epsilon)$. The map $y \to f(x, y)$ defines an increasing function on $J = (y_0 - \epsilon, y_0 + \epsilon)$. Since $f(x_0, y_0) = c$, $f(x_0, y_0 - \epsilon) < c$ and $f(x_0, y_0 + \epsilon) > c$, the Mean Value Theorem implies the existence of a unique $y \in (y_0 - \epsilon, y_0 + \epsilon)$ related to x. Now, we define $y = \xi(x)$, such that $f(x, \xi(x)) = c$, and so the function $\xi : B_\delta(x_0) \to (y_0 - \epsilon, y_0 + \epsilon), x \to \xi(x)$. Assume for now that ξ is continuous. Differentiability of ξ is achieved as shown next; let $h(x) = (x, \xi(x)), h : B_\delta(x_0) \to \mathbb{R}$,

$$
\begin{aligned}
0 = \frac{h(x + te_i) - h(x)}{t} &= \frac{f(x + te_i, \xi(x + te_i)) - f(x, \xi(x))}{t} \\
&= \frac{f(x + te_i, \xi(x + te_i)) - f(x, \xi(x + te_i))}{t} + \frac{f(x, \xi(x + te_i)) - f(x, \xi(x))}{t} \\
&= \frac{\partial f}{\partial x_i}(x + ce_i, \xi(x + te_i)) + \frac{\partial f}{\partial y}(x, d)\frac{\xi(x + te_i) - \xi(x)}{t},
\end{aligned}
$$

(19)

such that $0 < c < t$ and $\xi(x) < d < \xi(x + te_i)$. Since ξ is continuous, taking the limit $t \to 0$ in Eq. (19), we get

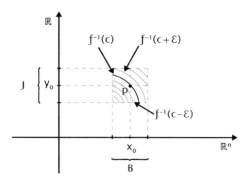

Fig. 9 $V = B \times J$

$$\frac{\partial \xi}{\partial x_i}(x) = \lim_{t \to 0} \frac{\xi(x + te_i) - \xi(x)}{t} = -\frac{\frac{\partial f}{\partial x_i}(x, \xi(x))}{\frac{\partial f}{\partial y}(x, \xi(x))}.$$

Therefore the identity (18) is verified. The continuity of ξ is a consequence of the following argument: given a compact set $K \subset J$, let $\{x_k\}_{k \in \mathbb{N}} \subset B$ be a sequence converging to $\bar{x}$, such that $\xi(x_k) \in K$ for all $k \in \mathbb{N}$. If the sequence $\{y_k = \xi(x_k)\}_{k \in \mathbb{N}}$ converges to $\alpha \in K$, then $f(\bar{x}, \alpha) = \lim_k f(x_k, y_k) = f(\bar{x}, \xi(\bar{x})) = c$. Since ξ is injective, it follows that $\xi(\bar{x}) = \alpha$, i.e, $\alpha = \lim y_k$. Hence ξ is continuous. □

In the above statement, we highlighted the importance of the hypothesis $\frac{\partial f}{\partial y}(x_0, y_0) \neq 0$ for the existence of a differentiable bijection between $B_\delta(x_0)$ and the image of the application $x \to (x, \xi(x))$, in which case the inverse is obviously differentiable. As a by-product, the set $f^{-1}(c)$ is locally a graph. This is an important remark when generalizing the Implicit Function Theorem for differentiable maps (Fig. 9).

Definition 11 Let $U \subset \mathbb{R}^n$ be an open subset and $f \in C^k(U)$, $k \geq 1$.
(i) $c \in \mathbb{R}$ is a regular value of f if there are no critical points in $f^{-1}(c)$; otherwise c is a critical value of f.
(ii) A subset $M \subset \mathbb{R}^{n+1}$ is a hypersurface of class C^k if it is locally a graph of a function, i.e., for all $p \in M$, we have an open neighborhood $V(p) \subset \mathbb{R}^{n+1}$ of p, an open set $U \subset \mathbb{R}^n$, and a C^k-function $\xi : U \to \mathbb{R}$, such that $V(p) \cap M = \{(x, \xi(x)) \mid x \in U\}$.

If c is a regular value of f, the ImFT implies that $M(c) = f^{-1}(c)$ is a hypersurface. A function $f : U \to \mathbb{R}$ is constant when restricted to M. Given a smooth curve $\gamma : (-\epsilon, \epsilon) \to M(c)$, we have $h'(t) = df_{\gamma(t)}.\gamma'(t) = <\nabla f(\gamma(t)), \gamma'(t)> = 0$. So $\gamma'(t)$ is orthogonal to $\nabla f(\gamma(t))$ for all $t \in (-\epsilon, \epsilon)$. Consequently, the tangent plane to $M(c)$ at p is

$$T_p M(c) = \{v \in \mathbb{R}^{n+1} \mid \exists \gamma : (-\epsilon, \epsilon) \to M, \ \gamma(0) = p, \ \gamma'(0) = v\} = \text{Ker}(df_p).$$

Therefore $T_p M(c) = (\nabla f(p))^\perp$ is a subspace of $\mathbb{R}^{n+1}$ of dimension n. M^c is also called an n-submanifold of $\mathbb{R}^{n+1}$, or equivalently, a hypersurface of $\mathbb{R}^{n+1}$. Since $M(c)$ is locally defined as the graph of $\xi : U \to \mathbb{R}$, let $V \subset M$ be an open set such that $M \cap V = \{(x, \xi(x)) \mid x \in U\}$. A curve in $M(c)$ is defined as follows: let $\gamma : (-\epsilon, \epsilon) \to M(c)$, $\gamma(0) = p$ and $\gamma'(t) = v$, and therefore $\gamma(t) = (x_1, \dots, x_n(t), \xi(x(t)))$ and

$$\gamma'(t) = \left(x_1'(t), \dots, x_n'(t), d\xi_{x(t)}.x'(t) \right) = \left(x_1'(t), \dots, x_n'(t), \sum_i \frac{\partial \xi}{\partial x_i}(x(t)).x_i'(t) \right)$$

$$= \sum_{i=1}^n x_i'(t)[e_i + \frac{\partial \xi}{\partial x_i}(x(t))e_{n+1}].$$

Defining the linearly independent vectors $\{v_1, \dots, v_n\}$, $v_i = e_i + \frac{\partial \xi}{\partial x_i}(x(0))e_{n+1}$, we get at $t = 0$ $v = \gamma'(0) = \sum_{i=1}^n x_i'(0)v_i$. Conversely, we claim that every linear combination $v = \sum_i \alpha_i v_i$ is the tangent vector $\gamma'(0)$ of some curve $\gamma : (-\epsilon, \epsilon) \to M(c)$ as shown by the following argument: let $\xi : B_\epsilon \to \mathbb{R}$ be the function given by ImFT and let $v = (\alpha_1, \dots, \alpha_n) \in \mathbb{R}^n$. Choose $\epsilon > 0$ so that $p + tv \in B_\epsilon$ and $\gamma(t) = (p + tv, \xi(p + tv))$. Hence $\gamma(0) = (p, \xi(p))$ and $\gamma'(0) = v$.

The following are necessary conditions for a hypersurface $M \subset \mathbb{R}^{n+1}$ to be a level set of the function f:

(i) there must be a non-null orthogonal field (∇f) over M,

(ii) in the neighborhood of any point $(x_0, y_0) \in M$, we have $\epsilon > 0$ and an open set, i.e., $V = B_\delta(x_0) \times (y_0 - \epsilon, y_0 + \epsilon) \subset \mathbb{R}^{n+1}$ with the property that $V \cap M = \{(x, \xi(x)) \mid x \in B_{\delta(x_0)}\}$.

This concept of hypersurface can be extended to an n-submanifold of $\mathbb{R}^{n+k}$, $1 \le k \le n$;

Definition 12 $M^n \subset \mathbb{R}^{n+k}$ is an n-submanifold of $\mathbb{R}^{n+k}$ of class C^l if for any point $p \in M^n$, there are open sets $\tilde{V} \subset \mathbb{R}^{n+k}$ ($p \in \tilde{V}$) and $U \subset \mathbb{R}^n$, such that $\tilde{V} \cap M^n$ is the graph of a map $f : U \to \mathbb{R}^k$, $f(x) = (f_1(x), \dots, f_k(x))$, $f_i \in C^l(U)$; i.e.,

$$V = \tilde{V} \cap M^n = \left\{ (x_1, \dots, x_n, f_1(x_1, \dots, x_n), \dots, f_k(x_1, \dots, x_n)) \right\}.$$

Example 3 We work out some examples;

(1) Let $U \subset \mathbb{R}^{n+1}$, $f : U \to \mathbb{R}$ and $M = \{(x, f(x)) \in \mathbb{R}^{n+1} \mid x \in U\}$. In this case, M is globally a graph, so it is trivially a hypersurface.

(2) $S^n = \{x \in \mathbb{R}^{n+1}; \mid x \mid = 1\}$.

Let $x = (x_1, \dots, x_{n+1}) \in \mathbb{R}^{n+1}$ be the coordinate system in $\mathbb{R}^{n+1}$ and let $S^n = \cup_i V_i^\pm$, $V_i^+ = \{x \in S^n \mid x_i > 0\}$ and $V_i^- = \{x \in S^n \mid x_i < 0\}$, $1 \le i \le n$, are the hemispheres. Let $D_i^n = \{x = (x_1, \dots, \hat{x}_i, \dots, x_{n+1}); \mid x \mid < 1\}$ be the unit open ball contained in $\mathbb{R}^n = \mathbb{R}^{i-1} \times \{0\} \times \mathbb{R}^{n+1-i} \subset \mathbb{R}^{n+1}$, and define the functions $f_i^\pm : D_i^n \to \mathbb{R}$, $f_i^\pm(x) = \pm\sqrt{1 - \sum_{l \ne i} x_l^2}$. So $V_i^\pm = \{(x, f_i^\pm(x)) \mid x \in D_i^n\}$. Therefore S^n is a hypersurface in $\mathbb{R}^{n+1}$.

(3) S^n being a hypersurface of $\mathbb{R}^{n+1}$ is an easy consequence of the ImFT. Considering the function $f : \mathbb{R}^n \to \mathbb{R}$, $f(x_1, \ldots, x_n) = \sum_i x_i^2$. Since any $a \in (0, \infty)$ is a regular value of f, it follows that $S^{n-1}(a) = f^{-1}(a)$ is a hypersurface. Let $p \in S^n$ and let $\vec{op}$ be the vector defined by p. The identity $df_p.v = 2 < \vec{op}, v \geq 0$ yields that the tangent plane at $p \in S^n$ is the orthogonal subspace to the vector $\vec{op}$. Indeed, $T_p S^n = \mathrm{Ker}(df_p)$.

(4) Consider the function $f : \mathbb{R}^3 \to \mathbb{R}$, $f(x, y, z) = -x^2 - y^2 + z^2 - 1$. The level surface $f^{-1}(0)$ is the non-connected 2-manifold of $\mathbb{R}^3$.

(5) Let $A : \mathbb{R}^n \to \mathbb{R}^n$ be a self-adjoint operator and let $f : \mathbb{R}^n \to \mathbb{R}$ be the bilinear form $f_A(x) = \frac{1}{2} < A(x), x >$. To find the gradient $\nabla f_A(x)$ at x, let $v \in \mathbb{R}^n$; therefore

$$d(f_A)_x.v = \lim_{t \to 0} \frac{f_A(x + tv) - f_A(x)}{t} = \frac{1}{2}[< A(x), v > + < A(v), x >] = < A(x), v > .$$

Therefore $\nabla f_A(x) = A(x)$ and $\mathrm{Cr}(f_A) = \mathrm{Ker}(A)$. So any $c \neq 0$ belonging to the image of f_A is a regular value of f_A. Hence $M_A^c = f_A^{-1}$ is a hypersurface. Indeed, M_A^c is a quadric in $\mathbb{R}^n$. The example in the last item is a quadric defined by the matrix

$$A = \begin{pmatrix} -1 & 0 & 0 \\ 0 & -1 & 0 \\ 0 & 0 & 1 \end{pmatrix}.$$

(6) The surface M in Fig. 10 is parametrized by rotating the curve $y = z^2 - 4$ around the z-axis. It is not a hypersurface of $\mathbb{R}^3$ since there are points with a local neighborhood that is not a local graph.

$$\begin{pmatrix} x \\ y \\ z \end{pmatrix} = \begin{pmatrix} \cos(v) & -\sin(v) & 0 \\ \sin(v) & \cos(v) & 0 \\ 0 & 0 & 1 \end{pmatrix} \begin{pmatrix} u^2 - 4 \\ 0 \\ u \end{pmatrix}.$$

(7) By the same argument in the last example, Whitney's umbrella illustrated in Fig. 11, with its parametrization given by $f(x, y) = (xy, x, y^2)$ is not a 2-submanifold of $\mathbb{R}^3$.

(8) Consider the functions $f(x, y, z) = xy - z$ and $g(x, y, z) = z$ and define the intersection of sets $M = (f)^{-1}(0) \cap (g)^{-1}(0)$. The gradients $\nabla f(x, y, z) = (y, x, -1)$ and $\nabla g(x, y, z) = (0, 0, 1)$ are non-zero, so 0 is a regular value of each of the functions. However, the vectors $\nabla f(0, 0, 0)$ and $\nabla g(0, 0, 0)$ are linearly dependent at $(0, 0, 0)$. The set $M = \{(x, y, 0) \in \mathbb{R}^3 \mid xy = 0\}$ is not a 1-submanifold of $\mathbb{R}^3$ since it is not a graph of a function at $(0, 0, 0)$. At $(0, 0, 0)$, M has no tangent plane.

(9) $SL_n(\mathbb{R})$.

The matrix set $M_n(\mathbb{R})$ is a vector space isomorphic to $\mathbb{R}^{n^2}$. The determinant $\det : M_n(\mathbb{R}) \to \mathbb{R}$ is a C^∞ function. Indeed, it is a homogeneous polynomial of degree n with unknowns the matrix entries. The subset of special matrices $SL_n(\mathbb{R}) = \{A \in M_n(\mathbb{R}) \mid \det(A) = 1\}$ is a subgroup of the group $GL_n(\mathbb{R})$ of invertible matrices. Considering $x = (x_1, \ldots, x_n)$, such that $x_i = (x_{i1}, \ldots, x_{in})$ is a column vector, the determinant is an alternating n-linear function satisfying the following properties:

(a) $\det(I) = 1$, $(I = (e_1, \ldots, e_n))$.

(b) (alternating) For all pairs (i, j), $1 \le i, j \le n$,

$$\det(x_1, \ldots, x_i, \ldots, x_j, \ldots, x_n) = -\det(x_1, \ldots, x_j, \ldots, x_i, \ldots, x_n);$$

(c) (n-linear) For all $1 \le i \le n$ and $a, b \in \mathbb{R}$,

$$\det(x_1, \ldots, ax_i + by_i, \ldots, x_n) = a \det(x_1, \ldots, x_i, \ldots, x_n) + b \det(x_1, \ldots, y_i, \ldots, x_n).$$

Let's compute to the case $n = 3$;

$$d(\det)_A.V = \lim_{t \to 0} \frac{\det(A + tV) - \det(A)}{t} = \det(A). \lim_{t \to 0} \frac{\det(I + tA^{-1}V) - 1}{t}.$$

Since A is invertible, we can define $B = A^{-1}V = (b_1, \ldots, b_n)$;

$\det(I + tB) = \det(e_1 + tb_1, e_2 + tb_2, e_3 + tb_3)$

$= \det(e_1, e_2 + tb_2, e_3 + tb_3) + t \det(b_1, e_2 + tb_2, e_3 + tb_3)$

$= \det(e_1, e_2, e_3 + tb_3) + t \det(e_1, b_2, e_3 + tb_3) + t \det(b_1, e_2, e_3 + tb_3)$

$+ t^2 \det(b_1, b_2, e_3 + tb_3) = \det(e_1, e_2, e_3) + t[\det(e_1, e_2, b_3) + \det(e_1, b_2, e_3) + \det(b_1, e_2, e_3)]$

$+ t^2[\det(b_1, e_2, b_3) + \det(e_1, b_2, b_3) + \det(b_1, b_2, e_3)] + t^3 \det(b_1, b_2, b_3)].$

Taking the limit $t \to 0$, we get

$$\lim_{t \to 0} \frac{\det(I + tB) - 1}{t} = b_{11} + b_{22} + b_{33},$$

that is,

$$d(\det)_A.V = \det(A).\mathrm{Tr}(A^{-1}.V).$$

Therefore 1 is a regular value of $\det$, since $\det(A) = 1$ and $d(\det)_A$ is surjective. Consequently, $SL_3(\mathbb{R}) = \det^{-1}(1)$ is a hypersurface of $M_3(\mathbb{R})$. The tangent plane at $A \in SL_3(\mathbb{R})$ is $T_A SL_3(\mathbb{R}) = \{V \in M_3(\mathbb{R}) \mid \mathrm{Tr}(A^{-1}V) = 0\}$. At the point $A = I$, the tangent plane $T_I SL_3(\mathbb{R}) = \{V \in M_3(\mathbb{R}) \mid \mathrm{Tr}(V) = 0\}$ is the subspace $\mathrm{End}_0(\mathbb{R}^3)$ of traceless matrices in $M(\mathbb{R}^3)$. The inner product $< A, B >= \frac{1}{3}\mathrm{Tr}(AB^t)$ defined on $M_3(\mathbb{R})$ induces the orthogonal decomposition $M_3(\mathbb{R}) =< I > \oplus \, \mathrm{End}_0(\mathbb{R}^3)$.

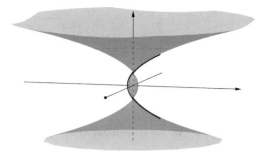

Fig. 10 Parametrized surface

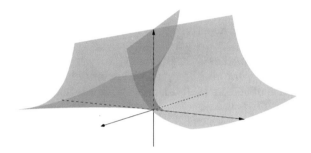

Fig. 11 Whitney's umbrella

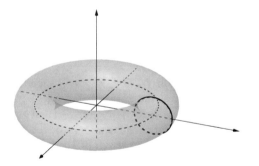

Fig. 12 Torus T^2

(10) The torus T^2 is the surface generated by moving the circumference $(y - a)^2 + z^2 = r^2$ along the circumference $x^2 + y^2 = R^2$, as illustrated in Fig. 12. Now T^2 can be defined as the level set $f^{-1}(0)$ of the function $f : \mathbb{R}^3 \to \mathbb{R}$,

$$f(x, y, z) = (\sqrt{x^2 + y^2} - a)^2 + z^2.$$

Since 0 is a regular value of f, T^2 is a 2-submanifold of $\mathbb{R}^3$.

Exercises

(1) Let $A \in GL_2(\mathbb{R})$ be a positive self-adjoint matrix and $f_A : \mathbb{R}^2 \to \mathbb{R}$, $f_A(x) =< A(x), x >$. Find the set of the regular values of f_A and describe the level curves of f_A.

(2) Suppose $A \in GL_3(\mathbb{R})$ is a self-adjoint matrix. Find the level surfaces of f_A : $\mathbb{R}^3 \to \mathbb{R}$, $f_A(x) =< A(x), x >$.

(3) Find the critical points of the function $f_A : \mathbb{R}^n \to \mathbb{R}$, $f_A(x) = \frac{<A(x),x>}{|x|^2}$.

(4) Prove that the solution set of the non-linear system below is a 1-submanifold of $\mathbb{R}^3$.

$$xy - z = 0$$
$$\ln(xy) + z^2 = 1.$$

(5) Prove that $SL_n(\mathbb{R})$ is a hypersurface of R^{n^2}. Is the level set $(\det)^{-1}(0)$ a hypersurface?

(6) (Non-Linear System) Let $U \subset \mathbb{R}^n$ and $f_1, \ldots, f_k \in C^l(U)$. Assume 0 is a regular value of f_i for all $1 \le i \le k$, and define $M_i = f_i^{-1}(0)$. Prove that if the set of gradient vectors $\{\nabla f_1(x), \ldots, \nabla f_k(x)\}$ is linearly independent, then $M = M_1 \cap \ldots \cap M_k$ is a $(n - k)$-submanifold of $\mathbb{R}^n$ (suppose $M \neq \varnothing$).

(7) Let $U \subset \mathbb{R}^n$ be an open set and $f_1, \ldots, f_k$ functions in $C^1(U)$. Find a condition such that the solution set S of the non-linear system below is a submanifold of $\mathbb{R}^n$ and find $\dim(S)$ (Fig. 15).

$$\begin{cases} f_1(x0 = 0, \\ \vdots \\ f_k(x) = 0. \end{cases}$$

(8) Prove that there is no function $f : U \subset \mathbb{R}^2 \to \mathbb{R}$ such that the Möebius band $\mathbb{M}$ in Fig. 13 can be defined as the level set of f.

(9) Let $f, g : \mathbb{R}^3 \to \mathbb{R}$, $f(x, y, z) = x^2 + y^2 + z^2 - 1$ and $g(x, y, z) = x^2 + (y - a)^2 - a^2$. For which values of the parameter a is the set $f^{-1}(0) \cap g^{-1}(0)$ a submanifold of $\mathbb{R}^3$? Fig. 14 shows the cases $a = 1$ and $a = \frac{1}{2}$).

(10) For which real values a is the intersection of the circumferences $x^2 + y^2 = 1$ and $(x - 2)^2 + y^2 = a^2$ not transversal?

(11) Let $M \subset \mathbb{R}^n$ be a compact, orientable hypersurface. Prove that there is an open subset $V \subset \mathbb{R}^n$ and a function $f : V \to \mathbb{R}$ such that $M = f^{-1}(0)$. (hint: the orientability allows us to assume the existence of vector field X, normal to M, such that $X(p) \neq 0$ for all $p \in M$. Compactness guarantees that a tubular neighborhood of M exists).

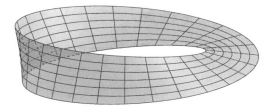

Fig. 13 Möebius band

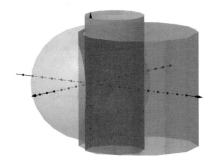

Fig. 14 Intersection of surfaces

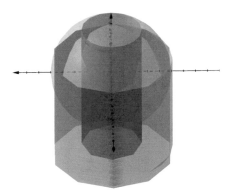

Fig. 15 Intersection of surfaces

5 Lagrange Multipliers

In this section the main issue is to optimize a function $f \in C^k(U), k \geq 1$, restricted to the hypersurface $M = \phi^{-1}(c) \subset U$, in which c is a regular value of the differentiable function $\phi : U \to \mathbb{R}$. An example is to determine the shortest distance from a point $p = (x_0, y_0, z_0)$ to the plane $\pi = \{(x, y, z) \in \mathbb{R}^3 \mid ax + by + cz + d = 0\}$. This is equivalent to finding the minimum value of the function $f : \mathbb{R}^3 \to \mathbb{R}$,

$$f(x, y, z) = \sqrt{(x - x_0)^2 + (y - y_0)^2 + (z - z_0)^2},$$

restricted to the plane π. Considering the function $\phi : \mathbb{R}^3 \to \mathbb{R}$, $\phi(x, y, z) = ax + by + cz + d$, then $\pi = \phi^{-1}(0)$.

To address the general case of optimizing a smooth function $f : \mathbb{R}^n \to \mathbb{R}$, restricted to a hypersurface $M = \phi^{-1}(0) \subset \mathbb{R}^n$, $\phi : \mathbb{R}^n \to \mathbb{R}$, we stress the following items:

(1) $\nabla\phi(p) \perp T_p M, \forall p \in M$.
(2) If p is a critical point of $f \mid_M$, then $df_p.v = < \nabla f(p), v > = 0$ for all $v \in T_p M$. Hence $\nabla f(p) \perp T_p M$.

Assuming $\nabla f(p) \neq 0$, the above items imply that $\nabla f(p)$ and $\nabla\phi(p)$ are linear dependent vectors, so we have $\lambda \neq 0 \in \mathbb{R}$ such that $\nabla f(p) = \lambda \nabla\phi(p)$. In this way, finding a critical point for $f \mid_M$ corresponds to finding $x \in M$ and $\lambda \neq 0$ satisfying the nonlinear system with $(n + 1)$-variables and $(n + 1)$-equations,

$$\begin{cases} \nabla f(x) - \lambda\nabla\phi(x) = 0, \\ \phi(x) = c. \end{cases} \tag{20}$$

It could happen that $p \in U$ is a critical point of f, so $\nabla f(p) = 0$. Now returning to the initial example, to find the shortest distance from p to π, we need to solve the system

$$\left(\frac{x - x_0}{f}, \frac{y - y_0}{f}, \frac{z - z_0}{f}\right) = \lambda.(a, b, c) \Rightarrow \begin{cases} x - x_0 = \lambda af, \\ y - y_0 = \lambda bf, \\ z - z_0 = \lambda cf, \\ ax + by + cz = d \end{cases} \Rightarrow f^2 = \lambda^2(a^2 + b^2 + c^2)f^2.$$

Since $f(x) \neq 0$ for all x, then $\lambda = \pm\frac{1}{\sqrt{a^2+b^2+c^2}}$. It is possible to solve the equations to find the point $p_m = (x_m, y_m, z_m)$ realizing the shortest distance; however, to find the minimum value of f, we take a short cut:

$$a(x_m - x_0) = \frac{a^2}{\sqrt{a^2 + b^2 + c^2}}f, \; b(y_m - y_0) = \frac{b^2}{\sqrt{a^2 + b^2 + c^2}}f, \; c(z_m - z_0) = \frac{c^2}{\sqrt{a^2 + b^2 + c^2}}f.$$

Using the identity $ax_m + by_m + cz_m + d = 0$, then

$$a(x_m - x_0) + b(y_m - y_0) + c(z_m - z_0) = f(p_m).\sqrt{a^2 + b^2 + c^2} \Rightarrow f(p_m) = \frac{|ax_0 + by_0 + cz_0 + d|}{\sqrt{a^2 + b^2 + c^2}}.$$

Of course, p_m is the absolute minimum since f restricted to the plane π grows as much as we wish by pushing $(x, y, z) \in \pi$ away from p_m.

An interesting interpretation of Eq. (20) is the following: the normal vector to the hypersurface $M = \phi^{-1}(c)$ at $x \in M$ is $\nabla\phi(x)$. By projecting the vector $\nabla f(x)$ over the tangent plane $T_p M = (\nabla\phi)^{\perp}$, we obtain the vector

$$\widetilde{\nabla} f(x) = \nabla f(x) - \frac{< \nabla f(x), \nabla \phi(x) >}{| \nabla \phi(x) |^2} \nabla \phi(x).$$ (21)

The vector $\widetilde{\nabla} f(x)$ is the projection of the gradient vector of $f : M \to \mathbb{R}$ over $T_x M$. If $x_0 \in M$ is a critical point, then $\widetilde{\nabla} f(x_0) = 0$; this is equivalent to Eq. (20).

Exercises

(1) Find the point p_m realizing the shortest distance from $p = (x_0, y_0, z_0)$ to the plane $\pi = \{(x, y, z) \in \mathbb{R}^3 \mid ax + by + cz + d = 0\}$.

(2) Find the shortest distance of a point $p = (x_0, y_0, z_0)$ to the sphere of radius R centered at the origin.

(3) Consider the ellipse $E : \frac{x^2}{a^2} + \frac{y^2}{b^2} = 1$ and the line $L : x + y = 1$. Find a necessary condition so that $E \cap L = \varnothing$ and determine the shortest distance between them.

(4) Let $A \in M_n(\mathbb{R})$ be a positive definite symmetric matrix and let $u_0 \neq 0$ in $\mathbb{R}^n$. Considering in $\mathbb{R}^n$ the ellipsoid $E = \{x \in \mathbb{R}^n \mid < Ax, x >= 1\}$ and the hyperplane defined by $L = \{x \in \mathbb{R}^n \mid < u_0, x >= 1\}$, prove the following:
(i) $E \cap L = \varnothing$ if and only if $< A^{-1}x, x > < 1$.
(ii) Find the distance

$$d(E, L) = \inf_{y \in E,\ x \in L} \| y - x \| .$$

(5) Let $Q_1 = \{(x_1, \ldots, x_n) \mid x_i > 0, \ \forall i = 1, \ldots, n\}$. Find the maximum value of

$$f(x_1, \ldots, x_n) = x_1 x_2 \ldots x_n$$

considering $(x_1, \ldots, x_n) \in Q_1$ and $x_1 + x_2 + \ldots + x_n = c$, c constant. Prove that

$$\sqrt[n]{x_1 x_2 \ldots x_n} \le \frac{x_1 + x_2 + \ldots + x_n}{n}.$$

(6) Let $A : \mathbb{R}^n \to \mathbb{R}^n$ be a linear self-adjoint operator and let $f_A : \mathbb{R}^n \to \mathbb{R}$ be the function $f_A(x) =< x, A(x) >$. Prove that the critical point set of $(f_A) |_{S^{n-1}}$ is equal to the set of eigenvectors of A. Prove that if $\sigma(A)$ is the spectrum of A, $\lambda_m = \min \sigma(A)$ and $\lambda_M = \max \sigma(A)$, then

$$\max_{x \in S^{n-1}} f_A(x) = \lambda_M, \ \min_{x \in S^{n-1}} f_A(x) = \lambda_m.$$

(7) Let S be the set defined by the equations $x^2 + y^2 + z^2 = 1$ and $x + y + z = 1$. Find the critical points of the function $f : S \to \mathbb{R}$ given by $f(x, y, z) = x^4 + y^4 + z^4$.

(8) Assume f, ϕ_1 and ϕ_2 are in $C^1(U)$. Assume that 0 is a regular value of ϕ_1 and ϕ_2 and define $M_1 = \phi^{-1}(0)$ and $M_2 = \phi^{-1}(0)$. Prove that if $p \in M = M_1 \cap M_2$ is a critical value of f, then there are constants λ_1, $\lambda_2 \in \mathbb{R}$ such that

$$\begin{cases} \nabla f(p) + \lambda_1 \nabla \phi_1(p) + \lambda_2 \phi_2(p) = 0. \\ \phi_1(p) = 0, \\ \phi_2(p) = 0. \end{cases} \qquad (22)$$

5.1   The Ultraviolet Catastrophe: The Dawn of Quantum Mechanics

The genesis of Quantum Mechanics lies in Max Planck's work on the radiation of a blackbody. Let's use the Lagrange multipliers to obtain the Maxwell-Boltzmann distribution used by Planck in his seminal work. This distribution is fundamental in the theory of gases and in many others topics in Statistical Physics.

The ultraviolet catastrophe was a mathematical prediction with respect to the spectral distribution of blackbody radiation when the temperature T is assumed to be large. The experimental evidences showed that the energy $\mathcal{E}(, \lambda, T)$ per unity volume of the radiation with wavelength between λ and $\lambda + d\lambda$ would have the following behavior as shown in Fig. 16:

(1) the short length cutoff advances toward the origin as the temperatures increase,
(2) raising the temperature increases the energy of all spectral components,
(3) the peak of the curve shifts to a shorter wavelength as the temperature increases.

Indeed, the ultraviolet catastrophe is the divergence of the density of energy

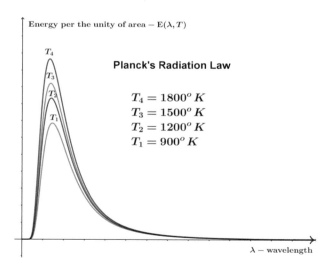

Fig. 16 Plancks' radiation law

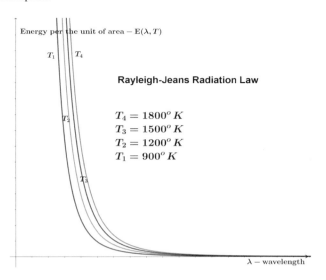

Fig. 17 Rayleigh-Jeans radiation law

$$\int_0^\infty \mathcal{E}(\lambda, T)d\lambda. \tag{23}$$

Considering c as the velocity of light, the energy density $\mathcal{E}(\lambda, T) = \frac{c}{4}I(\lambda, T)$ emitted is the energy radiated per the unit of area at a given wavelength λ and at temperature T. The Rayleigh-Jeans law obtained from the classical theory of equipartition of energy gives us the equation (plotted in Fig. 17)

$$\mathcal{E}_{RJ}(\lambda, T) = \frac{8\pi}{\lambda^4}kT. \tag{24}$$

in which k is the Boltzmann constant. Consequently we have that $E(\lambda, T) \to \infty$ when $\lambda \to 0$. This inconsistency with the experimental reality, a failure of the model, implied that there was something fundamentally wrong with either the equipartition theorem or the theory of electromagnetic radiation. Max Planck was aware of the shortcomings of the Rayleigh-Jeans law, so he proposed a condition that later became one of the axioms of Quantum Mechanics. Planck's model for a radiating body is to regard the body as a collection of $N_0 = 6,022 \times 10^{23}$ (Avogadro constant) of linear oscillators performing harmonic motion, which then radiates electromagnetic waves. Planck considered the blackbody as a set of N_0 harmonic oscillators, the energy levels of each oscillator with wavelength λ were also discretized in energy levels $\{E_n \mid n \in \mathbb{N}\}$, and applied the Maxwell-Boltzmann distribution function $N(n) = n_0 e^{-E_n/kT}$. The fundamental assumption was to consider the density energy irradiated by an oscillator at frequency ν to assume values in the discrete set $h\nu\mathbb{N} = \{E_n = nh\nu \mid n \in \mathbb{N}\}$, in which $\nu = c/\lambda$ and $h = 6.6261 \times 10^{-34}$ m^{2}kg/s is the Planck constant. The number of oscillators per the unit of volume at each wavelength λ, which is known as Jeans' number, is $n(\lambda) = 8\pi/\lambda^4$. Planck was able to obtain the energy density $\mathcal{E}(\lambda, T)$ emitted by the oscillators at wavelength λ, to be (see Fig. 16)

$$\mathcal{E}(\lambda, T) = \frac{8\pi hc}{\lambda^5} \frac{1}{e^{hc/\lambda kT} - 1}.$$

This model made it possible for Planck to eliminate the ultraviolet catastrophe.

Next we apply the Lagrange multipliers to show how the Maxwell-Boltzmann distribution function is obtained.

Given a volume V divided into l cells of volume V_i $(1 \le i \le l)$, N identical particles are to be distributed among the l cells. The number n_i of particles occupying the ith-cell should be proportional to the volume of the cell. Let $g_i = V_i/V$ be the fractional volume. In this way, we have

$$\sum_{i=1}^{l} g_i = 1, \quad \text{and} \quad \sum_{i=1}^{l} n_i = N.$$

The probability of a given distribution with n_1 particles in V_1, n_2 particles in V_2 and so on is

$$W = \frac{N!}{n_1! n_2! \cdots n_K!} (g_1)^{n_1} (g_2)^{n_2} \cdots (g_K)^{n_K}. \tag{25}$$

For example, consider the case of distributing 12 particles over 4 cells such that $n_1 = 1, n_2 = 2, n_3 = 3$ and $n_4 = 6$ is

$$W(1, 2, 3, 6) = \frac{12!}{1! 2! 3! 6!} (g_1)^1 (g_2)^2 (g_3)^3 (g_4)^6.$$

If we assume that the cells are equal with the same volume, then we have $g_i = 1/4$ and $W(1, 2, 3, 6) = 0,33045 \times 10^{-2}$.

We would like to find the most probable distribution, so we must maximize[2] the function $W(n_1, \ldots, n_l)$. Let's assume that the variables n_i, $1 \le i \le l$, take values in $\mathbb{R}$. The method of Lagrange's multipliers is applied as follows: we wish to find the extreme value of $W(n_1, \ldots, n_l)$ subject to the constraint $\phi = N - \sum_{i=1}^{l} n_i = 0$. It is convenient to take the logarithm in Eq. (25), so

$$\ln(W) = \ln(N!) - \sum_{i=1}^{l} \ln(n_i!) + \sum_{i=1}^{l} n_i \ln(g_i).$$

When N is large, we can use Stirling's approximation $\ln(n!) \sim n \ln(n) - n$. So we get

$$\ln(W) = N \ln(N) + \sum_{i=1}^{l} n_i \ln\left(\frac{g_i}{n_i}\right).$$

---

[2]It is easily seem that a critical point must be a upper limit.

To apply the method of Lagrange's multipliers, we must solve Eq. (20), i.e., we must solve the system of equations $\frac{\partial W}{\partial n_i} + \alpha \frac{\partial \phi}{\partial n_i} = 0$, $1 \leq i \leq l$. It follows that $n_i = g_i e^{-(\alpha+1)}$, and so

$$g_1 + \ldots + g_K = (n_1 + \ldots + n_K)e^{\alpha+1} = 1 \implies e^{\alpha+1} = \frac{1}{N}.$$

Consequently we get $n_i = g_i N$. This means that the most probable distribution is the uniform distribution, that is when the number of particles in a cell is proportional to the size of the cell.

Now we consider the problem of distributing N identical particles into l cells with the additional constraint on the energy of the system. The cells are regarded as a discrete set of the energy states E_i, such that the total energy is $E = \sum_{i=1}^{l} E_i$. In this case, we have two constraints given by

$$\phi_1(n_1, \ldots, n_l) = N - \sum_{i}^{l} n_i = 0,$$

$$\phi_2(n_1, \ldots, n_l) = E - \sum_{i}^{l} n_i E_i = 0.$$

In Exercise 8 of this section, the reader was left to prove that due to having two constraints, we have to solve the system of Eq. (22). Then we get

$$\ln\left(\frac{g_i}{n_i}\right) = 1 + \alpha_1 + E_i \alpha_2 \implies n_i = g_i e^{-(\alpha_1+1)} e^{-E_i \alpha_2}.$$

Let's define[3] $Z = \sum_i g_i e^{-E_i \alpha_2}$, so we then have $N = \sum_{i=1}^{K} n_i = e^{-(\alpha_1+1)} Z$. Therefore the number n_i of particles in the ith cell is $n_i = \frac{N g_i e^{-E_i \alpha_2}}{Z}$.

The Maxwell-Boltzmann distribution is obtained when we consider that $\lambda_2 = \frac{1}{kT}$ and $N = N_0$ is the Avogadro constant, then

$$n_i = \frac{N_0 g_i}{Z} e^{-E_i/kT}.$$

Assuming that $g_i = g_0$ for all $1 \leq i \leq l$, and defining $n_0 = \frac{N_0 g_i}{Z}$, we have

$$n_i = n_0 e^{-E_i/kT}.$$

Therefore if the energy E_i increases, then the number n_i of particles placed in the ith cell becomes small. So the higher energy levels are less likely to contain particles. It is completely different from the classical situation when $n_i = g_i N$.

---

[3]Z is the Partition Function.

Now we return to using oscillators instead of cells. Planck's hypothesis was that the energy emitted by an oscillator at frequency ν can only take value in the discrete set $h\nu\mathbb{N} = \{h\nu, 2h\nu, 3h\nu, \ldots, nh\nu, \ldots\}$, and not in a continuous set of values. As show next, this has a profound effect on the average energy calculated using the Maxwell-Boltzmann distribution. The probability distribution in a system that is in equilibrium at temperature T, and partitioned into equal oscillators, is

$$P(E_i) = \frac{n_i}{N} = \frac{e^{-E_i/kT}}{\sum_i e^{-E_i/kT}}.$$

Here the fact that the ith energy level at frequency ν is given by $E_i(\nu) = ih\nu$, $i \in \mathbb{N}$, comes into play. The number of oscillators is large, so we can assume $l \to \infty$. In this way, the average energy $\overline{E}$ per oscillator with frequency ν (wavelength λ) is

$$\overline{E}(\nu, T) = \sum_{i=1}^{\infty} E_i(\nu) P(E_i(\nu)) = \frac{\sum_{i=1}^{\infty} n_0 e^{-ih\nu/kT} ih\nu}{\sum_{i=1}^{\infty} n_0 e^{-ih\nu/kT}} = \frac{h\nu}{e^{h\nu/kT} - 1}.$$

Therefore Planck's radiation law describing the energy density per wavelength λ then reads as

$$\mathcal{E}(\lambda, T) = n(\lambda)\overline{E}(\lambda, T) = \frac{8\pi hc}{\lambda^5} \frac{1}{(e^{hc/\lambda kT} - 1)}. \tag{26}$$

Equation (26) obtained by Planck explained the experimental facts about black-body radiation to overcome the crises due to ultraviolet catastrophe. The assumption that the energy emitted by a tiny oscillator, as small as an atom, assumes discrete values that are multiples of $E = h\nu$ called *quantas* was a fundamental breakthrough to start understanding Physical systems in the atomic scale. Later the *quanta* hypothesis became an axiom of Quantum Theory.

Ultraviolet catastrophe arose because in the classical model, it was considered that all energy levels were equally likely to be occupied by particles.

Exercises

(1) Prove Eq. (26).
(2) Show that when $\lambda \to \infty$, the behavior of the energy density obtained by Planck in (26) is similar to the classical energy density (24).
(3) Assuming that the energy takes continuous values, show that

$$\overline{E}(\lambda) = \frac{\int_0^{\infty} E e^{-E/kT} dE}{\int_0^{\infty} e^{-E/kT} dE} = kT.$$

and conclude that $\mathcal{E}(\lambda, T) \propto \frac{kT}{\lambda^4}$.

6 Differentiable Maps I

Let $U \subset \mathbb{R}^n$ and $V \subset \mathbb{R}^m$ be open subsets. This section is the first step in studying the differentiability of maps $f : U \rightarrow V$. The basic concepts and the Inverse Function Theorem are introduced and some applications are worked out. The proof is postponed until Chap. 3.

6.1 Basics Concepts

Using the usual orthogonal systems, a point $x \in U$ is represented as $x = (x_1, \ldots, x_n) \in \mathbb{R}^n$ and a point $y \in V$ is represented as $y = (y_1, \ldots, y_m) \in \mathbb{R}^m$.

Definition 13 A map $f : U \rightarrow V$ is differentiable at $p \in U$ if we have a linear transformation $df_p : T_p U \rightarrow T_{f(p)} V$ such that

$$f(p + v) - f(p) = df_p.v + r(v),$$

for all $v \in T_p U$, and the map $r : \mathbb{R}^n \rightarrow \mathbb{R}^m$ satisfies the condition $\lim_{v \to 0} \frac{r(v)}{|v|} = 0$. f is differentiable in U if it is differentiable for all $p \in U$.

In the local coordinates of U and V, a map $f : U \rightarrow V$ is given by

$$f(x_1, \ldots, x_n) = \big(f_1(x_1, \ldots, x_n), \ldots, f_m(x_1, \ldots, x_n)\big).$$

Now assume that for every component $f_i \in C^1(U)$, there are m functions r_i, $1 \le i \le m$, satisfying $\lim_{v \to 0} \frac{r_i(v)}{|v|} = 0$. Defining $r(v) = (r_1(v), \ldots, r_m(v))$, it follows that $\lim_{v \to 0} \frac{r(v)}{|v|} = 0$. Then at every $p \in U$,

$$f(p + v) - f(x) = \big(f_1(p + v) - f_1(p), \ldots, f_m(p + v) - f_m(p)\big)$$
$$= \big(d(f_1)_p.v + r_1(v), \ldots, d(f_m)_p.v + r_m(v)\big) = \big(d(f_1)_p.v, \ldots, d(f_m)_p.v\big) + r(v)$$
$$= \big(< \nabla f_1(p), v >, \ldots, < \nabla f_m(p), v > \big) + r(v).$$

Taking $v = (v_1, \ldots, v_n)$ in the local coordinate systems defined on U, df_p is the linear map $df_p : \mathbb{R}^n \rightarrow \mathbb{R}^m$ given by

$$df_p.v = \begin{pmatrix} \frac{\partial f_1}{\partial x_1} & \frac{\partial f_1}{\partial x_2} & \cdots & \frac{\partial f_1}{\partial x_n} \\ \frac{\partial f_2}{\partial x_1} & \frac{\partial f_2}{\partial x_2} & \cdots & \frac{\partial f_2}{\partial x_n} \\ \vdots & \vdots & \cdots & \vdots \\ \frac{\partial f_m}{\partial x_1} & \frac{\partial f_m}{\partial x_2} & \cdots & \frac{\partial f_m}{\partial x_n} \end{pmatrix} \cdot \begin{pmatrix} v_1 \\ v_2 \\ \vdots \\ v_n \end{pmatrix}.$$

The column vectors $\{\frac{\partial f}{\partial x_1}, \ldots, \frac{\partial f}{\partial x_n}\}$ generate the image of df_p in $\mathbb{R}^m$; let $v = \sum_i v_k e_i \in \mathbb{R}^n$, so $df_p.v = \sum_{i=1}^m v_i \frac{\partial f}{\partial x_i}(x)$. The differential of f at p is represented by the matrix $df_p \in \mathcal{L}(\mathbb{R}^n, \mathbb{R}^m)$; it is also called the Jacobian matrix of f.

Let $C^k(U, V)$ be the set of differentiable maps $f : U \to V$ of class C^k. The next result is straightforward based on a previous discussion.

Theorem 7 *Let $k \geq 1$. If $f_i \in C^k(U)$ for all $i \in \{1, \ldots, m\}$, then the map $f : U \to V$, $f = (f_1, \ldots, f_m)$ belongs to $C^k(U, V)$.*

Recalling that $\mathcal{L}(\mathbb{R}^n, \mathbb{R}^m) = \mathbb{R}^{nm}$, the condition $f \in C^1(U, V)$ is equivalent to the continuity of the map $df : U \to \mathcal{L}(\mathbb{R}^n, \mathbb{R}^m) = \mathbb{R}^{nm}$, $x \to df_x$.

The derivative of a map satisfies the following properties: let $x \in U$, $f, g \in C^1(U, V)$, $t \in C^1(U)$ and $a, b \in \mathbb{R}$;

(1) $d(af + bg)_x = adf_x + bdg_x$ ($\mathbb{R}$-linear),
(2) $d(t.f)_x = dt_x.f(x) + t(x).df_x$ (Leibniz's rule).

Definition 14 A map $f : U \to V$ is a diffeomorphism if $f : U \to f(U)$ is bijective, differentiable, and the inverse map $f^{-1} : f(U) \to U$ is also differentiable. $f : U \to V$ is a local diffeomorphism when it is the diffeomorphism $U_p \to V_p$, restricted to neighborhoods $U_p \subset U$ and $V_{f(p)} \subset V$ of p and $f(p)$, respectively for all $p \in U$.

Several references define the Jacobian of f at p as the determinant $\det(df_p)$ and it is denoted by $\det(df_p) = \frac{\partial(f_1, \ldots, f_n)}{\partial(x_1, \ldots, x_n)}(p)$. If f is a diffeomorphism, the linear map $df_p : T_p U \to T_{f(p)} V$ is an isomorphism for all $p \in U$, since the inverse $df_p^{-1} = [df_p]^{-1}$ is well-defined.

Example 4 The next examples reflect the importance of coordinate systems;

(1) The function $f : (0, \infty) \to \mathbb{R}$, $f(x) = \ln(x)$, is bijective and differentiable in $(0, \infty)$. The inverse $f^{-1} : \mathbb{R} \to (0, \infty)$, $(f^{-1})(x) = e^x$ is also differentiable. Therefore f is a diffeomorphism.
(2) The function $f : \mathbb{R} \to \mathbb{R}$, $f(x) = x^3$ is bijective and differentiable; its derivative is $f'(x) = 3x^2$. The inverse $f^{-1} : \mathbb{R} \to \mathbb{R}$, $(f^{-1})(x) = \frac{1}{3\sqrt[3]{x^2}}$ is non-differentiable at $x = 0$, so f is not a diffeomorphism. Let $y_0 = f(x_0)$; if $f'(x_0) = 0$, then $(f^{-1})'(y_0) = \frac{1}{f'(f^{-1}(y_0))}$ is not defined at x_0.
(3) Let $f : [0, 2\pi) \to S^1$, $f(t) = (\cos(2\pi t), \sin(2\pi t))$. Then $f'(t) \neq 0$ for all $t \in [0, 2\pi)$. The function f cannot be a diffeomorphism since S^1 is compact and $[0, 2\pi)$ is not. Both functions f and f^{-1} are well-defined, however, f^{-1} is not continuous.
(4) The polar coordinate system in $\mathbb{R}^2$ defines a map $P : \mathbb{R}^2 \to \mathbb{R}^2$,

$$P(r, \theta) = (r \cos(\theta), r \sin(\theta)) \tag{27}$$

with a derivative at (r, θ) that is

$$dP_{(r,\theta)} = \begin{pmatrix} \cos(\theta) & -r\sin(\theta) \\ \sin(\theta) & r\cos(\theta) \end{pmatrix}.$$

By restricting the map P to $\overline{P} : (0, \infty) \times (0, 2\pi) \to \mathbb{R}^2 \setminus \{(x, 0) \mid x \geq 0\}$, it becomes a diffeomorphism when we notice that $\overline{P}$ is a bijection and $\det(dP_{(r,\theta)}) = r \neq 0$.

(5) The spherical coordinate system in $\mathbb{R}^3$ defines the map $F : \mathbb{R}^3 \to \mathbb{R}^3$,

$$F(\rho, \theta, \psi) = (\rho.\cos(\theta)\sin(\psi), \rho.\sin(\theta)\sin(\psi), \rho.\cos(\psi)), \tag{28}$$

with derivative at (ρ, θ, ψ) given by the linear map

$$d(F)_{(\rho,\theta,\psi)} = \begin{pmatrix} \cos(\theta)\sin(\psi) & -\rho\sin(\theta)\sin(\psi) & \rho\cos(\theta)\cos(\psi) \\ \sin(\theta)\sin(\psi) & \rho\cos(\theta)\sin(\psi) & \rho\sin(\theta)\cos(\psi) \\ \cos(\psi) & 0 & -\rho\sin(\psi) \end{pmatrix}.$$

So, $\det(d(F)_{(\rho,\theta,\psi)}) = -\rho^2\sin(\psi)$. By restricting F to the domain $U = (0, \infty) \times (0, 2\pi) \times (0, \pi)$, we get a diffeomorphism $F : U \to F(U)$.

(6) The map $f : \mathbb{R}^n \to B = \{x \in \mathbb{R}^n; \mid x \mid < 1\}$, $f(x) = \dfrac{x}{\sqrt{1+|x|^2}}$ is a diffeomorphism. The inverse is $f^{-1}(y) = \dfrac{y}{\sqrt{1-|y|^2}}$ and both are C^1.

(7) (Cauchy-Riemann Equations) Let $\mathbb{C} = \{z = x + iy \mid x, y \in \mathbb{R}\}$ and let $f : \mathbb{C} \to \mathbb{C}$ be a function of a one complex variable. Assume that $f(z) = u(z) + iv(z)$ and $u, v : \mathbb{C} \to \mathbb{R}$. Identifying $\mathbb{C}$ with $\mathbb{R}^2$, and using the map $\mathcal{I} : \mathbb{C} \to \mathbb{R}^2$, $\mathcal{I}(x + iy) = (x, y)$, the function f induces the map $f_{\mathbb{R}}(x, y) = (u(x, y), v(x, y))$ and the following diagram commutes:

$$\begin{array}{ccc} \mathbb{C} & \xrightarrow{f} & \mathbb{C} \\ \downarrow{\scriptstyle\mathcal{I}} & & \downarrow{\scriptstyle\mathcal{I}} \\ \mathbb{R}^2 & \xrightarrow{f_{\mathbb{R}}} & \mathbb{R}^2. \end{array} \tag{29}$$

The function f is holomorphic (differentiable with respect to z) if we have functions $f'(z) : \mathbb{C} \to \mathbb{C}$ and $r : \mathbb{C} \to \mathbb{C}$ such that for all $v \in \mathbb{C}$,

$$f(z + v) - f(z) = f'(z).v + r(v),$$

and $\lim_{v \to} \frac{r(v)}{v} = 0$. Consider $f'(z) = \alpha + i\beta$, $v = v_1 + iv_2$ and $r(v) = r_1(v) + ir_2(v)$. The equation above implies that

$$f(z + v) - f(z) = [(\alpha v_1 - \beta v_2)] + i(\alpha v_2 + \beta v_1)] + r_1(z) + ir_2(z).$$

So,

$$u((x, y) + (v_1, v_2)) - u(x, y) = (\alpha v_1 - \beta v_2) + r_1(v) = < (\alpha, -\beta), (v_1, v_2) > + r_1(v)$$
$$v((x, y) + (v_1, v_2)) - v(x, y) = (\alpha v_2 + \beta v_1) + r_2(v) = < (\beta, \alpha), (v_1, v_2) > + r_2(v).$$

Assuming $\lim_{v \to 0} \frac{r_1}{|v|} = \lim_{v \to 0} \frac{r_2}{|v|} = 0$, it follows that the u, v are differentiable and the gradients are $\nabla u = (\alpha, -\beta)$ and $\nabla v = (\beta, \alpha)$. Consequently, f is holomorphic if and only if u and v satisfy the Cauchy-Riemann equations

$$\frac{\partial u}{\partial x} = \frac{\partial v}{\partial y}, \quad \frac{\partial u}{\partial y} = -\frac{\partial v}{\partial x}. \tag{30}$$

The Jacobian of $f_{\mathbb{R}} : \mathbb{R}^2 \to \mathbb{R}^2$ is

$$df_{\mathbb{R}} = \begin{pmatrix} \frac{\partial u}{\partial x} & -\frac{\partial u}{\partial x} \\ \frac{\partial u}{\partial x} & \frac{\partial u}{\partial x} \end{pmatrix} \quad \Rightarrow \quad \det(df_{\mathbb{R}}) = \left(\frac{\partial u}{\partial x} \right)^2 + \left(\frac{\partial u}{\partial y} \right)^2.$$

Therefore f is holomorphic and this implies that $f_{\mathbb{R}}$ is a conformal map (preserve angles), since

$$< df_{\mathbb{R}}.u, df_{\mathbb{R}}.v > = \det(df_{\mathbb{R}}). < u, v > .$$

To finish the argument, $\lim_{v \to} \frac{r(v)}{v} = 0$ implies that $\lim_{v \to 0} \frac{r_1}{|v|} = \lim_{v \to 0} \frac{r_2}{|v|} = 0$.

Exercises

Prove that the following maps are C^1-maps and find their derivatives;

(1) $T : \mathbb{R}^n \to \mathbb{R}^m$, $T(x) = A.x$, and $A \in M(m \times n, \mathbb{R})$.
(2) $S : \mathbb{R}^n \times \mathbb{R}^n \to \mathbb{R}^n$, $S(x, y) = x + y$.
(3) $B : \mathbb{R}^n \times \mathbb{R}^m \to \mathbb{R}^p$ a bilinear map.
(4) $f : \mathbb{R}^n \to B = \{x \in \mathbb{R}^n; \mid x \mid < 1\}$, given by $f(x) = \frac{x}{\sqrt{1+|x|^2}}$.
(5) Let $f, g \in C^1(U, \mathbb{R}^n)$ and consider the function $< f, g >: U \times U \to \mathbb{R}$, given by $< f, g > (x) = < f(x), g(x) >$. Prove that

$$d(< f, g >)_p.v = < df_p.v, g(p) > + < f(p), dg_p.v >, \quad v \in T_pU \tag{31}$$

Show that if $\mid f \mid$ is constant, then $f(p) \perp df_p.v$ for all $p \in U$ and $v \in T_pU$.

6.2 Coordinate Systems

Coordinate systems are often used to simplify different kinds of problems. For example, using spherical coordinates instead of Cartesian coordinates simplifies problems with spherical symmetry, and similarly for cylindrical coordinates when there is cylindrical symmetry.

To change a coordinate system, it is necessary to be able to reverse the transformation, so it must be a diffeomorphism. For example, we have learned how to change the integration variable to solve an integral; this change has to be a diffeomorphism. By changing the coordinate system, the derivative of a map changes as well, as shown in the Chain Rule below.

Proposition 2 (Chain Rule). *Let $U \subset \mathbb{R}^n$ and $V \subset \mathbb{R}^m$ be open subsets and let $f : U \to \mathbb{R}^p$ and $T : V \to U$ be C^1. Taking $p \in V$ and $q = T(p) \in V$, the composition $f \circ T : V \to \mathbb{R}^p$ is C^1 and*

$$d(f \circ T)_p = df_{T(p)}.dT_p : T_p\mathbb{R}^n \to T_{f(q)}\mathbb{R}^p. \tag{32}$$

Proof If f and T are C^1, then we have maps ρ_1 and ρ_2 such that

$$f(q + w) - f(q) = df_p.w + \rho_1(w). \mid w \mid$$
$$T(p + v) - T(p) = dT_p.v + \rho_2(v). \mid v \mid$$

and $\lim_{w \to 0} \rho_1(w) = \lim_{v \to 0} \rho_2(v) = 0$. Therefore

$$(f \circ T)(p + v) = f(T(p + v)) = f(\overbrace{T(p) + dT_p.v + \rho_2(v) \mid v \mid}^{w})$$
$$= f(T(p)) + df_q(dT_p.v + \rho_2(v) \mid v \mid) + \rho_1(dT_p.v + \rho_2(v) \mid v \mid) \mid (dT_p.v + \rho_2(v) \mid v \mid) \mid$$
$$= (f \circ T)(p) + df_q.dT_p.v + R(v) \mid v \mid$$

and $R(v) = df_q.\rho_2(v) + \rho_1(dT_p.v + \rho_2(v) \mid v \mid). \mid (dT_p.\frac{v}{|v|} + \rho_2(v) \mid v \mid) \mid$. Since the term $\frac{v}{|v|}$ is bounded, by passing to the limit $v \to 0$, we get $R(v) \to 0$ verifying the identity (32). $\qquad\square$

Corollary 1 *Let $U \subset \mathbb{R}^n$ and $V \subset \mathbb{R}^m$ be open subsets, and let $f : U \to \mathbb{R}^p$ and $T : V \to U$ be C^k. So $f \circ T : V \to \mathbb{R}^p$ is C^k.*

Now, we work out the gradient and Laplacian formulas using polar coordinates. Considering $\beta = \{e_1, e_2\}$ the canonical basis of $\mathbb{R}^2$, let's compute the gradient $\nabla f = \frac{\partial f}{\partial x}e_1 + \frac{\partial f}{\partial y}e_2$ and the Laplacian $\triangle f = \frac{\partial^2 f}{\partial x^2} + \frac{\partial^2 f}{\partial y^2}$ using the polar coordinates defined in (27). Since $r = \sqrt{x^2 + y^2}$ and $\theta = \text{arctg}\left(\frac{y}{x}\right)$, we have the partial derivatives

$$\frac{\partial f}{\partial x} = \frac{\partial f}{\partial r}.\frac{\partial r}{\partial x} + \frac{\partial f}{\partial \theta}.\frac{\partial \theta}{\partial x} = \frac{\partial f}{\partial r}.\cos(\theta) + \frac{\partial f}{\partial \theta}\left(-\frac{\sin(\theta)}{r}\right),$$
$$\frac{\partial f}{\partial y} = \frac{\partial f}{\partial r}.\frac{\partial r}{\partial y} + \frac{\partial f}{\partial \theta}.\frac{\partial \theta}{\partial y} = \frac{\partial f}{\partial r}.\sin(\theta) + \frac{\partial f}{\partial \theta}\left(\frac{\cos(\theta)}{r}\right).$$

Taking the orthogonal unit vectors $\hat{r} = (\cos(\theta), \sin(\theta))$ and $\hat{\theta} = (-\sin(\theta), \cos(\theta))$, we get

$$\nabla f = \frac{\partial f}{\partial r}\hat{r} + \frac{1}{r}\frac{\partial f}{\partial \theta}\hat{\theta}. \tag{33}$$

Analogously,

$$\frac{\partial^2 f}{\partial x^2} = \frac{\partial}{\partial r}\left(\frac{\partial f}{\partial x}\right)\cdot\frac{\partial r}{\partial x} + \frac{\partial}{\partial\theta}\left(\frac{\partial f}{\partial x}\right)\cdot\frac{\partial\theta}{\partial x} = \frac{\partial}{\partial r}\left[\frac{\partial f}{\partial r}\cdot\cos(\theta) + \frac{\partial f}{\partial\theta}\left(-\frac{\sin(\theta)}{r}\right)\right]\cos(\theta)$$
$$+ \frac{\partial}{\partial\theta}\left[\frac{\partial f}{\partial r}\cdot\cos(\theta) + \frac{\partial f}{\partial\theta}\left(-\frac{\sin(\theta)}{r}\right)\right]\left(-\frac{\sin(\theta)}{r}\right),$$

$$\frac{\partial^2 f}{\partial y^2} = \frac{\partial}{\partial r}\left(\frac{\partial f}{\partial y}\right)\cdot\frac{\partial r}{\partial y} + \frac{\partial}{\partial\theta}\left(\frac{\partial f}{\partial y}\right)\cdot\frac{\partial\theta}{\partial y} = \frac{\partial}{\partial r}\left[\frac{\partial f}{\partial r}\cdot\sin(\theta) + \frac{\partial f}{\partial\theta}\left(\frac{\cos(\theta)}{r}\right)\right]\sin(\theta)$$
$$+ \frac{\partial}{\partial\theta}\left[\frac{\partial f}{\partial r}\cdot\sin(\theta) + \frac{\partial f}{\partial\theta}\left(\frac{\cos(\theta)}{r}\right)\right]\left(\frac{\cos(\theta)}{r}\right).$$

Therefore

$$\triangle f = \frac{1}{r}\frac{\partial f}{\partial r} + \frac{\partial^2 f}{\partial r^2} + \frac{1}{r^2}\frac{\partial^2 f}{\partial\theta^2}. \tag{34}$$

Applying the formula (33) to the function $f(x, y) = (x^2 + y^2)^{-3/2}$, we get $\nabla f(r, \theta) = -\frac{3}{r^4}\hat{r}$.

Exercises

(1) Let $f = f(r, \theta)$ and $g(x, y) = f \circ P^{-1}(x, y)$. Use the chain rule to prove the Eq. (33).
(2) The cylindrical coordinates are defined using the map

$$C : \mathbb{R}^3 \to \mathbb{R}^3,$$
$$C(r, \theta, z) = \big(r\cos(\theta), r\sin(\theta), z\big). \tag{35}$$

Find an open set $U \subset \mathbb{R}^3$ such that $C : U \to C$ is a diffeomorphism. Using the cylindrical coordinates, prove that the gradient and the Laplacian of a function are given by

$$\nabla f = \frac{\partial f}{\partial r}\hat{r} + \frac{1}{r}\frac{\partial f}{\partial\theta}\hat{\theta} + \frac{\partial f}{\partial z}\hat{z},$$
$$\triangle f = \frac{1}{r}\frac{\partial f}{\partial r} + \frac{\partial^2 f}{\partial r^2} + \frac{1}{r^2}\frac{\partial^2 f}{\partial\theta^2} + \frac{\partial^2 f}{\partial z^2}. \tag{36}$$

(3) Using spherical coordinates, prove that the gradient and the Laplacian of a function $f \in C^1(\mathbb{R}^3)$ are

$$\nabla f = \frac{\partial f}{\partial r}\hat{r} + \frac{1}{r\sin(\psi)}\frac{\partial f}{\partial\theta}\hat{\theta} + \frac{1}{r}\frac{\partial f}{\partial\psi}\hat{\psi},$$
$$\triangle f = \frac{\partial^2 f}{\partial r^2} + \frac{2}{r}\frac{\partial f}{\partial r} + \frac{1}{r^2\sin^2(\psi)}\frac{\partial^2 f}{\partial\theta^2} + \frac{\cot(\psi)}{r^2}\frac{\partial f}{\partial\psi} + \frac{1}{r^2}\frac{\partial^2 f}{\partial\psi^2}. \tag{37}$$

(4) Using spherical coordinates in R^n, prove that the Laplacian is given by

$$\Delta f = \frac{\partial^2 f}{\partial \rho^2} + \frac{n-1}{\rho} \frac{\partial f}{\partial \rho} + \frac{1}{\rho^2} \Delta_{S^{n-1}} f, \tag{38}$$

such that $\Delta_{S^{n-1}} f$ is an operator involving only the 1st and 2nd partial derivatives of f with respect to the angles $\theta_1, \ldots, \theta_{n-1}$. The operator $\Delta_{S^{n-1}} f$ is the Laplace-Beltrami operator defined on S^{n-1}.

One of the most important results in the theory of differentiable maps is the Inverse Function Theorem (InFT), our next result.

Theorem 8 (InFT). *Let U, V be subsets of $\mathbb{R}^n$ and let $f : U \to V$ be a C^1-map. Take $p \in U$ and assume that $df_p : T_p U \to T_{f(p)} V$ is an isomorphism of vector spaces. So we have an open neighborhood $W \subset U$ containing p, such that $f : W \to f(W)$ is a diffeomorphism.*

Now, to verify whether a map is a diffeomorphism in the neighborhood of a point p, it is sufficient to check if the linear transformation df_p is an isomorphism. The proof is postponed to Chap. 4. Next, we will apply the InFT to understand some standard cases of differentiable maps. To this end, we will need the following definitions;

Definition 15 Let $U \subset \mathbb{R}^n$ and let $V \subset \mathbb{R}^m$ be open subsets. Let $f \in C^1(U, V)$ and $p \in U$. Thus, f is:

(1) an immersion at p if $n < m$ and rank$(df_p) = n$. f is an immersion on U if it is an immersion for all $p \in U$.
(2) a submersion at p if $n > m$ and rank$(df_p) = m$. f is a submersion on U if it is a submersion for all $p \in U$.

Example 5 In the examples below, we show standard examples of immersions and submersions;

(1) The map $f : (0, 2\pi) \times (0, \pi) \to \mathbb{R}^3$, given by

$$f(\theta, \psi) = \big(\cos(\theta) \sin(\psi), \sin(\theta) \sin(\psi), \cos(\psi) \big)$$

is an immersion.
(2) Let $B_1 = \{(x, y) \in \mathbb{R}^2 \mid x^2 + y^2 < 1\}$. The map $\phi_1 : B_1 \to \mathbb{R}^3$ given by

$$\phi_1(x, y) = (x, y, \sqrt{1 - x^2 - y^2})$$

is an immersion.
(3) A map $F : \mathbb{R}^n \to \mathbb{R}^{n+m}$, $F(x) = \big(x, f(x) \big)$, defining the graph of a differentiable map $f : \mathbb{R}^n \to \mathbb{R}^m$ is the standard model of a local immersion. It is proved in the sequel that all immersions can be described locally with the map F.
(4) The function $f : \mathbb{R}^3 \to \mathbb{R}$, $f(x, y, z) = x^2 + y^2 + z^2$ is a submersion in $\mathbb{R}^3 - \{0\}$.

(5) Let $n > m$; the projection $\pi : \mathbb{R}^n \to \mathbb{R}^m$, $\pi(x_1, \ldots, x_n) = (x_1, \ldots, x_m)$, is a submersion.

(6) (Whitney's umbrella) The map $f : \mathbb{R}^2 \to \mathbb{R}^3$, $f(x, y) = (xy, x, y^2)$ is not an immersion at $p = (0, 0)$.

(7) $M \subset \mathbb{R}^m$ is a parametrized n-dimensional submanifold of $\mathbb{R}^m$ if the tangent plane $T_p M$ is a well-defined n-dimensional vector subspace of $\mathbb{R}^m$ for all $p \in M$.
If $f : U \subset \mathbb{R}^n \to \mathbb{R}^m$ is an immersion, then $M = f(U)$ is a parametrized n-dimensional submanifold of $\mathbb{R}^m$.

(8) (Boy's Surface) David Hilbert asked if there is an immersion of the Projective Plane $\mathbb{R}P^2$ into $\mathbb{R}^3$. In 1901, Werner Boy answered positively displaying the following immersion (Figs. 18 and 19):

$$x = \frac{\sqrt{2}\cos^2(v)\cos(2u) + \cos(u)\sin(2v)}{2 - \sqrt{2}\sin(3u)\sin(2v)}, \quad y = \frac{\sqrt{2}\cos^2(v)\sin(2u) - \sin(u)\sin(2v)}{2 - \sqrt{2}\sin(3u)\sin(2v)},$$
$$z = \frac{3\cos^2(v)}{2 - \sqrt{2}\sin(3u)\sin(2v)}.$$

$$(39)$$

Fig. 18 Boy's surface

Fig. 19 Boy's surface

Definition 16 Let $U \subset \mathbb{R}^n$ and $V \subset \mathbb{R}^m$ be open subsets and $f \in C^1(U, V)$. A point $c \in \mathbb{R}^m$ is a regular value of f if the linear map $df_p : T_p U \to T_c V$ has maximum rank for all $p \in f^{-1}(c)$.

The values taken by immersions and submersions are always regular. A map $f : \mathbb{R}^n \to \mathbb{R}^m$ is an immersion if $\mathrm{rank}(df_p) = n$ and f is a submersion if $\mathrm{rank}(df_p) = m$.

6.3   The Local Form of an Immersion

We start by studying the easiest example of an immersion. Let $U \subset \mathbb{R}^2$ be an open subset and let $\phi : U \to \mathbb{R}$ be a differentiable function such that $\nabla \phi(x) \neq 0$ for all $x = (x_1, x_2) \in U$. The map $\iota : \mathbb{R}^2 \to \mathbb{R}^3$ given by

$$\iota(x_1, x_2) = \left(x_1, x_2, \phi(x_1.x_2)\right) \tag{40}$$

is an immersion. Indeed, $\iota(U)$ is the graph of $\phi : U \to \mathbb{R}$ and the rank of the matrix

$$d\iota_{(x_1, x_2)} = \begin{pmatrix} 1 & 0 \\ 0 & 1 \\ \frac{\partial \phi}{\partial x_1} & \frac{\partial \phi}{\partial x_2} \end{pmatrix}$$

is equal to 2. We claim $\iota : U \to f(U)$ is a diffeomorphism. ι is obviously bijective, so it is sufficient to prove that the inverse ι^{-1} is differentiable. Extending ι to a map $\Phi : U \times \mathbb{R} \to \mathbb{R}^3$,

$$\Phi(x_1, x_2, x_3) = \left(x_1, x_2, \phi(x_1, x_2) + x_3\right),$$

we have the linear map $d\Phi_p$ at $p = (x_1, x_2, x_3)$ given by

$$d\Phi_p = \begin{pmatrix} 1 & 0 & 0 \\ 0 & 1 & 0 \\ \frac{\partial \phi}{\partial x_1} & \frac{\partial \phi}{\partial x_2} & 1 \end{pmatrix}.$$

So $d\phi_p : T_p(U \times \mathbb{R}) \to T_{\Phi(p)}\mathbb{R}^3$ is an isomorphism of vector spaces since $\mathrm{rank}(d\Phi_p) = 3$. By the InFT, we have an open set $W \subset U \times \mathbb{R}$ such that $\Phi : W \to \Phi(W)$ is a diffeomorphism. Consequently, the restriction $\Phi : W \cap (U \times \{0\}) \to \Phi(W \cap (U \times \{0\}))$, $\Phi(x_1, x_2, 0) = \iota(x_1, x_2)$, is a diffeomorphism. The inverse map is given by

$$\iota^{-1}\left(u, v, \phi(u, v)\right) = \Phi^{-1}\left(u, v, \phi(u, v)\right) = (u, v, 0).$$

Theorem 9 (Local Form of Immersions). *Let $U \subset \mathbb{R}^n$ and $V \subset \mathbb{R}^m$ be open subsets and let $f \in C^1(U, V)$ be an immersion. So there are open subsets $U' \subset U$ and $W \subset \mathbb{R}^n$, and a diffeomorphism $\Phi : W \to U'$ such that the composition $f \circ \Phi : W \to V$ is given by*

$$f \circ \Phi(x) = \big(x, \xi_1(x), \ldots, \xi_{m-n}(x)\big), \tag{41}$$

$x = (x_1, \ldots, x_n) \in W$ and the functions $\xi_1, \ldots, \xi_{m-n} : W \to \mathbb{R}$ are all differentiable.

Proof The proof is carried out for the case $n = 2$ and $m = 3$; the general case is analogous. Let $f(x_1, x_2) = \big(f_1(x_1, x_2), f_2(x_1, x_2), f_3(x_1, x_2)\big)$; then

$$df_{(x_1,x_2)} = \begin{pmatrix} \frac{\partial f_1}{\partial x_1} & \frac{\partial f_1}{\partial x_2} \\ \frac{\partial f_2}{\partial x_1} & \frac{\partial f_2}{\partial x_2} \\ \frac{\partial f_3}{\partial x_1} & \frac{\partial f_3}{\partial x_2} \end{pmatrix}.$$

Since $\operatorname{rank}(df_p) = 2$, we assume the matrix

$$\begin{pmatrix} \frac{\partial f_1}{\partial x_1} & \frac{\partial f_1}{\partial x_2} \\ \frac{\partial f_2}{\partial x_1} & \frac{\partial f_2}{\partial x_2} \end{pmatrix}$$

defines an isomorphism $\forall (x_1, x_2) \in U$. Taking the map $g : U \to \mathbb{R}^2$, $g(x_1, x_2) = (f_1(x_1, x_2), f_2(x_1, x_2))$, from the InFT, there are open neighborhoods $U' \subset U$ and W of $f(p)$, such that $g : U' \to W$ is a diffeomorphism. Let $\Phi = g^{-1} : W \to U'$ and set $u = f_1(x_1, x_2)$ (Fig. 20), $v = f_2(x_1, x_2)$ and $\xi_1(u, v) = f_3 \circ \Phi(u, v)$, so

$$f \circ \Phi : W \to U' \quad f \circ \Phi(u, v) = (u, v, \xi_1(u, v)).$$

$\square$

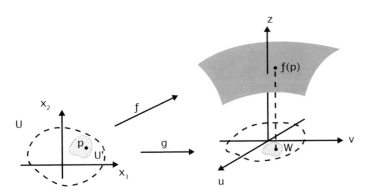

Fig. 20 Local form of immersions

Fig. 21 Klein Bottle

Figure 21 illustrates the immersion of Klein's Bottle; defined as follows: let u, $v \in [0, 2\pi] \times [0, 2\pi)$;

$$x = 6\cos(u)\big(1 + \sin(u)\big) + 4\left(1 - \frac{\cos(u)}{2}\right)\cos(u)\cos(v),$$

$$y = 4\left(1 - \frac{\cos(u)}{2}\right)\sin(v),$$

$$z = 16\sin(u) + 5\left(1 - \frac{\cos(u)}{2}\right)\sin(u)\cos(v).$$

The Projective Plane $\mathbb{R}P^2$ and Klein's Bottle $\mathbb{K}^2$ cannot be embedded into $\mathbb{R}^3$ without self-intersections. Since both are non-orientable, inside they contain an embedded Möbius Band. Suppose that $\mathbb{K}^2$ could be embedded into $\mathbb{R}^3$ without self-intersections, so there would be a tubular neighborhood $N = \mathbb{K}^2 \times (-\epsilon, \epsilon)$ such that an arbitrary closed curve $C \subset N$ intersecting the surface $\mathbb{K}^2 \times \{0\}$ would have two points of intersection with $\mathbb{K}^2 \times \{0\}$; however, we can prove that there is a curve with only one such point, as proved in Hatcher [22].

Definition 17 Consider $n < m$, $U \subset \mathbb{R}^n$ and let $V \subset \mathbb{R}^m$ be open subsets. Let $f : U \to V$ be a differentiable map;

(1) f is an open map if for all open subsets $W \subset U$, the image $f(W)$ is an open subset of $f(U)$, i.e., there is an open set $V \subset \mathbb{R}^m$ such that $f(W) = f(U) \cap V$.
(2) f is a differentiable embedding, or just an embedding, if f is an open immersion.

Theorem 10 *Consider $n < m$, $U \subset \mathbb{R}^n$ and $V \subset \mathbb{R}^m$ open subsets, and assume $f : U \to V$ is an embedding. So $M = f(U)$ is an n-submanifold of V.*

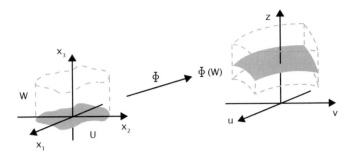

Fig. 22 Embedding

Proof The proof is carried out to the case $n = 2$ and $m = 3$; the general case is analogous. From the Local Form of an Immersion (9), for all $p \in U$ we have an open neighborhood $U'_p \subset U$ such that $f : U'_p \to f(U'_p)$ is a diffeomorphism given by $f(x_1, x_2) = (x_1, x_2, f_1(x_1, x_2))$. Since f is an open map, we have an open set $V \subset \mathbb{R}^3$ such that $f(U'_p) = V \cap f(U)$ is open in $f(U)$. Indeed, given $\epsilon > 0$, take the open set $V = f(U'_p) \times (-\epsilon, \epsilon)]$ (Fig. 22).   □

Example 6 There are examples of immersions that are not an embedding; this is a subtle issue. The following sheds some light on what may occur. Consider the embedding of the torus T^2 given by $\phi : [0, 2\pi] \times [0, 2\pi] \to \mathbb{R}^3$,

$$\phi(u, v) = \big((6 + 2\cos(u)) \cos(v), (6 + 2\cos(u)) \sin(v), 2\sin(u)\big), \qquad (42)$$

and the curve $\alpha_{a,b} : \mathbb{R} \to \mathbb{R}^2$, $\alpha_{a,b}(t) = (at, bt)$. For $r = \frac{a}{b}$, define the curve $\gamma_r : \mathbb{R} \to T^2$ given by the composition $\gamma_r = \phi \circ \alpha_{a,b}$ and consider the following families of curves;
(i) $r \in \mathbb{Q}$; illustrated in Fig. 23,
(ii) $r \in \mathbb{R} \backslash \mathbb{Q}$; illustrated in Fig. 24.
Both families have distinct topological properties as a subset of T^2. For every $r \in \mathbb{Q}$ the curve α_r is closed in T^2; so it is an embedding of $\mathbb{R}$ into T^2. However, if $r \notin \mathbb{Q}$, then the curve α_r is not closed; indeed it is dense in T^2 (see a proof in Lima [30]). Consequently, the case $r \notin \mathbb{Q}$ defines a curve $\alpha_r \subset T^2$ which is not an embedding of $\mathbb{R}$ into T^2, even though it is an immersion.

6.4 The Local Form of Submersions

Let $V \subset \mathbb{R}^3$ be an open subset and $p \in V$. The projection $\pi_x : V \to \mathbb{R}, \pi_x(x, y, z) = x$, is a submersion since the linear functional

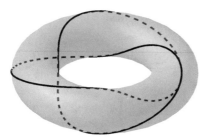

Fig. 23 $\gamma_{2/3}$ is closed

Fig. 24 $\gamma_{\frac{\sqrt{3}}{2}}$ is not closed

$$d(\pi_x)_p : T_p V \to \mathbb{R}, \ d(\pi_x)_p.v = \langle (1, 0, 0), (v_1, v_2, v_3) \rangle = v_1$$

is surjective. For every $c \in \mathrm{Im}(\beta_x)$, the set $\pi_x^{-1}(c)$ is the intersection of the plane $x = c$ with the open set V. Another example is the submersion defined by $\pi_{xy} : V \to \mathbb{R}^2$, $\pi_{xy}(x, y, z) = (x, y)$. The projections are standard models of submersions.

Theorem 11 (Local Form of Submersions). *Let $U \subset \mathbb{R}^{m+k}$ and $V \subset \mathbb{R}^m$ be open subsets and $p \in U$. If $f \in C^1(U, V)$, $f = (f_1, \ldots, f_m)$, is a submersion such that $\frac{\partial(f_1, \ldots, f_m)}{\partial(x_1, \ldots, x_m)} \neq 0$, then we have an open neighborhood $U'(\subset U)$ of p, an open subset $W \subset \mathbb{R}^{m+k}$ and a diffeomorphism $\Phi : W \to U'$ such that the composition $f \circ \Phi : W \to V$ is the projection*

$$f \circ \Phi(U_1, \ldots, u_m, \ldots, u_{m+k}) = (U_1, \ldots, u_m).$$

Proof The proof is carried out for the case $n = 3$ and $m = 2$. The general case follows using the same reasoning. Let $f : \mathbb{R}^3 \to \mathbb{R}^2$ be given by

$$f(x, y, z) = \big(f_1(x, y, z), f_2(x, y, z)\big).$$

Therefore

$$df_{(x,y,z)} = \begin{pmatrix} \frac{\partial f_1}{\partial x} & \frac{\partial f_1}{\partial y} & \frac{\partial f_1}{\partial z} \\ \frac{\partial f_2}{\partial x} & \frac{\partial f_2}{\partial y} & \frac{\partial f_2}{\partial z} \end{pmatrix}.$$

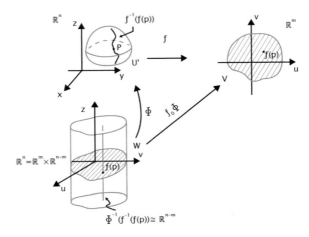

Fig. 25 Local form of submersions

From the hypothesis $\mathrm{rank}(df_p) = 2$. Without loss of generality, we can assume the matrix

$$\begin{pmatrix} \frac{\partial f_1}{\partial x} & \frac{\partial f_1}{\partial y} \\ \frac{\partial f_2}{\partial x} & \frac{\partial f_2}{\partial y} \end{pmatrix}(p)$$

is invertible for all $p \in U$. The InFT implies that the map $F : U \to V \times \mathbb{R}$, $F(x, y, z) = (f_1(x, y, z), f_2(x, y, z), z)$ is a diffeomorphism since

$$dF_p = \begin{pmatrix} \frac{\partial f_1}{\partial x} & \frac{\partial f_1}{\partial y} & \frac{\partial f_1}{\partial z} \\ \frac{\partial f_2}{\partial x} & \frac{\partial f_2}{\partial y} & \frac{\partial f_2}{\partial z} \\ 0 & 0 & 1 \end{pmatrix}$$

is invertible. Now we have open neighborhoods $U'(\subset U)$ containing p, and $W(\subset V \times \mathbb{R})$ containing $f(p)$, such that $F : U' \to W$ is a diffeomorphism. Let $\Phi = F^{-1} : W \to U'$; considering $u = f_1(x, y, z)$ and $v = f_2(x, y, z)$, we get

$$f \circ \Phi(u, v, z) = f(x, y, z) = (u, v)$$

$\square$

The InFT ensures that locally $\Phi(u, v, z) = \big(\phi_1(u, v, z), \phi_2(u, v, z), z\big)$ is a diffeomorphism. In this way for $c = (c_1, c_2) \in \mathrm{Im}(f)$, $f^{-1}(c) = \{(\phi_1(c_1, c_2, z), \phi_2(c_1, c_2, z), z) \mid (c_1, c_2, z) \in W\}$ is the local graph of $\widehat{\phi} : \mathbb{R} \to \mathbb{R}^2$, $\widehat{\phi} = (\phi_1, \phi_2)$ (Fig. 25).

Theorem 12 *Let $U \subset \mathbb{R}^n$, $V \subset \mathbb{R}^m$ be open subsets, let $f : U \to V$ be a differentiable map and let $c \in \mathrm{Im}(f)$ be a regular value of f. So the set $M^c = f^{-1}(c)$ is an $(n\text{-}m)$-submanifold of $\mathbb{R}^n$ and $T_p M^c = \mathrm{Ker}(df_p)$ for all $p \in M^c$, .*

Proof The proof is carried out for the case $n = 3$ and $m = 2$. Let $c = (c_1, c_2) \in \mathrm{Im}(f)$ and $p = (x, y, z) \in f^{-1}(c)$. From the proof of Theorem 11, there are open neighborhoods $U'_p \ni p$ and $W \ni f(p)$ such that U'_p is diffeomorphic to $W \subset f(U'_p) \times \mathbb{R}$ and a map $F : U' \to W$. Therefore $f^{-1}(c) \cap U'_p$ is diffeomorphic to $\{(c_1, c_2, z) \mid z \in \mathbb{R}\}$ since

$$f(x, y, z) = c \quad \Longleftrightarrow \quad F(x, y, z) = (c_1, c_2, z).$$

$\square$

Example 7 Examples of submanifolds.

(1) Orthogonal Group $O_n = \{A \in GL_n(\mathbb{R}) \mid A.A^t = A^t.A = I\}$.
   Consider $S_n = \{A \in M_n(\mathbb{R}) \mid A^t = A\}$ and $A_n = \{A \in M_n(\mathbb{R}) \mid A^t = -A\}$ to be the subspaces of symmetric and skew-symmetric matrices, respectively. Since any matrix $B \in M_n(\mathbb{R})$ decomposes as $B = \frac{B+B^t}{2} + \frac{B-B^t}{2}$, we take the direct sum $M_n(\mathbb{R}) = S_n \oplus A_n$. As vector spaces, $S_n = \mathbb{R}^{\frac{n(n+1)}{2}}$ and $A_n = \mathbb{R}^{\frac{n(n-1)}{2}}$. Defining the map $f : M_n(\mathbb{R}) \to S_n$, $f(A) = A.A^t$, the orthogonal group is given by $O_n = f^{-1}(I)$. To show that O_n is a submanifold, it is sufficient to prove that $I \in S_n$ is a regular value of f. Taking $A \in O_n$, the derivative of f at A is given by

$$df_A.V = \lim_{t \to 0} \frac{(A + tV)(A + tV)^t - AA^t}{t} = \lim_{t \to 0} \frac{t[AV^t + VA^t] + t^2 VV^t}{t}$$
$$= AV^t + VA^t \in S_n.$$

Since the equation $AV^t + VA^t = B$ admits a solution for all $B \in S_n$, the derivative df_A is surjective. Applying the decomposition $B = \frac{B+B^t}{2} + \frac{B-B^t}{2}$, it is easy to verify that $V = \frac{1}{2}BA$ is a solution. Once I is a regular value of f, it follows that O_n is a differentiable submanifold of $\mathbb{R}^{n^2}$. The tangent plane at A is $T_A O_n = \mathrm{Ker}(df_A)$, so $\dim(O_n) = \frac{n(n-1)}{2}$. At $A = I$, the tangent plane is $T_I O_n = A_n$. The determinant of a matrix $A \in O_n$ is either 1 or -1, so the group O_n has two connected components. The connected component containing the identity is the subgroup $SO_n = \{A \in O_n \mid \det(A) = 1\}$ of special orthogonal matrices.

(2) Unitary Group $U_n = \{A \in M_n(\mathbb{C}) \mid A^* = A\}$ $(A^* = \overline{A}^t)$.

Let $M_n(\mathbb{C})$ be the algebra of matrices with complex entries; let $H_n = \{A \in M_n(\mathbb{C}) \mid A^* = A\}$ be the subspace of Hermitian matrices and let $A_n^h = \{A \in M_n(\mathbb{C}) \mid A^* = -A\}$ be the subspace of skew-Hermitian matrices. For any unitary matrix $A \in U_n$, the following hold:

(i) Given two complex vectors x and y, multiplication by A preserves their inner product; that is, $< Ax, Ay > = < x, y >$.

(ii) A is normal.

(iii) A is diagonalizable; that is, A is unitarily similar to a diagonal matrix, as a consequence of the spectral theorem. A now has a decomposition of the form $A = PDP^*$, with P unitary, and D is diagonal and unitary.

(iv) $\mid \det(A) \mid = 1$.

(v) Its eigenspaces are orthogonal.

(vi) A can be written as $A = e^S$ (e indicates the exponential), i is the imaginary unit and S is a skew-Hermitian matrix.

The decomposition $A = \frac{A+A^*}{2} + \frac{A-A^*}{2}$ induces the direct sum $M_n(\mathbb{C}) = H_n \oplus A_n^h$. The real dimension of each space is given by $\dim_\mathbb{R}(H_n) = n^2$ and $\dim_\mathbb{R}(A_n^h) = n^2$. Considering the map $f : M_n(\mathbb{C}) \to H_n$, $f(X) = X.X^*$, the space of unitary matrices is given by $U_n = f^{-1}(I)$. The vector space $M_n(\mathbb{C})$ is isomorphic to $\mathbb{R}^{2n^2}$ and H_n is a vector subspace (over $\mathbb{R}$) of $\mathbb{R}^{2n^2}$. We prove that I is a regular value of f. Fix $A \in U_n$, the derivative of f at A is given by

$$df_A.V = \lim_{t \to 0} \frac{(A + tV)(A + tV)^* - AA^*}{t} = \lim_{t \to 0} \frac{t[AV^* + VA^*] + t^2 VV^*}{t}$$
$$= AV^* + VA^* \in H_n.$$

Since the equation $AV^* + VA^* = B$ admits a solution $V = \frac{1}{2}B^*A$, df_A is surjective and I is a regular value of f. The tangent plane at $A = I$ is $T_I U_n = A_n^h$. Hence $U_n = f^{-1}(I)$ is a submanifold of $\mathbb{R}^{2n^2}$ with dimension $\dim_\mathbb{R}(U_n) = n^2$. To prove that U_n is connected, we use the property (iii) that any unitary matrix A, being diagonalizable, can be written as $A = PDP^*$, with P unitary and $D = \mathrm{diag}(e^{i\cdot 1}, \ldots, e^{i\cdot n})$. So by considering $D(t) = \mathrm{diag}(e^{it\cdot 1}, \ldots, e^{it\cdot n})$, the path $\gamma : [0, 1] \to U_n, \gamma(t) = PD(t)P^*$, connects A to I. Hence the group U_n is connected. A second proof that U_n is a submanifold uses the determinant function $\det : M_n(\mathbb{C}) \to \mathbb{C}$. Since the determinant of a unitary matrix is a complex number with norm 1, the determinant gives a group homomorphism $\det : U_n \to U_1$. However, since the determinant restricted to the case of U_n takes values in U_1, and not in a Euclidean space, the proof that $U_n = \det^{-1}(U_1)$ is a submanifold requires concepts of transversality beyond the scope of this text.

(3) Unitary Special Group $SU_n = \{A \in U_n \mid \det(A) = 1\}$.

To prove that SU_n is a submanifold, we must use the property $\det(A) = 1$ for all $A \in SU_n$. As mentioned before, this situation requires further concepts.

Exercise

Show that the groups O_n, SO_n, U_n and SU_n are compact spaces.

6.5 *Generalization of the Implicit Function Theorem*

In $\mathbb{R}^{n+k} = \mathbb{R}^n \times \mathbb{R}^k$, consider the coordinates (x, y), $x \in \mathbb{R}^n$ and $y \in \mathbb{R}^k$. Let $U \subset \mathbb{R}^{n+k}$ be an open set and assume that the functions $f_1, \ldots, f_k$ are differentiable; then consider the non-linear system of equations

$$\begin{cases} f_1(x, y) = c_1, \\ \quad \vdots \\ f_k(x, y) = c_k. \end{cases} \tag{43}$$

Next, define the map $F : U \to \mathbb{R}^k$, $F(x, y) = (f_1(x, y), \ldots, f_k(x, y))$ and value $c = (c_1, \ldots, c_k) \in \mathbb{R}^k$. Suppose the point $(x_0, y_0) \in F^{-1}(c)$ is known and $\frac{\partial F}{\partial y}(x_0, y_0) = \left(\frac{\partial f_i}{\partial y_j}(x_0, y_0) \right)$ is an invertible matrix. Then the InFT can be applied to the map $\tilde{F} : U \to \mathbb{R}^{n+k}$, $\tilde{F}(x, y) = (x, F(x, y))$ in a neighborhood of (x_0, y_0). So we have open sets U', $V \in \mathbb{R}^{n+k}$, $(x_0, y_0) \in U' \subset U$, such that $\tilde{F} : U' \to V$ is a diffeomorphism. Consider the coordinates in V given by $(x, u) \in \mathbb{R}^n \times \mathbb{R}^k$ and let $\tilde{\xi} = \tilde{F}^{-1}$, and we get

$$(x, y_1, \ldots, y_k) = \tilde{\xi}(x, u) = (x, \xi_1(x, u), \ldots, \xi_1(x, u)) = (x, \xi(x, u)).$$

Therefore if $u = c \in \mathbb{R}^k$ is a fixed value, then $F^{-1}(c) = (x, \xi(x, c))$; i.e., $F^{-1}(c)$ restricted to $U' \ni (x_0, y_0)$ is the graph of the map $\xi : \mathbb{R}^n \to \mathbb{R}^k$.

Exercises

(1) Let $U \subset \mathbb{R}^n$, $V \subset \mathbb{R}^m$ be open sets and $f \in C^1(U, V)$. Prove the following;

   (a) The local form of immersions $(n < m)$.
   (b) The local form of submersions $(n > m)$.

(2) Prove that the set of matrices 2×2 having rank-1 is a 3-submanifold of $\mathbb{R}^4$.
(3) Consider $p : \mathbb{R}^n \to \mathbb{R}$ to be a homogeneous polynomial of degree k and $c \neq 0$. Prove that $p^{-1}(c)$ is a (n-1)-submanifold of $\mathbb{R}^n$.
(4) Prove that any submersion is an open map. Conclude that if K is a compact set, then there is no submersion $f : K \to \mathbb{R}^m$.
(5) Let $U_1 \subset \mathbb{R}^n$, $U_2 \subset \mathbb{R}^m$ be open subsets and $U = \{(x, y) \in U \mid x \in U_1, y \in U_2\} = U_1 \times U_2$. Consider $f : U_1 \times U_2 \to \mathbb{R}^m$ as the map $f(x, y) = (f_1(x, y), \ldots, f_m(x, y))$, such that $f_1, \ldots, f_m : \mathbb{R}^n \to \mathbb{R}$ are C^1-functions. Let $d_{(x,y)} f = d_1 f_{(x,y)} + d_2 f_{(x,y)} : \mathbb{R}^n \oplus \mathbb{R}^m \to \mathbb{R}^m$ be the derivative of f and

assume that $d_2 f_{(x_0, y_0)}$ is invertible. Using this setting, write a statement and prove a generalization of the Implicit Function Theorem, and also prove that the set $M = f^{-1}(0)$ is locally the graph of a map $\xi : \mathbb{R}^n \to \mathbb{R}^m$.

(6) Find the critical values set of the function $f : \mathbb{R}^3 \to \mathbb{R}$, $f(x, y, z) = x^2 + y^2 - z^2$. Show that if c_1 and c_2 are regular values of f having the same sign, then $f^{-1}(c_1)$ and $f^{-1}(c_2)$ are diffeomorphic.

(7) Let $Q : \mathbb{R}^n \to \mathbb{R}$ be a non-degenerate quadratic form. The spectrum $\sigma(Q)$ of Q admits a decomposition $\sigma(Q) = \sigma^+ \cup \sigma^-$, in which $\sigma^+ = \sigma(Q) \cap (0, \infty)$ and $\sigma^- = \sigma(Q) \cap (-\infty, 0)$. Since Q is diagonalizable, $\mathbb{R}^n$ has a basis $\beta = \{u_1^+, \ldots, u_p^+, u_1^-, \ldots, u_q^-\}$ of eigenvectors of Q, and $Q(u_i^\pm) = \lambda_i^\pm u_i^\pm$, $\lambda_i^+ > 0$ and $\lambda_i^- < 0$. By Sylvester's theorem, quadratic forms are classified by their rank $n = p + q$ and by their signature $\sigma(Q) = p - q$. Show the set $SO(p, q) = \{A \in GL_n(\mathbb{R}) \mid A.Q.A^t = Q\}$ is a group and a submanifold. Find its dimension.

(8) Let I_n be the $n \times n$ identity matrix and $J_0 = \begin{pmatrix} 0 & -I_n \\ I_n & 0 \end{pmatrix}$. The symplectic group is defined as $\mathrm{Sp}_{2n}(\mathbb{R}) = \{A \in GL_n(\mathbb{R}) \mid A^t J_0 A = J_0\}$. Show $\mathrm{Sp}_{2n}(\mathbb{R})$ is a submanifold.

(9) Consider the linear transformation $J_0 : \mathbb{R}^{2n} \to \mathbb{R}^{2n}$ in which the matrix matrix J_0 is defined in the last item. The real vector space $\mathbb{R}^{2n}$ and the complex vector space $\mathbb{C}^n$ can be identified by taking the canonical basis $\beta = \{e_1, e_2, \ldots, e_i, e_{i+1}, \ldots, e_{2n-1}, e_n\}$ and defining the complex orthonormal basis $\beta^{\mathbb{C}} = \{e_1 + J_0 e_2, \ldots, e_i + J_0 e_{i+1}, \ldots, e_{2n-1} + J_0 e_{2n}\}$. Show that $GL_n(\mathbb{C}) \subset GL_{2n}(\mathbb{R})$ and $U_n \subset \mathrm{Sp}(2n)$ are subgroups; indeed, they are submanifolds $(A \in GL_n(\mathbb{C}) \Leftrightarrow AJ_0 = J_0 A)$.

(10) Using the Gram-Schmidt procedure to obtain an orthonormal basis, show that any compact subset of $GL_n(\mathbb{R})$ can be continuously deformed into a compact subset of O_n. Indeed, given a compact $K \subset GL_n(\mathbb{R})$, we have a continuous map $F : [0, 1] \times K \to GL_n(\mathbb{R})$ such that $F(0, x) = x$ e $F(1, x) \in O_n$.

(11) Let (x_0, y_0) be a solution to the non-linear system (43). Prove that if the gradient vectors $\nabla f_1(x_0, y_0), \ldots, \nabla f_k(x_0, y_0)$ are linearly independent in a neighborhood of (x_0, y_0), then the solution set is an n-submanifold of $\mathbb{R}^{n+k}$. Conclude that the solution of the system below is not a 1-submanifold of $\mathbb{R}^3$.

$$\begin{cases} x^3 - y^2 + z = 0, \\ xy - z = 0. \end{cases}$$

7 Fundamental Theorem of Algebra

This section is devoted to applying the theory of differentiable maps to prove the Fundamental Theorem of Algebra (FTA) on $\mathbb{R}$. The proof relies only on the properties of differentiable maps; the condition that the field be algebraically closed is not necessary. The complex case will be left as an exercise. This proof is due to Pukhlikov [37].

Polynomials of one real variable define the ring $\mathbb{R}[x]$. Let $\widehat{\mathcal{P}_d} \subset \mathbb{R}[x]$ be the vector subspace of polynomials of degree d. An element $P \in \widehat{\mathcal{P}_d}$ is represented by

$$P(x) = a_d x^d + a_{d-1} x^{d-1} + \cdots + a_1 x + a_0,$$

and $a_d \neq 0, a_{d-1}, \ldots, a_1, a_0$ are real numbers. $\widehat{\mathcal{P}_d}$ is an $\mathbb{R}$-vector space identified with $\mathbb{R}^{d+1}$ by the vector space isomorphism

$$\widehat{\mathcal{P}_d} \to \mathbb{R}^{d+1}$$
$$a_d x^d + \cdots + a_0 \to (a_d, \ldots, a_0).$$

We introduce in $\widehat{\mathcal{P}_d}$ the equivalent relation "$P(x) \sim Q(x) \Leftrightarrow$ if we have $\lambda \in \mathbb{R}^*$ such that $Q(x) = \lambda P(x)$", and every polynomial $P(x) \in \widehat{\mathcal{P}_d}$ can be identified with a monic polynomial

$$P(x) = x^d + \frac{a_{d-1}}{a_d} x^{d-1} + \cdots + \frac{a_1}{a_d} x + \frac{a_0}{a_d}.$$

The space $\mathcal{P}_d$ of monic polynomials in $\widehat{\mathcal{P}_d}$ can be identified with $\mathbb{R}^d = \{(1, 0, \ldots, 0)\} \times \mathbb{R}^d$. The quotient space $(\widehat{\mathcal{P}_d}/\sim)$ is exactly the projective space $\mathbb{R}\mathbf{P}^d = (\mathbb{R}^{d+1}/\sim)$.

Remarks (1) As described in Appendix B, the real projective spaces $\mathbb{R}\mathbf{P}^n$, $n \in \mathbb{N}$, can be assembled with the inductive process

$$\mathbb{R}\mathbf{P}^1 = S^1,$$
$$\mathbb{R}\mathbf{P}^n = \mathbb{R}^n \bigcup \mathbb{R}\mathbf{P}^{n-1}. \tag{44}$$

(2) The space $\mathcal{P}_d$ of monic polynomials with degree d is identified as $\mathbb{R}^d \subset \mathbb{R}\mathbf{P}^d$, and the space $\mathcal{P}_{d-1}$ of monic polynomials of degree $\leq (d-1)$ is identified as $\mathbb{R}\mathbf{P}^{d-1} = \{(0 : a_{d-1} : \ldots : a_0)\} \subset \mathbb{R}\mathbf{P}^d$.

Theorem 13 (FTA). *Every non-constant polynomial $P(x) \in \mathbb{R}[x]$ can be factored into a product of linear and quadratic factors.*

Example Let $P(x) \in \mathcal{P}_4$. Factoring an arbitrary monic polynomial $P(x) = x^4 + Ax^3 + Bx^2 + Cx + D$ as

$$x^4 + Ax^3 + Bx^2 + Cx + D = (x^2 + ac + b)(x^2 + cx + d),$$

we get a system of four equations in four unknowns:

$$a + c = A, \qquad ad + bc = C,$$
$$b + d + ac = B, \quad bd = D.$$

Rewriting the system in the variables x, y, z and w, we have

$$f : \mathbb{R}^4 \to \mathbb{R}^4,$$
$$(x, y, z, w) \to (x + z, y + w + xz, xw + yz, yw).$$

We would like to know if the map f above is surjective. The main ingredients to understand the topology of the solution set for the equation

$$f(x, y, z, w) = (A, B, C, D)$$

are the Inverse Function Theorem and the fact that f is a proper map.

Let X and Y be non-compact locally compact Hausdorff spaces. Take compact subsets $K_X \subset X$ and $K_Y \subset Y$ and let $\widehat{X} = X \cup K_X$ and $\widehat{Y} = Y \cup K_Y$ be the compactifications, respectively. If a map $f : X \to Y$ admits a continuous extension $\widehat{f} : \widehat{X} \to \widehat{Y}$, then f is proper.[4] There are two important cases for our purposes. The first is used in the next section regarding the Jacobian Conjecture and the second is to prove the FTA.

(i) Let $\widehat{\mathbb{R}^n} = \mathbb{R}^n \cup \{\infty\}$ be the one point compactification by stereographic projection. We know that $\widehat{\mathbb{R}^n} = S^n$.

(ii) The real projective plane $\mathbb{R}P^n = \mathbb{R}^n \cup \mathbb{R}P^{n-1}$ is a compactification of $\mathbb{R}^n$ by adding a copy of $\mathbb{R}P^{n-1}$ at infinity.

The image $f(X)$ with a proper map $f : X \to Y$ between two topological Hausdorff spaces is closed[5] in Y. Furthermore, if Y is connected and $f(X) \subset Y$ is open, then f is surjective.

The polynomial ring $\mathbb{R}[x]$ is composed of a chain of linear subspaces, each corresponding to a space of monic polynomials:

$$\mathbb{R} = \mathcal{P}_0 \subset \mathcal{P}_1 \subset \ldots \subset \mathcal{P}_n \subset \ldots \subset \mathbb{R}[x].$$

For $d \geq 3$ and $1 \leq k \leq d - 1$, we define the multiplication map

$$\mu_k : \mathcal{P}_k \times \mathcal{P}_{d-k} \to \mathcal{P}_d, \tag{45}$$
$$(P, Q) \to P(x) \cdot Q(x).$$

The image of μ_k corresponds to the polynomials that can be factored as a product of a polynomial of degree k with another polynomial of degree $(d - k)$.

---

[4] See Appendix B.
[5] See Appendix A.

Proposition 3 *For $1 \leq k \leq d - 1$, the mapping $\mu_k : \mathcal{P}_k \times \mathcal{P}_{d-k} \to \mathcal{P}_d$ is proper and differentiable.*

Proof f is a proper map since μ_k extends to a map $\mu_k : \mathbb{RP}^k \times \mathbb{RP}^{d-k} \to \mathbb{RP}^d$, and $\mathcal{P}_k = \mathbb{R}^k \subset \mathbb{RP}^d$. To differentiate, we consider a map $\mu_k : \mathbb{R}^k \times \mathbb{R}^{d-k} \to \mathbb{R}^d$. For the polynomials $P = (P_1, P_2) \in \mathcal{P}_k \times \mathcal{P}_{d-k}$ and $T = (f, g) \in \mathcal{P}_{k-1} \times \mathcal{P}_{d-k-1}$, we have

$$d\left(\mu_k\right)_P . T = \lim_{\epsilon \to 0} \frac{\mu_k(P + \epsilon T) - \mu_k(R)}{\epsilon} = g\,P_1 + f\,P_2. \tag{46}$$

$\square$

Corollary 2 *If $d = d_1 + \ldots + d_k$, $d_i \in \mathbb{N}$, then the map $\mu : \mathcal{P}_{d_1} \times \mathcal{P}_{d_2} \times \ldots \times \mathcal{P}_{d_K}$ given by*

$$\mu(P_1, \ldots, P_k) = P_1 \cdot P_2 \cdots P_{k-1} \cdot P_k$$

is proper.

Next, we will continue to demonstrate TFA.

Proof (FTA) The argument is by induction on d. The theorem is true for $d = 1$ and $d = 2$. For $d = 3$, it is also true since every $P(x) \in \mathcal{P}_3$ admits at least one real root. Let $W_k = \mu_k(\mathcal{P}_k \times \mathcal{P}_{d-k}) \subset \mathcal{P}_d$ be the subset of reducible polynomials and define

$$W(d) = \bigcup_{k=1}^{d-1} W_k.$$

The proof is completed if the induction principle leads us to the identity $\mathcal{P}_d = W(d)$. The proof is divided into four statements:

$1st$ $-$ Claim : $W(d) \subset \mathcal{P}_d$ is closed in $\mathcal{P}_d$.
Pf: Each map μ_k is proper, so its image is closed. $W(d)$ is a finite union of closed subsets, and hence is closed.

The subset $W(d)$ would be an open subset in $\mathcal{P}_d$ if for every polynomial $P = (P_1, P_2) \in W(d)$ there was an open neighborhood $U \subset \mathcal{P}_d$. Here the InFT comes into play. The derivative of $\mu_K : \mathcal{P}_k \times \mathcal{P}_{d-k} \to \mathcal{P}_d$ defined by Eq. (46) is

$$d(\mu_k)_P(f, g) = g\,P_1 + f\,P_2.$$

Recall that the pair (P_1, P_2) is relatively prime if and only if the greatest common divisor $(P_1, P_2) = gcd(P_1, P_2) = 1$. In this way, the kernel of $d(\mu_k)_P$ is

$$\mathrm{Ker}\, d(\mu_k)_P = \begin{cases} 0, & \text{if } (P_1, P_2) = 1, \\ \neq 0, & \text{otherwise.} \end{cases}$$

Let $\mathcal{D} = \{P = (P_1, P_2) \mid (P_1, P_2) \neq 1\} \subset \mathcal{P}_k \times \mathcal{P}_{d-k}$ be the space of degenerate monic polynomials. $\mathcal{D}$ is closed in $W(d)$. Consequently, if $P = (P_1, P_2) \in W(d) \backslash \mathcal{D}$, then there are open neighborhoods $U \subset \mathcal{P}_k \times \mathcal{P}_{d-k}$ of P and $V \subset \mathcal{P}_d$ of $P_1 \cdot P_2$, such that $\mu_k : U \to V$ is a diffeomorphism. Therefore every element in U can be factored into the product of nonconstant relatively prime polynomials.

2nd $-$ Claim : $W(d) \backslash \mathcal{D}$ is open in $\mathcal{P}_d \backslash \mathcal{D}$.
Pf: By the InFT, $W(d) \backslash \mathcal{D}$ is open in $\mathcal{P}_d$ for all $1 \leq k \leq d-1$. In particular, $W(d) \backslash \mathcal{D}$ is open in $\mathcal{P}_d \backslash \mathcal{D}$.
3rd $-$ Claim : $W(d) \backslash \mathcal{D}$ is closed in $\mathcal{P}_d \backslash \mathcal{D}$.
Pf: Since $W(d)$ is closed in $\mathcal{P}_d$, it follows that $W(d) \backslash \mathcal{D}$ is closed in $\mathcal{P}_d \backslash \mathcal{D}$.
4th $-$ Claim : $\mathcal{P}_d \backslash \mathcal{D} = W(d) \backslash \mathcal{D}$.
Pf: The claim follows if $\mathcal{P}_d \backslash \mathcal{D}$ is connected. Let's describe the set $\mathcal{D}$ of monic polynomials that cannot be written as a product of nonconstant relatively prime polynomials. Given an element $P = (P_1, P_2) \in \mathcal{D}$, by the induction hypothesis P_1 and P_2 can be factored into linear or quadratic polynomials and they have a common factor. There are two possibilities:

(1) d is odd.
In this case, the only possibility is $P(x) = (x + a)^d$. So

$$\mathcal{D} = \{(x + a)^d \mid a \in \mathbb{R}\}$$

is an embedded smooth curve in $\mathbb{R}^d$, i.e.,

$$(x + a)^d \to \left(\binom{d}{1} a, \binom{d}{2} a^2, \ldots, \binom{d}{d} a^d \right).$$

(2) d is even.
In this case, the possibilities are either $P(x) = (x + a)^d$ or $P(x) = (x^2 + bx + c)^{d/2}$. Therefore

$$\mathcal{D} = \{(x + a)^d \mid a \in \mathbb{R}\} \bigcup \{(x^2 + bx + c)^{d/2} \mid b, c \in \mathbb{R}\}$$

is the union of an embedded curve in $\mathbb{R}^d$ with an embedded surface parametrized by (b, c). If $d = 6$, the surface is given by

$$(b, c) \xrightarrow{\phi} (3b, 3b^2 + 3c, b^3 + 6bc, 3b^2c + 3c^2, c^3).$$

The embedding is differentiable since the matrix

$$d\phi_{(b,c)} = \begin{pmatrix} 3 & 6b & 3b^3 + 6c & 6bc & 3c^2 & 0 \\ 0 & 3 & 6b & 3b^2 + 6c & 6bc & 3c^2 \end{pmatrix}$$

has rank 2. Of course, the embedding and the smoothness of the surface can be checked for all $d \in \mathbb{N}$.

If $d \geq 4$, the codimension of $\mathcal{D}$ is greater or equal to 2. It is well-known from topology that a smooth subspace with codimension ≥ 2 does not disconnect $\mathbb{R}^n$. Hence $\mathcal{P}_d \setminus \mathcal{D}$ is connected.

Since $W(d) \setminus \mathcal{D} = \mathcal{P}_d \setminus \mathcal{D}$ and $\mathcal{D} \subset W(d)$, it follows that $\mathcal{P}_d = W(d)$. Therefore every polynomial $P(x) \in \mathcal{P}_d$ can be factored into the product of linear and quadratic polynomials. Applying the Principle of Induction, we conclude the proof of the Fundamental Theorem of Algebra on field $\mathbb{R}$. $\qquad\qquad\qquad\square$

Exercises

(1) Show the Fundamental Theorem of Algebra when the field is $\mathbb{C}$. In this case, prove that every polynomial can be factored into linear irreducible polynomials.

8 Jacobian Conjecture

In the preceding sections, the study of differentiable maps defined on $\mathbb{R}^n$ was carried out in great detail. The main concepts and theorems were carefully introduced. To conclude this chapter, we describe an open problem since 1939, which is very easy to enunciate and basic enough considering the elements addressed in the question.

In this section, we consider the field being either $\mathbb{K} = \mathbb{R}$ or $\mathbb{K} = \mathbb{C}$. The Jacobian of a differentiable map $f : \mathbb{K}^n \to \mathbb{K}^n$ at $x \in \mathbb{K}^n$ we define to be the determinant $J(f) = \det(df_x)$.

Let $\mathbb{K}[x_1, \ldots, x_n]$ be the polynomial ring of $\mathbb{K}^n$. A polynomial mapping $P : \mathbb{K}^n \to \mathbb{K}^n$ is a map given by

$$P(x_1, \ldots, x_n) = \big(p_1(x_1, \ldots, x_n), \ldots, p_n(x_1, \ldots, x_n)\big),$$

and $p_i \in \mathbb{K}[x_1, \ldots, x_n]$, $1 \leq i \leq n$. The degree of P is defined as $\deg(P) = \sup_i \deg(p_i)$.

We may use the bold notation $\mathbf{x} = (x_1, \ldots, x_n)$ when referring to a point $\mathbf{x} \in \mathbb{K}^n$.

Jacobian Conjecture $\mathfrak{JC}_0^n$: Let $P : \mathbb{K}^n \to \mathbb{K}^n$ be a polynomial mapping. Then P is invertible and its inverse is also a polynomial mapping if and only if the Jacobian $J(P)$ is constant and nonzero ($J(P) \in \mathbb{K}^*$).

Remarks

(1) A polynomial automorphism is a polynomial map with an inverse that is well-defined and is also a polynomial map.

(2) Let $P : \mathbb{K}^n \to \mathbb{K}^n$ be a polynomial automorphism and let Q be its inverse. So $Q(P(x)) = x$ implies $dQ_{P(x)}.dP_x = I$. The Jacobians $J(P)$ and $J(Q)$ are also polynomials. Therefore

$$J(Q \circ P) = J(Q).J(P) = 1 \implies \deg(J(P)) = \deg(J(Q)) = 0.$$

Hence $J(P) \in \mathbb{K}^*$.

(3) The $\mathfrak{J}\mathfrak{C}_0^n$-Conjecture on $\mathbb{R}$ is sometimes called the Strong $\mathfrak{J}\mathfrak{C}_0^n$-Conjecture. This is because if the conjecture is true for real polynomial maps $P : \mathbb{R}^m \to \mathbb{R}^m$, then it is also true for complex polynomial maps $\widehat{P} : \mathbb{C}^n \to \mathbb{C}^n$. By identifying $\mathbb{C}^n$ with $\mathbb{R}^{2n}$, such that $\mathbf{z} = \mathbf{x} + i\mathbf{y} \to (\mathbf{x}, \mathbf{y})$, a complex polynomial map $\widehat{P} : \mathbb{C}^n \to \mathbb{C}^n$ induces a real polynomial map $P : \mathbb{R}^{2n} \to \mathbb{R}^{2n}$, and

$$J(P)(\mathbf{x}) = \det(dP_x) = \left| \det(d\widehat{P}_{\mathbf{z}}) \right|^2 = \left(J(P)(\mathbf{z}) \right)^2. \tag{47}$$

Therefore we have $J(\widehat{P}) \in \mathbb{C}^*$ if and only if $J(P) \in \mathbb{R}^*$.

The Jacobian Conjecture is attributed to O.H. Keller, who first posted it in [26]. So far, the only case in which the conjecture has been settled is in dimension $n = 1$. There are variants of the conjecture. They are important in order to understand the issues involved in the question. So, before diving into the question, let's introduce some notation and concepts.

Consider X and Y two subsets of $\mathbb{K}$ and $\mathcal{F}$ a category of maps. Let $\mathcal{F}_X$ and $\mathcal{F}_Y$ be the category atributed to X and Y, respectively. A map $f : X \to Y$ induces a ring morphism $f^* : \mathcal{F}_Y \to \mathcal{F}_X$ by $f^*(g) = g \circ f$. Let $\mathbb{K}[X]$ be the class of polynomial mappings on X. The $\mathfrak{J}\mathfrak{C}_0^n$-Conjecture can be rephrased in this setting.

$\mathfrak{J}\mathfrak{C}_0^n$-**Conjecture:** Let $P : X \to Y$ be a polynomial mapping. The morphism $P^* : \mathbb{K}[Y] \to \mathbb{K}[X]$ is a ring automorphism if and only if the Jacobian $J(P) = \det(dP_x) \in \mathbb{K}^*$.

Originally, the question was raised in Algebraic Geometry, however, it can be extended to larger categories of maps. We can define extensions to the conjecture by considering other categories of maps: given a category $\mathcal{F}$, let $\mathfrak{J}\mathfrak{C}_{\mathcal{F}}$ be the corresponding Jacobian Conjecture.

(i) C^∞ = the category of smooth maps $\rightsquigarrow \mathfrak{J}\mathfrak{C}_\infty^n$.
(ii) C^ω = the category of analytic maps $\rightsquigarrow \mathfrak{J}\mathfrak{C}_\omega^n$.
(iii) $\mathfrak{R}$ = the category of rational maps $\rightsquigarrow \mathfrak{J}\mathfrak{C}_\mathfrak{R}^n$.

It is natural to consider the following weaker form of the Jacobian Conjecture when the field is $\mathbb{K} = \mathbb{R}$.

$\mathfrak{W}\mathfrak{J}\mathfrak{C}$-**Conjecture** - Let $F : X \to Y$ be a map in $\mathcal{F}$. The induced morphism $F^* : \mathbb{K}[Y] \to \mathbb{K}[X]$ is a ring automorphism if and only if $J(F) \neq 0$.

We mentioned different extensions to the $\mathfrak{J}\mathfrak{C}$-Conjecture in order to broaden the reader's knowledge regarding the question, since all of them are addressed in the references. The sub-index 0 is used exclusively for the original conjecture formulated for polynomials.

A local diffeomorphism $f : \mathbb{K}^n \to \mathbb{K}^n$ induces a local change of coordinates. If the conjecture is true, then it would imply f is a global change of coordinates.

The conjecture can be formulated using a general field $\mathbb{K}$. Our interest is restricted to the cases either $\mathbb{K} = \mathbb{C}$ or $\mathbb{K} = \mathbb{R}$, but it can be addressed for fields with arbitrary characteristic $k \geq 0$.

8.1 Case $n = 1$

We intend to shed light where there is darkness; it means we would like to help the reader become sensitive to the inherent difficulties present in the problem. So we start considering the simplest case:

Question 1: ($\mathfrak{WJC}^1_\infty$) Suppose $f : \mathbb{R} \to \mathbb{R}$ is a local diffeomorphism. Is it true that f is a global diffeomorphism?

Assuming f is a C^1-map such that $f'(x) \neq 0$ for all $x \in \mathbb{R}$, the question is whether f is injective and surjective. This case is rather exceptional, since in dimension $n = 1$ we can count on the following theorems.

Theorem 14 (Intermediate Value). *Let $f : [a, b] \to$ be a continuous function. Then f attains all values between $f(a)$ and $f(b)$.*

Theorem 15 (Mean Value). *Let $f : \mathbb{R} \to \mathbb{R}$ be a C^1-function. For every pair $a, b \in \mathbb{R}$, such that $a < b$, there is $c \in (a, b)$ such that*

$$f(b) - f(a) = f'(c)(b - a). \tag{48}$$

Corollary 3 *If $n = 1$, then a local diffeomorphism $f : \mathbb{R} \to \mathbb{R}$ is injective.*

Proof It follows straighforward from the Mean Value Theorem. Assuming we have $a, b \in \mathbb{R}$ such that $f(a) = f(b)$, then by Eq. (48), we have $c \in (a, b)$ such that $f'(c) = 0$ contradicting the hypothesis. Hence f is injective. □

The surjectivity is more subtle, as we can see from the next examples.

Examples

(1) The function $f : \mathbb{R} \to \mathbb{R}$, $f(x) = \arctan(x)$ satisfies the following conditions:
   (1i) $f'(x) = \frac{1}{1+x^2} > 0$ for all $x \in \mathbb{R}$,
   (1ii) $\lim_{x \to \infty} f'(x) = 0$,
   (1iii) $\lim_{x \to -\infty} f'(x) = 0$,
   (1iv) f is not surjective since $f(\mathbb{R}) \subset (-\frac{\pi}{2}, \frac{\pi}{2})$.
(2) The function $\tanh : \mathbb{R} \to \mathbb{R}$ defined by

$$\tanh(x) = \frac{e^x - e^{-x}}{e^x + e^{-x}} \Rightarrow (\tanh)'(x) = \frac{4}{(e^x + e^{-x})^2} > 0,$$

satisfies conditions (1i)-(1iv) in item (1). It is not surjective since $\tanh(\mathbb{R}) = (-1, 1)$.

Remark From the examples above, we learn that the conditions $\lim_{x \to \pm\infty} f'(x) = 0$ are not desirable to prove the surjectivity of f.

(3) The function $f : \mathbb{R} \to \mathbb{R}$, $f(x) = e^x$ satisfies $f'(x) = e^x > 0$ and defines a diffeomorphism with an inverse $f^{-1} : \mathbb{R} \to (0, \infty)$ is given by $f^{-1}(x) = \ln(x)$.
Therefore f satisfies both conditions (1i) and (1iii). f is not surjective since
$f(\mathbb{R}) = (0, \infty)$.
(4) The function $f : (0, \infty) \to (0, \infty)$, $f(x) = \sqrt{x}$ with a derivative $f'(x) = \frac{1}{2\sqrt{x}}$
satisfies conditions (1i) and (1ii). However f is surjective.

Therefore the Conjecture $\mathfrak{WJC}_\infty^1$ is false. We need extra conditions on f.
Remark Example (4) above suggests we could construct a diffeomorphism $h : \mathbb{R} \to$
$\mathbb{R}$ by composing the following:

$$\mathbb{R} \xrightarrow{e^x} (0, \infty) \xrightarrow{\sqrt{x}} (0, \infty) \xrightarrow{\ln(x)} \mathbb{R}.$$

$$x \longrightarrow h(x) = \ln(\sqrt{e^x}) = \frac{1}{2}x. \tag{49}$$

It turned out that h is a polynomial of degree 1.

Without lost of generality, fix $a \in \mathbb{R}$ and let $f : \mathbb{R} \to \mathbb{R}$ be a C^1-function such that
$f'(x) > 0$ for all $x \in \mathbb{R}$. So f is monotone increasing. Given a sequence $\{x_n\}_{n \in \mathbb{N}} \subset$
$\mathbb{R}$, such that $\lim_{n \to \infty} x_n = \infty$, we have a sequence $\{c_n\}_{n \in \mathbb{N}} \subset \mathbb{R}$ such that $c_n \in (a, x_n)$
and

$$f(x_n) - f(a) = f'(c_n)(x_n - a) > 0. \tag{50}$$

From the examples above, we learn that the right-hand side term of Eq. (50) reveals
the behavior of the function when $n \to \infty$. The surjectivity does depend on the decay
of f, which is determined by $\lim_{n \to \infty} | f'(c_n) |$. There are two possibilities:
(1) $\lim_{n \to \infty} | f'(c_n)(x_n - a) |= L < \infty$, as in examples (1) and (2);
(2) $\lim_{x \to \infty} | f'(c_n)(x_n - a) |= \infty$, as in examples (3) and (4).

Examples (1), (2) and (3) above are not proper functions, while example (4)
is proper. Examples (3) and (4) are in the twilight zone, as we can see assuming
the existence of $r > 0$ such that $| f'(x) |> r$ for all $x \in \mathbb{R}$. In this case the limit
$\lim_{n \to \infty} | f'(c_n)(x_n - a) |= \infty$ yields $\lim_{n \to \infty} f(x_n) = \infty$. Therefore the surjectivity of f follows from the Intermediate Value Theorem. The following theorem
closes Question 8.1 when $n = 1$.

Theorem 16 *Let $f : \mathbb{R} \to \mathbb{R}$ be a local diffeomorphism such that*

$$| f'(x) |> 0$$

*for all $x \in \mathbb{R}$. Then f is a global diffeomorphism if and only if for any fixed $a \in \mathbb{R}$,
we get*

$$\lim_{x \to \pm\infty} | f'(x)(x - a) |= \infty. \tag{51}$$

Finally, for the case $n = 1$, we list the followings facts:

(1) The $\mathfrak{JC}_0^1$ conjecture is true.

Let $P(x) = \sum_{i=0}^{k} p_k x^k$ be an invertible polynomial and let $Q(x) = \sum_{l=0}^{l} q_l x^l$ be the inverse polynomial, i.e., $(Q \circ P)(x) = x$. If $Q'(P(x)) \cdot P'(x) = 1$ for all $x \in \mathbb{R}$, then $\deg(P) = \deg(Q) = 0$ and $\deg(P) = \deg(Q) = 1$. Therefore

$$P(x) = p_0 + p_1 x \ \Rightarrow \ Q(x) = q_0 + p_1^{-1} x.$$

(2) The $\mathfrak{WJC}_0^1$-Conjecture is false. The polynomial $p(x) = x^3 + x$ satisfies $p'(x) = 3x^2 + 1 > 0$ and $p(\mathbb{R}) = \mathbb{R}$. So p is a global diffeomorphism; neverthless, $p^{-1}(x)$ is not a polynomial.[6]

8.2 Case $n \geq 2$

As before, $\mathbb{K} = \mathbb{R}$ or $\mathbb{K} = \mathbb{C}$. When $n \geq 2$, a new novel starts. The simplest case $\mathfrak{JC}_0^2$ remains unproven. These cases are far more difficult since there are no results similar to the Intermediate Value and Mean Value Theorems. Now injectivity becomes a hard issue and surjectivity becomes more difficult to be achieved.

The main ingredients to achieve surjectivity are the InFT and to verify if the map is proper.

Examples Below we give some simple examples to enlighten our understanding about polynomial and differentiable maps.

(1) Let $p : \mathbb{R} \to \mathbb{R}$ be a polynomial. A very simple example of a polynomial auto-morphism is the polynomial map $P : \mathbb{R}^2 \to \mathbb{R}$ given by

$$P(x, y) = \left(x + p(y), y\right),$$

$$dP_{(x,y)} = \begin{pmatrix} 1 & \frac{dp}{dy} \\ 0 & 1 \end{pmatrix} \ \Rightarrow \ J(P)(x, y) = 1.$$

P is a proper map.

(2) Let's consider the following question: "is every quadratic polynomial defined on $\mathbb{K}$ the product of two linear polynomials?" Given a quadratic polynomial $f(x) = x^2 + \alpha x + \beta$, we would like to prove that

$$x^2 - \alpha x + \beta = (x - a)(x - b).$$

The decomposition above is equivalent to finding solutions to the system

$$\begin{cases} a + b = \alpha, \\ ab = \beta, \end{cases} \tag{52}$$

---

[6] Solve the cubic equation $x^3 + x - \alpha = 0$ to obtain the inverse.

or to solve the quadratic equation $x^2 - \alpha x + \beta = 0$. The existence of the solution to the system (52) is equivalent to proving that the polynomial map $P : \mathbb{R}^2 \to \mathbb{R}^2$, $P(x, y) = (x + y, xy)$, is surjective. Suppose there are two solutions x_1 and x_2 to $P(x, y) = (\alpha, \beta)$, then $P^{-1}(\alpha, \beta) = \{(x_1, x_2), (x_2, x_1)\}$. Two possibilities can occur according to the field $\mathbb{K}$:

(i) If $\mathbb{K} = \mathbb{C}$, then equation $P(x, y)$ is surjective but it is not injective.

(ii) If $\mathbb{K} = \mathbb{R}$, then $P(x, y)$ is not surjective and admits a solution if and only if $\alpha^2 \geq 4\beta$.

P is a proper map. This example corresponds to the Fundamental Theorem of Algebra in its more humble case. Observe that

$$d P_{(x,y)} = \begin{pmatrix} 1 & 1 \\ y & x \end{pmatrix} \quad \Rightarrow \quad J(P)(x, y) = x - y.$$

The singular set corresponds to the cases when the linear factors are identical, i.e., $f(x) = (x - a)^2$.

(3) Let $P : \mathbb{R}^2 \to \mathbb{R}^2$ be the polynomial map $P(x, y) = (xy, -x^2 + y^2)$. Then

$$d P_{(x,y)} = \begin{pmatrix} y & x \\ -2x & 2y \end{pmatrix} \quad \text{and} \quad J(P)(x, y) = 2(x^2 + y^2). \tag{53}$$

The map P shares the following properties:

(i) $J(P)(x, y) = 0$ if and only if $(x, y) = (0, 0)$.

(ii) P is surjective.

(iii) For any point $(a, b) \neq (0, 0)$, we have $\#P^{-1}(a, b) = 2$ and $\#P^{-1}(0, 0) = 1$. So P is not injective.

(iv) P is a proper map.

Fix a point $(a, b) \in \mathbb{R}^2$ and consider the equations

$$\begin{matrix} xy = a \\ -x^2 + y^2 = b \end{matrix} \quad \Rightarrow \quad \begin{cases} y = \frac{a}{x} \\ x^4 + bx^2 - a^2 = 0. \end{cases}$$

Solving the quadratic equation $z^2 + bz - a^2 = 0$, with the discriminant $\triangle = b^2 + 4a^2 \geq 0$, we get

$$z_1 = \frac{-b + \sqrt{\triangle}}{2}, \quad z_2 = \frac{-b - \sqrt{\triangle}}{2}.$$

So the biquadratic equation admits only two[7] real solutions, namely $\alpha = \sqrt{z_1}$ and $-\alpha$. We get $P(\alpha, \frac{a}{\alpha}) = P(-\alpha, -\frac{a}{\alpha})$. So the items (ii) and (iii) are verified. Thus for all $(a, b) \in \mathbb{R}^2$, the solution set $P^{-1}(a, b)$ is the intersection of level curves, i.e.,

---

[7]If $\mathbb{K} = \mathbb{C}$, then $\#P^{-1}(a, b) = 4$ for all $(a, b) \neq 0$.

$$P^{-1}(a, b) = \{(x, y) \in \mathbb{R}^2 \mid xy = a\} \bigcap \{(x, y) \in \mathbb{R}^2 \mid -x^2 + y^2 = b\}.$$

The origin is a ramification point. Indeed, the behavior of P near the origin is similar to a rotation. Furthermore, $P\big((0, \infty) \times \mathbb{R}\big) = \mathbb{R}^2$.
So that a polynomial map $P : \mathbb{R}^2 \to \mathbb{R}^2$, $P(x, y) = \big(f(x, y)g(x, y)\big)$ defines a diffeomorphism, we need to find polynomials $f(x)$ and $g(x)$ that level curves define a coordinate system, i.e., for every point $(a, b) \in \mathbb{R}^2$, the solution set $P^{-1}(a, b) = f^{-1}(a) \cap g^{-1}(b)$ must satisfy $\# P^{-1}(a, b) = 1$. It is a difficult task to find two families of algebraic curves defining transversal foliations whose leaves $f^{-1}(a)$ and $g^{-1}(b)$ intersect at most one point for all $a, b \in \mathbb{R}$. It is even harder to find families in which infinitesimal parallelograms defined by the level curves have a constant area. The tangent vector to the curves $f^{-1}(a)$ and $g^{-1}(b)$ define the vector fields X_f and X_g, respectively, as to which Lie bracket is $[X_f, X_g] = 0$.

(4) Let $F : \mathbb{R}^2 \to \mathbb{R}^2$ be the map

$$F(x, y) = \big(e^x(1 - y^2), e^x y\big).$$

So $J(F)(x, y) = e^{2x}(1 + y^2) > 0$. It is an interesting example; note F is not surjective since $(-\infty, 0) \times \{y = 0\} \not\subset \mathrm{Im}(F)$, however, F is injective. So it is clear that the $\mathfrak{WIC}^2_\infty$-Conjecture is false.

(5) Let B^2 be the open 2-disk centered at the origin with radius 1 and let $F : \mathbb{R}^2 \to \mathbb{R}^2$ be the C^∞-map

$$F(x, y) = \left(\frac{x}{\sqrt{1 + x^2 + y^2}}, \frac{y}{\sqrt{1 + x^2 + y^2}} \right).$$

Therefore $\mid F(x, y) \mid < 1 \Rightarrow \mathrm{Im}(F) = B^2$. We have

$$dF_{(x,y)} = \frac{1}{(1 + x^2 + y^2)^{3/2}} \begin{pmatrix} 1 + y^2 & -xy \\ -xy & 1 + x^2 \end{pmatrix},$$

$$J(F)(x, y) = \frac{1}{(1 + x^2 + y^2)^{1/2}}.$$

$F : \mathbb{R}^2 \to B^2$ is a diffeomorphism. The map $F : \mathbb{R}^2 \to \mathbb{R}^2$ is injective and not proper.

(6) Let $F : \mathbb{R}^2 \setminus \{y = 0\} \to \mathbb{R}^2$ be the C^∞-map given by $F(x, y) = (x^2 y^6 + 2xy^2,$ $xy^3 + \frac{1}{y})$. So $J(F) = -2$. F is not injective since $F(-1, 1) = F(-1, 1) = (-1, 0)$ and is not surjective since $(a, 0) \notin \mathrm{Im}(F)$ for all $a > 0$. The images of the horizontal and vertical curves tell us a little bit about how F transforms the plane:

(i) For every constant b, the curves $(t, b) \overset{F_b}{\to} F(t, b)$ describe the parabola $x = y^2 - \frac{1}{b^2}$. The curve $F(t, -b)$ describes the same parabola as $F(t, b)$.

(ii) For a constant a, the curves $(a, t) \overset{F_a}{\to} F(a, t)$ show the following behaviors:
    (ii.1) If $a < 0$, we have $\mid a \mid = -a$ and

$$F_a \left((\mid a \mid)^{-1/4} \right) = F_a \left(-(\mid a \mid)^{-1/4} \right) = \left(-\sqrt{\mid a \mid}, 0 \right).$$

(ii.2) If $a > 0$, then $\text{Im}(F_a) \cap \{y = 0\} = \varnothing$.

This example is not a counterexample since its domain is a disconnected subset in $\mathbb{R}^2$, and it shows that the condition on the Jacobian is far from being sufficient.
(7) The examples above are showing that the $\mathfrak{WIC}^n_\infty$-Conjecture fails and they are all in the C^∞-category. Pinchuk [36] gave a counterexample to the $\mathfrak{WIC}^2_0$ on $\mathbb{R}$. The outline of the counterexample is described as follows:

(a) Define the polynomials

$$t = xy - 1, \; h = t(xt + 1), \; f = \frac{(xt + 1)^2 (h + 1)}{x} = (xt + 1)(t^2 + y).$$

(b) Define the polynomial $p = h + f$.
(c) Show that

$$J \left(p, -t^2 + 6th(h + 1) \right) = t^2 + \left(t + f(13 + 15h) \right)^2 + f^2 - fv,$$

and $v = v(f, h) = f + f(13 + 15h)^2 + 12h + 12fh$.
(d) Find a polynomial $u(f, h)$ such that $J(p, u) = -fv$. This is equivalent to finding a solution to the PDE

$$\frac{\partial u}{\partial f} - \frac{\partial u}{\partial h} = v.$$

(e) Define the polynomial $q = -t^2 - 6th(h + 1) - u(f, h)$. Then

$$J(p, q) = t^2 + f^2 + \left(t + f(13 + 15h) \right)^2 \geq 0.$$

So $J(p, q) = 0$ if and only if $t = 0$ and $f = 0$. But this is impossible, since if $t = 0$, then $f = \frac{1}{x}$. Therefore $J(p, q)(x, y) > 0$ for all (x, y). The polynomial map $F : \mathbb{R}^2 \to \mathbb{R}^2$, $F(x, y) = \left(p(x, y), q(x, y) \right)$, cannot be injective because the set $p^{-1}(0)$ is the disconnected algebraic curve $x^2 y - x + 1 = 0$. Indeed, the reasoning follows from the next claim:
Claim: Let $f, g \in C^\infty$ and consider the map $F = (f, g)$ such that $J(F)(p) \neq 0$ for all $p \in \mathbb{R}^2$. If $f^{-1}(c)$ is connected for all $c \in \mathbb{R}$, then F is injective. Moreover, if F is a polynomial map, then the reciprocal is true.
Pf: Let's assume $J(F)(p) > 0$ for all $p \in \mathbb{R}^2$. Suppose we have points $A, B \in \mathbb{R}^2$ such that $F(A) = F(B) = c$. Solving[8] the ODE

---

[8] In Chap. 4, we prove the existence and uniqueness of the solution for this ODE (IVP).

$$\gamma'(t) = (\nabla F)^{\perp}(\gamma(t)) \iff \begin{cases} \gamma_1'(t) &= -\frac{\partial f}{\partial y}(\gamma(t)), \\ \gamma_2'(t) &= \frac{\partial f}{\partial x}(\gamma(t)), \end{cases}$$

with the initial condition $\gamma(0) = A$, we get a parametrization of the level curve $f^{-1}(c)$ given by $\gamma(t) = (\gamma_1(t), \gamma_2(t))$. For all t, we have $F(\gamma(t)) = (c, g(\gamma(t)))$, and so

$$\frac{d}{dt}[F(\gamma(t))] = dF_{\gamma(t)}.\gamma'(t) = \begin{pmatrix} \frac{\partial f}{\partial x} & \frac{\partial f}{\partial y} \\ \frac{\partial g}{\partial x} & \frac{\partial g}{\partial y} \end{pmatrix} \cdot \begin{pmatrix} -\frac{\partial f}{\partial y} \\ \frac{\partial f}{\partial x} \end{pmatrix} = \begin{pmatrix} 0 \\ J(F)(\gamma(t)) \end{pmatrix},$$

$$\frac{d}{dt}[F(\gamma(t))] = (0, \nabla g(\gamma(t)).\gamma'(t)).$$

Considering the function $h(t) = g(\gamma(t))$, we have $h'(t) = J(F)(\gamma(t)) > 0$. Therefore h is increasing along the curve $f^{-1}(c)$ contradicting the condition $g(A) = g(B)$. If F is a polynomial map, the proof of the reciprocal is left as an exercise.

Exercises

(1) Let $f : \mathbb{C} \to \mathbb{C}$ be the entire analytic function[9]

$$f(z) = \sum_{n \geq 0} \frac{z^{2n+1}}{n!(2n+1)}.$$

The real map induced by f is $F : \mathbb{R}^2 \to \mathbb{R}^2$ given by $F(x, y) = (\mathrm{Re}(f(z)), \mathrm{Im}(f(z)))$.

   (i) Show that $f'(z) = e^{z^2}$. Deduce that $J(F)(x, y) > 0$ for all $(x, y) \in \mathbb{R}^2$.
   (ii) Show that f is surjective (hint: assume it is not surjective and apply the little Picard theorem).
   (iii) Show that f is not injective (hint: use the Big Picard theorem).

So far, a better understanding of the relationships between the conditions given by the hypothesis and the assertion of the $\mathfrak{JC}_0^n$-Conjecture is given in the equivalences proved in Bass-Connell-Wright [2].

Theorem 17 (Bass, Connell, Wright). *Let $p : \mathbb{C}^n \to \mathbb{C}^n$ be a polynomial mapping with constant nonzero Jacobian. Then the following statements are equivalent:*
(i) p is invertible and p^{-1} is a polynomial mapping.
(ii) p is injective.
(iii) p is proper.

Remark The theorem above considers the complex case. The hypothesis of being a polynomial map is essential.

---

[9]See Ref. [16].

As n increases, the question becomes much more difficult to tackle using methods based on properties of the determinant, since computing determinants grow factorially as n grows. We hope for a general property to ensure that the conjecture is true. To hope might be wishful, neverthless we live for it.

Theorem 17 suggests that to solve the $\mathfrak{J}\mathfrak{C}_0^n$-Conjecture, we need to prove that a map F satisfying the condition $J(F) \in \mathbb{K}^*$ is proper or injective.

The fact that it is not a proper map might be a problem for polynomial maps of many variables since the inverse image of a compact set may not be compact. Consider the example $P(x, y) = (xy, x^2 - xy)$; so we have $p^{-1}(0) = \{(0, y) \mid y \in \mathbb{R}\}$, i.e., the inverse image of a compact set is unbounded.

If we assume $F : \mathbb{R}^n \to \mathbb{R}^n$ is proper, since the image $F(\mathbb{R}^n)$ using a proper map is always[10] closed, it remains to prove that $F(\mathbb{R}^n)$ is also open to achieve surjectivity.

Let's discuss the cases assuming the polynomials are proper.

Proposition 4 *Let $P : \mathbb{K}^n \to \mathbb{K}^n$ be a polynomial mapping such that $J(P)(\mathbf{x}) \neq 0$ for all $\mathbf{x} \in \mathbb{K}^n$. Then for each $\mathbf{y} \in \mathbb{K}^n$, $P^{-1}(\mathbf{y})$ is finite.*

Proof Since dP_x has maximal rank, the map $P : \mathbb{K}^n \to \mathbb{K}^n$ defines a submersion with fibers over $y \in \text{Im}(P)$ that must be a discrete set of points. Let $P = (p_1, \ldots, p_n)$, such that $p_i : \mathbb{K}^n \to \mathbb{K}$ is a polynomial for all $1 \leq i \leq n$, and $\mathbf{y} = (y_1, \ldots, y_n)$. So

$$P^{-1}(\mathbf{y}) = p_1^{-1}(y_1) \bigcap \cdots \bigcap p_n^{-1}(y_n).$$

The derivative dP_x being non-singular implies that the set of gradient vectors

$$\{\nabla p_1(x), \ldots, \nabla p_n(x)\}$$

is linearly independent for all $x \in \mathbb{K}^n$; therefore the intersection is a 0-dimensional submanifold. Hence $P^{-1}(\mathbf{y})$ is a discrete set of points. The claim follows from the fact that a discrete algebraic set over $\mathbb{K} = C$ or $\mathbb{K} = \mathbb{R}$ is finite. $\square$

Remark Proposition 4 does not extend to smooth maps.

Once $\dim(P^{-1}(\mathbf{y})) = 0$, we can assign to each element $\mathbf{x} \in P^{-1}(\mathbf{y})$ the number

$$\text{sgn}\big(J(P)(\mathbf{x})\big) = \begin{cases} +1, & \text{if } J(P)(\mathbf{x}) > 0, \\ -1, & \text{if } J(P)(\mathbf{x}) < 0, \end{cases} \tag{54}$$

and so, we can define the Index of P at $\mathbf{y} \in \text{Im}(P)$ to be

$$\text{I}(P; \mathbf{y}) = \sum_{\mathbf{x} \in P^{-1}(\mathbf{y})} \text{sgn}\big(J(P)(\mathbf{x})\big). \tag{55}$$

The map $\mathbf{y} \to \text{I}(P; \mathbf{y})$ may not be continuous, as shown in Example 3 by the polynomial map $P(x, y) = (xy, -x^2 + y^2)$. In this way, we cannot guarantee $P^{-1}(\mathbf{y})$ is

---

[10]See Appendix A.

locally constant. However, if we assume $P^{-1}(\mathbf{y}) \in \mathbb{Z}$ is locally constant, then we get a well-defined Index since $\mathbb{K}^n$ is connected.

Definition 18 Let $P : \mathbb{K}^n \to \mathbb{K}^n$ be a proper polynomial mapping such that $J(P)(\mathbf{x}) \neq 0$ for all $\mathbf{x} \in \mathbb{K}^n$. If the Index $I : \mathbb{K} \to \mathbb{Z}$ is a locally constant function, the Index of P is

$$I(P) = I(P; \mathbf{y}), \qquad (56)$$

for all $\mathbf{y} \in \mathrm{Im}(P)$.

An interpretation can be given for the Index of P.
(1) Let the map $P : \mathbb{R}^{2n} \to \mathbb{R}^{2n}$ be induced by $\widehat{P} : \mathbb{C}^n \to \mathbb{C}^n$. So we have $J(P) = | J(\widehat{P}) |^2$. Therefore

$$I(P) = \#P^{-1}(\mathbf{y}). \qquad (57)$$

(2) Fix an orientation on $\mathbb{R}^n$ and let $P : \mathbb{R}^n \to \mathbb{R}^n$ be a polynomial map defining a local diffeomorphism preserving orientation. So $J(P)(\mathbf{x}) > 0$. In this case, we also have $I(P) = \#P^{-1}(\mathbf{y})$.

In the presence of a singularity, the Index may change. If the Index changes, the map must have a singularity.

Proposition 5 *Let $r \in \mathbb{N}$. Let $P : \mathbb{K}^n \to \mathbb{K}^n$ be a polynomial mapping with $J(P)(\mathbf{x}) \neq 0$ for all $\mathbf{x} \in K^n$ and such that $\mathrm{I}(P) = r$. Then $P : \mathbb{K}^n \to \mathrm{Im}(P)$ is a proper map onto its image.*

Proof Let $K \subset \mathrm{Im}(P)$ be a compact subset; we need to establish the compactness of $P^{-1}(K)$. Let $\{x_n\}_{n \in \mathbb{N}} \subset P^{-1}(K)$ be an infinite sequence. The compactness will be achieved if $\{x_n\}_{n \in \mathbb{N}}$ has a convergent subsequence with limit in $\mathrm{Im}(P)$. Consider the sequence $\{y_n\}_{n \in \mathbb{N}} \subset K$ given by $y_n = P(x_n)$. We can consider $\{y_n\}_{n \in \mathbb{N}}$ convergent, so let $y_\infty = \lim_{n \to \infty} y_n$ be the limit. Let's consider

$$P^{-1}(y_\infty) = \{a_1, \ldots, a_r\}.$$

Given the InFT, we have open subsets $U_i \ni a_i$ and $V_i \ni y_\infty$ such that $P : U_i \to V_i$ is a diffeomorphism. Taking $V = \bigcap_i V_i$, let $Q_i : V \to U_i$ be the inverse, i.e., $P \circ Q_i = \mathrm{id}$. Let's assume $\{y_n\}_{n \in \mathbb{N}} \subset V$, so $P(Q_i(y_n)) = y_n$ for all $n \in \mathbb{N}$. Therefore $Q_i(y_n) \in P^{-1}(y_n)$ and

$$P^{-1}(\{y_n \mid n \in N\}) \subset P^{-1}(V) = \bigcup_{i=1}^{r} U_i.$$

By construction, we have $x_n \in P^{-1}(y_n)$, so $x_n = Q_i(y_n)$ in which the index $i \in \{1, \ldots, r\}$ can be changing randomly. Nevertheless, for some $i \in \{1, \ldots, r\}$, the subsequence

$$\{x_{n_i}\} = \{x_n\} \bigcap \mathrm{Im}(Q_i)$$

has an infinite number of elements. Since $x_{n_i} = Q_i(y_n)$, it follows that

$$\lim_{n_i \to \infty} x_{n_i} = \lim_{n \to \infty} Q_i(y_n) = Q_i(y_\infty) = a_i.$$

Hence $P^{-1}(K)$ is compact.   □

Remark Proposition 5 extends to non-singular smooth maps.

8.3   Covering Spaces

The polynomial P in Proposition 5 is not injective when $r > 1$ and may not be surjective. To proceed, we will apply the theory of Covering Spaces. For further reading about Covering Spaces we recommend [33].

Definition 19 Let $\widetilde{X}$ and X be path-connected Hausdorff spaces. A map $f : \widetilde{X} \to X$ is a covering map if the following conditions are satisfied:
(i) f is surjective,
(ii) f is locally a homeomorphism.
In this case, $\widetilde{X}$ is a cover space for X.

A local homeomorphism may not be a covering map, since it may not be surjective. However, a proper map that is locally a homeomorphism is a covering map.
The subset $f^{-1}(\mathbf{x}_0)$ is called the fiber over $\mathbf{x}_0$. The fibers are discrete subsets of $\widetilde{X}$. When f is a polynomial map, the fiber is a finite set. Once an element $p_0 \in f^{-1}(\mathbf{x}_0)$ has been fixed, we can establish an order such that

$$f^{-1}(\mathbf{x}_0) = \{p_0, p_1, \dots, p_r\}.$$

This can be done for all points in $\widetilde{X}$. In this framework, we have a group G and an action $G \times \widetilde{X} \to \widetilde{X}$, $(g, x) \to g \cdot x$, such that:
(i) G acts freely on $\widetilde{X}$.
(ii) $G \cdot f^{-1}(\mathbf{x}_0) = f^{-1}(\mathbf{x}_0)$.
(iii) $X = \widetilde{X}/G$.
(iv) The order of G is $o(G) = \# f^{-1}(\mathbf{x}_0)$.
(v) If $\pi_1(\widetilde{X}) = 0$, then $\pi_1(X) = G$.

Lemma 3 *Let $P : \mathbb{K}^n \to \mathbb{K}^n$ be a polynomial mapping with $J(P)(\mathbf{x}) \neq 0$ for all $\mathbf{x} \in K^n$ and such that $I(P) = r \in \mathbb{N}$. Then $r = 1$ and $P : \mathbb{K}^n \to P(\mathbb{K}^n)$ is injective.*

Proof Let's sketch the proof, since it requires a short stroll along the banks of Algebraic Topology. Let's assume $r > 1$; then we can write $r = k.p$, p prime. Take an element $g \in G$, such that $g^k \neq e$, and consider the cyclic subgroup $H = \{e, (g^k), \dots, (g^k)^{r-1}\}$. So $o(H) = p$. The group H acts freely on $\mathbb{K}^n$; consequently it acts freely on the sphere $S^{n-1} \subset \mathbb{K}^n$. Taking the stereographic projection, we get an action on the n-sphere S^n with one fixed point. A corollary of Smith theory [41] claims this cannot happen, since the fixed point set by a finite group action on S^n must be a homology m-sphere with $m = -1, 0, 1, \ldots, n$. Here, $m = -1$ denotes the empty set and $m = 1$ is $S^0 = \{N, S\}$. N is the North Pole and S is the South Pole. Therefore for $m \geq 0$, there are at least two fixed points, one of which belongs to $\mathbb{K}^n$. Hence $r = 1$ and $P : \mathbb{K}^n \to P(\mathbb{K}^n)$ is injective. □

From now on, the hypothesis of being a polynomial map is essential. The demonstration of the next lemmas requires that we delve into algebraic techniques that are not within the scope of this textbook. Therefore we present an outline and leave the reader with the references.

Lemma 4 *Let $P : \mathbb{K}^n \to \mathbb{K}^n$ be an injective polynomial mapping with $J(P)(x) \neq 0$ for all $x \in K^n$. Then P is a global diffeomorphism.*

Proof (sketch) Given the theorem of Bialynicki-Birula and Rosenlicht [3], an injective polynomial map of affine spaces of the same dimension is surjective, so $P : \mathbb{K}^n \to \mathbb{K}^n$ is a bijection. It follows from the InFT that P is a diffeomorphism. □

Lemma 5 *Let $P : \mathbb{K}^n \to \mathbb{K}^n$ be a bijective polynomial mapping with $J(P)(x) \neq 0$ for all $x \in K^n$. Then P is biregular, i.e, $P^*\mathbb{K}[x] = \mathbb{K}[x]$.*

Proof Let $\mathbb{K}(x)$ be the field of rational functions on $\mathbb{K}^n$. A bijection $P : \mathbb{K}^n \to \mathbb{K}^n$ induces a monomorphism $P^* : \mathbb{K}(x) \to \mathbb{K}(x)$, and so $\mathbb{K}(x)$ may be regarded as a finite dimensional vector space over the field $P^*\mathbb{K}(x)$ such that

$$\# f^{-1}(y) = \dim_{P^*\mathbb{K}(x)} \mathbb{K}(x).$$

Since $\# f^{-1}(y) = 1$, we have $P^*\mathbb{K}(x) = \mathbb{K}(x)$. P is a regular bijection between two smooth affine varieties, so by Zariski's main theorem then P is biregular. □

Proof Lemmas 4 and 5 conclude the proof of Theorem 17. □

8.4 *Degree Reduction*

Another strategy to tackle the $\mathfrak{J}\mathfrak{C}_0^n$-Conjecture was to analyze the question for polynomials of fixed degree. Stuart Wang proved in [44] that the $\mathfrak{J}\mathfrak{C}_0^n$ is true if $\deg(P) = 2$. The very simple proof of Wang's theorem given below is due to Oda-Yoshida [35].

Proposition 6 (Wang [44]). *If $\deg(P) \leq 2$, then the real $\mathfrak{J}\mathfrak{C}_0^n$ is true.*

Proof By Theorem 17 it suffices to prove that P is injective. Let's suppose $P(a) = P(b)$. Define the polynomial $Q(x) = P(x + a) - P(a)$ and let $c = b - a \neq 0$. So Q satisfies the following properties:

(i) $Q(c) = Q(0) = 0$,

(ii) $\deg(Q) \leq 2$,

(ii) $dQ_x = dP_{x+a} \Rightarrow J(Q)(x) = J(P)(x+a) \in \mathbb{R}^*$.

Now, decomposing $Q = Q_1 + Q_2$ into its homogeneous components Q_1 and Q_2 with degree 1 and 2, respectively, then we get

$$Q(tc) = tQ_1(c) + t^2 Q_2(c).$$

Differentiation gives

$$Q_1(c) + 2tQ_2(c) = \frac{d}{dt}(Q(tc)) = dQ_{tc} \cdot c \neq 0,$$

for all $t \in \mathbb{R}$. Substituting $t = \frac{1}{2}$ gives $Q(c) \neq 0$, a contradiction. Hence P is injective. $\qquad\square$

Proposition 6 improves the result for polynomials of degree 1. In $\mathbb{R}^2$, we know from [5] that any local diffeomorphism $F : \mathbb{R}^2 \to \mathbb{R}^2$ with $\deg(F) \leq 4$ satisfies the real Jacobian conjecture. The next theorem says we don't have to check the $\mathfrak{JC}_0^n$-Conjecture for all polynomials, only tfor the polynomials of degree 3. However, the task is not easy since we must check the conjecture for all polynomials of degree 3 and for all dimensions n.

Theorem 18 (Bass-Connell-Wright [2]). *If the Jacobian Conjecture $\mathfrak{JC}_0^n$ holds for all polynomial $P \in \mathbb{C}[x_1, \ldots, x_n]$ with $\deg(P) \leq 3$ and for all $n \in \mathbb{N}$, then the Jacobian Conjecture $\mathfrak{JC}_0^n$ holds for all polynomials $P \in \mathbb{C}[x_1, \ldots, x_n]$ such that $J(P) \in \mathbb{R}^*$.*

Proof (sketch) They proved it suffices to prove the Jacobian Conjecture for all polynomial P of the form

$$P(x) = (x_1 + H_1(x), \ldots, x_n + H_n(x)), \tag{58}$$

suh that each H_i is either zero or homogenous of degree 3. $\qquad\square$

Improvements: The following results improved the theorem above.

(1) Drużkowski proved in [11] that the polynomials H_i, $1 \leq i \leq n$, in (58) can be taken as

$$H_i(x_1, \ldots, x_n) = \left(\sum_{j=1}^{n} a_{ji} x_j\right)^3. \tag{59}$$

(2) If $n \leq 7$, Hubbers proved in [25] that the Jacobian Conjecture $\mathfrak{JC}_0^n$ holds for all polynomial P of the form $P(x) = x + H(x)$, and H is a polynomial given by (59).

Chapter 2
Linear Operators in Banach Spaces

In this chapter we present a brief introduction to basic concepts of Operator Theory, and some relevant classes of operators are introduced to what follows thereafter. The most explored Banach spaces in the text are the spaces $E = (C^k(K; \mathbb{R}^m), \| f \|_{C^k})$, as defined in Appendix A. Eventually, the spaces L^p are used, but we avoid them since more care is required with the analysis. Our larger goal is to study the differentiable maps; for this purpose the spaces C^k are enough.

1 Bounded Linear Operators on Normed Spaces

The concept of differentiable maps defined on a Banach space requires prior knowledge of examples and properties of linear operators between Banach spaces. Due to the complexity, we will cover only enough as needed for our purposes; more details are found in books on Functional Analysis. Let $\mathbb{K} = \mathbb{R}$ or $\mathbb{C}$ and let E, F be normed $\mathbb{K}$-vector spaces. A function $T : E \to F$ is a $\mathbb{K}$-linear operator if, for all $u, v \in E$ and $a, b \in \mathbb{K}$,

$$T(au + bv) = aT(u) + bT(v).$$

When $F = \mathbb{K}$, the operator T is a linear functional. The space of linear functions on E is denoted by E^*. We note that the concept is defined for normed spaces; the need to be a Banach space will arise according to the context whenever completeness is necessary.

Example 1 Examples of linear operators.

(1) Consider E, F vector spaces such that $\dim(E) = n$ and $\dim(F) = m$. A $\mathbb{K}$-linear operator $T : E \to F$ can be represented by a matrix with coefficients in $\mathbb{K}$. By fixing the bases $\beta_E = \{e_1, \ldots, e_n\}$ and $\beta_F = \{f_1, \ldots, f_m\}$ of E and F,

© Springer Nature Switzerland AG 2021
C. M. Doria, *Differentiability in Banach Spaces, Differential Forms and Applications*,
https://doi.org/10.1007/978-3-030-77834-7_2

respectively, we get $T(x) = A.x$. The matrix $A = (a_{ij}) \in M_{m \times n}(\mathbb{K})$ is defined by $T(e_i) = \sum_k a_{ki} f_k$, $1 \le i \le n$.

(2) On the functional spaces $E = C^1([a, b])$ and $F = C^0([a, b])$, we consider the $\mathbb{R}$-linear differential operator $\mathcal{D} : E \to F$, $(\mathcal{D}f)(x) = f'(x)$.

(3) On the functional spaces $E = C^0([a, b])$ and $F = C^1([a, b])$, we have the integral operator $\mathcal{I} : E \to F$, $\mathcal{I}(f)(x) = \int_a^x f(t)dt$, which is an $\mathbb{R}$-linear operator.

(4) Given a function $f \in E = C^0([a, b]; \mathbb{C})$, we define the linear functional $L_f :$ $E \to \mathbb{R}$, $L_f(g) = \int_a^b (f.\overline{g})(x)dx$. Consider the Hermitian product on E as

$$< f, g >= \int_a^b (f.\overline{g})(x)dx.$$

The space E, endowed with the induced norm by the Hermitian product, is not necessarily a Banach space. A fundamental question is to determine the continuous linear functionals defined over E. When E has finite dimension, every linear functional $f : E \to \mathbb{R}$ can be written as $f(v) =< u_f, v >$, with $u_f \in \mathbb{R}^n$, as proved in Proposition (2 in Appendix A). Similarly, when E is an infinite dimensional Hilbert space, we can apply the Representation Theorem of Riesz (12 in Appendix A) to represent any linear functional $f : E \to \mathbb{R}$ as $f(v) =< u_f, v >$.

(5) Given a continuous function $k : [a, b] \times [a, b] \to \mathbb{R}$, we define the $\mathbb{R}$-linear integral operator $K : C^0([a, b]) \to C^0([a, b])$ as

$$(Kf)(x) = \int_a^b k(x, y) f(y) dy.$$

The function k is called the kernel of K.

(6) Let E, F be normed spaces. Consider $a_{ij}, b_j, c \in C^k(E; \mathbb{R})$. The differentiable operator $\mathcal{D} : C^k(E, F) \to C^{k-2}(E, F)$ given by

$$\mathcal{D}f = \sum_{i,j=1}^n a_{ij}(x) \frac{\partial^2 f}{\partial x_i \partial x_j} + \sum_{i=1}^n b_i(x) \frac{\partial f}{\partial x_i} + c(x)$$

is $\mathbb{R}$-linear.

Definition 1 Let $(E, || \cdot ||_E)$, $(F, || \cdot ||_F)$ be normed spaces. A linear operator $T : E \to F$ is bounded if there is $M > 0$ such that

$$|| T(u) ||_F \le M || u ||_E,$$

for all $u \in E$. Let $\mathcal{L}(E, F)$ be the space of bounded linear operators $T : E \to F$; when $E = F$, let $\mathcal{L}(E) = \mathcal{L}(E, E)$.

Proposition 1 *Let $(E, || \cdot ||_E)$ and $(F, || \cdot ||_F)$ be normed spaces. For any $T \in \mathcal{L}(E, F)$, the following items are equivalent;*

  (i) T is continuous.
 (ii) T is continuous at the origin.
(iii) T is a bounded operator.

Proof We have $T(0) = 0$. Let's check the following directions;
(i) $\Rightarrow$ **(ii):** straightforward.
(ii) $\Rightarrow$ **(iii):** for any $x \neq 0 \in E$, consider $y = \delta \frac{x}{||x||}$. Taking $\epsilon = 1$, we have $\delta > 0$ such that, if $|| \, y \, || < \delta$, then $|| \, T(y) \, || < 1$. In this way, we get

$$|| \, T(y) \, || = \delta \frac{T(x)}{|| \, x \, ||} < 1 \;\Rightarrow\; || \, T(x) \, || < \frac{1}{\delta} \, || \, x \, || \,.$$

(iii) $\Rightarrow$ **(i):** Let $\{x_n\}_{n \in \mathbb{N}}$ be a sequence such that $x_n \to x \in E$. Therefore

$$|| \, T(x_n) - T(x) \, || = || \, T(x_n - x) \, || < M \, || \, x_n - x \, || \to 0.$$

$\square$

The space $\mathcal{L}(E, F)$ is a vector space with the following operations: take $T, S \in \mathcal{L}(E, F)$ and $a \in \mathbb{K}$

$$(i)\ (T + S)(x) = T(x) + S(x),$$
$$(ii)\ (aT)(x) = aT(x).$$

The operator norm of an operator $T \in \mathcal{L}(E, F)$ is defined as

$$| \, T \, | = \inf \left\{ M \in \mathbb{R};\ || \, T(x) \, ||_F \leq M \, || \, x \, ||_E \right\}.$$

It is straightforward that $|| \, T(x) \, || \leq | \, T \, | \, . \, || \, x \, ||$, for all $x \in E$. Moreover,

$$| \, T \, | = \sup_{||x||=1} || \, T(x) \, ||_F = \sup_{x \neq 0} \frac{|| \, T(x) \, ||_F}{|| \, x \, ||_E}$$

and $| \, S \circ T \, | \leq | \, S \, | \, . \, | \, T \, |$. Now we consider the normed space $\big(\mathcal{L}(E, F), | \, . \, |\big)$.

Proposition 2 *Let E and F be Banach spaces and let $| \, . \, |$ be the operator norm. So the normed space $\big(\mathcal{L}(E, F), | \, . \, |\big)$ is a Banach space.*

Proof For all $x \in E$, $|| \, T_n(x) - T_m(x) \, ||_F \leq | \, T_n - T_m \, | \, . \, || \, x \, ||_E$. Consider $\{T_n\}_{n \in \mathbb{N}}$ a Cauchy sequence in $\mathcal{L}(E, F)$. Since $| \, T_n - T_m \, | \to 0$, when $m, n \to \infty$, the sequence $\{T_n(x)\}_{n \in \mathbb{N}} \subset F$ is a Cauchy sequence for all $x \in E$. Since F is complete, define $T(x) = \lim_n T_n(x) \in F$. The linearity of T follows from

$$T(ax + by) = \lim_{n \to \infty} T_n(ax + by) = a \lim_{n \to \infty} T_n(x) + b \lim_{n \to \infty} T_n(y) = aT(x) + bT(y).$$

We have to check that T is a bounded operator and the sequence $\{T_n\} \subset \big(\mathcal{L}(E, F), |\,.\,|\big)$ converges to T. Since $\{T_n\}_{n \in \mathbb{N}}$ is Cauchy, given $\epsilon > 0$, we have $n(\epsilon)$ such that if $m, n > n(\epsilon)$, then

$$|| T_n(x) - T_m(x) ||_F \leq | T_n - T_m | \cdot || x ||_E < \epsilon || x ||_E . \tag{1}$$

Taking the limit $m \to \infty$, it follows that $|| T_n(x) - T_m(x) ||_F < \epsilon || x ||_E$, for all $x \in E$. Then, if $n_0 > n(\epsilon)$, we have

$$|| T(x) ||_F \leq || T_{n_0}(x) - T(x) ||_F + || T_{n_0}(x) ||_F < \epsilon || x ||_E + | T_{n_0} | \cdot || x ||_E$$

for all $x \in E$. Therefore $T \in \mathcal{L}(E, F)$ since $|| T(x) ||_F \leq (\epsilon + | T_{n_0} |) || x ||_E$. From (1) above, we have

$$| T_n - T | = \sup_{x \neq 0} \frac{|| T_n(x) - T(x) ||_F}{|| x ||_E} \leq \epsilon, \text{ for every } n > n(\epsilon).$$

Hence $\lim_{n \to \infty} T_n = T$.   □

Henceforth we will always consider $\mathcal{L}(E, F)$ to be endowed with the topology induced by the operator norm $| T | = \sup_{||x||=1} || T(x) ||$.

The theory of Linear Algebra for infinite dimensional vector spaces requires much more attention to topological issues. Several results proved for finite dimensional vector spaces are false in the general Banach space settings. Basic questions concerning continuity, the closedness of the graph and the closed range property of a linear operator require special attention. Continuity is understood in Proposition 1 and the closed range is fully answered in the next section.

Example 2 In what follows, we will give two examples of bounded linear operators defined on $C^0([a, b])$ considering different norms. Let $k \in C^0([a, b])$ and consider the integral linear operator

$$(Kf)(x) = \int_a^b k(x, y) f(y) dy.$$

(1) Let $E = \big(C^0([a, b]), || \,.\, ||_0 \big)$, so $K \in \mathcal{L}(E)$.
    First, let's check that K is well-defined, i.e., if $f \in E$, then $Kf \in E$. By Theorem 4 in Appendix A, $k(x, y)$ is absolutely continuous, so given $\epsilon > 0$, we have $\delta(\epsilon) > 0$ such that if $(x, y), (\tilde{x}, \tilde{y}) \in [a, b] \times [a, b], | \tilde{x} - x | < \delta(\epsilon)$ and $| \tilde{y} - y | < \delta(\epsilon)$, then

$$\left|(Kf)(\tilde{x}) - (Kf)(x)\right| = \left| \int_a^b (k(\tilde{x}, y) - k(x, y)) f(y) dy \right| \leq$$

$$\leq \int_a^b | k(\tilde{x}, y) - k(x, y) | \cdot | f(y) | dy < \epsilon \int_a^b | f(y) | dy \implies Kf \in C^0.$$

Since k is continuous, we have $M > 0$ such that $|k(x, y)| \le M$ in $[a, b] \times [a, b]$. Then for all $x \in \mathbb{R}$,

$$|(Kf)(x)| \le \int_a^b |k(x, y)| . |f(y)| \, dy \le M \int_a^b |f(y)| \, dy \le M(b - a) \|f\|_0 .$$

Therefore $\|K\|_0 \le M(b - a)$ and $K \in \mathcal{L}(E)$.

(2) Let $F = (C^0([a, b]), \|.\|_2)$, so $K \in \mathcal{L}(F)$.

The norm $\|f\|_{L^2} = \left(\int_a^b |f(x)|^2 \, dx \right)^{1/2}$ is induced by the inner product

$$< f, g > = \int_a^b f(x)\overline{g(x)}dx,$$

and satisfies the Cauchy-Schwarz inequality

$$|< f, g >| \le \|f\|_{L^2} . \|g\|_{L^2} .$$

Applying the inequality to $(Kf)(x)$, we get

$$|(Kf)(x)| \le \int_a^b |k(x, y)| . |f(y)| \, dy \le \left(\int_a^b |k(x, y)|^2 \, dy \right)^{1/2} . \left(\int_a^b |f(y)|^2 \, dy \right)^{1/2} .$$

Let $C^2 = \int_a^b \int_a^b |k(x, y)|^2 \, dydx$, so

$$\left(\|(Kf)\|_{L^2} \right)^2 \le \int_a^b \left(\int_a^b |k(x, y)|^2 \, dy \right) . \|f\|_{L^2}^2 \, dx \le$$

$$\le \left(\int_a^b \int_a^b |k(x, y)|^2 \, dydx \right) . \|f\|_{L^2}^2 \le C^2 . \|f\|_{L^2}^2 .$$

Therefore $\|(Kf)\|_2 \le C . \|f\|_{L^2}$ and $K \in L(F)$.

2 Closed Operators and Closed Range Operators

Let E and F be Banach spaces and let $T : E \to F$ be a linear operator. In Appendix A, Proposition 4, we give necessary and sufficient conditions for having a complementary closed subspace or, equivalently, a topological complement. They are useful to understand when the Kernel or the Range (Image) of a linear operator admits a complementary closed subspace. In the infinite dimension, several technical difficulties make the analysis harder than in the finite dimension. For the remainder of the text, the following notations are used;

- $\mathcal{D}(T)$ denotes the domain of T.
- $\mathcal{R}(T)$ denotes the range of T. When referring to the range, we may also use the symbols Im(T) or $T(E)$.
- Ker(T) $= \{x \in \mathcal{D}(T) \mid T(x) = 0\}$ denotes the Kernel of T.

The most elementary issues are:

(i)   A linear operator may not be continuous.
(ii) The graph Gr(T) $= \{(x, T(x)) \mid x \in \mathcal{D}(T)\}$ may not be closed.
(iii) The range $\mathcal{R}(T)$ may not be closed.

Definition 2 Let E, F be Banach spaces and let $T : E \to F$ be a linear operator with the domain $\mathcal{D}(T) \subset E$. T is a closed operator if the graph Gr(T) is a closed subset of $E \times F$.

Indeed, T is a closed operator if given any sequence $\{x_n\}_{n \in \mathbb{N}} \subset \mathcal{D}(T)$ such that $\lim x_n = x \in E$ and $\lim T(x_n) = y \in F$, then $x \in D(T)$ and $T(x) = y$. Considering the norm

$$\| x \|_{\text{Gr(T)}} = \| x \|_E + \| T(x) \|_F,$$

T is closed if $\big(\mathcal{D}(T), \| . \|_{\text{Gr(T)}}\big)$ is a Banach space. T is a closable operator when T admits an extension $S : \mathcal{D}(S) \to F$ such that $\mathcal{D}(T) \subset \mathcal{D}(S)$ and $S(x) = T(x)$ for all $x \in \mathcal{D}(T)$. Later, in Section 2.8, when treating unbounded linear operators, we will give an example of a non-closable operator. There we will also show more details concerning the closable operators defined on Hilbert spaces. For a full treatment of the topic, we recommend [20, 39].

The third issue concerns the closed range, which is a real pain in the neck. Given a linear operator $T : E \to F$, let $T(E)$ be the range of T.

Theorem 1 *Let $T : E \to F$ be a bounded linear operator. If the range $T(E)$ is closed in F, then given $\epsilon > 0$, there is a constant $N \in \mathbb{R}$, $N > 0$, such that for all $x \in E$ there is a point $x' \in E$ such that:*

(i) $\| T(x) - T(x') \|_E \le \epsilon$.
(ii) $\| x' \|_E \le N \| T(x) \|_F$.

Proof The image-set $T(E)$ is a Banach space. Let $B_E = \{x \in E; \mid x \mid < 1\}$ be the unit open ball. We split the proof into some steps:

Step 1: $T(E) = \cup_{n=1}^{\infty} n\overline{T(B_E)}$.

Let $x \in E$ and $n \in \mathbb{N}$ be such that $\| x \|_E \le n$. So $\frac{1}{n} \| x \|_E \in B_E$ and

$$T(x) = nT\left(\frac{1}{n}x\right) \in nT(B_E) \Rightarrow T(E) \subseteq \bigcup_{n=1}^{\infty} n\overline{T(B_E)}.$$

Since $T(E)$ is closed, $n\overline{T(B_E)} \subseteq \cup_{n=1}^{\infty} T(E)$. Thus, $T(E) = \cup_{n=1}^{\infty} n\overline{T(B_E)}$.
Let $B_{T(E)} = \{y \in T(E); \mid y \mid < 1\}$.

Step 2: We have $r > 0$ such that $r\,B_{T(E)} \subseteq 2n\overline{T(B_E)}$.

Given the Baire Category Theorem, $T(E)$ has a non-empty interior, so there must be an $m \in \mathbb{N}$ such that $m\overline{T(B_E)}$ has a non-empty interior. Taking $y_0 \in \mathrm{int}\left(m\overline{T(B_E)}\right)$, let $r > 0$ be such that

$$B_r(y_0) \subseteq \mathrm{int}\left(m\overline{T(B_E)}\right).$$

Therefore whenever $\|\,y_0 - T(x)\,\| < r$, we have $T(x) \in \mathrm{int}\left(m\overline{T(B_E)}\right)$. In particular, if $\|\,T(x)\,\| \le 1$, then

$$y_0 + rT(x) \in \mathrm{int}\left(m\overline{T(B_E)}\right).$$

Let $\{\alpha_n\}_{n \in \mathbb{N}}$ and $\{\beta_n\}_{n \in \mathbb{N}}$ be sequences in B_E such that

$$\lim_n mT(\alpha_n) = y_0, \quad \lim_n mT(\beta_n) = y_0 + rT(x).$$

Having the above sequences is a consequence of $T(E)$ being a Banach space. Define the sequence $\gamma_n = \frac{\beta_n}{2} - \frac{\alpha_n}{2}$ and note that $\{\gamma_n\}_{n \in \mathbb{N}}$ is a sequence in B_E since

$$\|\,\gamma_n\,\| \le \frac{1}{2}\,\|\,\alpha_n\,\| + \frac{1}{2}\,\|\,\beta_n\,\| \le 1.$$

Furthermore, $\lim 2mT(\gamma_n) = rT(x)$. Therefore $rT(x) \in 2m\overline{T(B_E)}$ for every $x \in B_E$. Hence $r\,B_{T(E)} \subseteq 2m\overline{T(B_E)}$.

Step 3: Final;

Let $x \in E$ be such that $T(x) \ne 0$. In the last step we have

$$r \cdot \frac{T(x)}{\|\,T(x)\,\|} \in \overline{B_{T(E)}} \subseteq 2m\overline{T(B_E)}.$$

So given $\epsilon > 0$, there is $z \in B_E$ such that

$$\left\|\, r \cdot \frac{T(x)}{\|\,T(x)\,\|} - 2mT(z)\right\| < \epsilon \frac{r}{\|\,T(x)\,\|},$$

and so

$$\left\|\,T(x) - \frac{2m\,\|\,T(x)\,\|}{r}T(z)\right\| < \epsilon.$$

Taking $x' = \frac{2m\|T(x)\|}{r}z$, we have $\|\,T(x) - T(x')\,\| < \epsilon$. Moreover, since $z \in B_E$,

$$\|\,x'\,\| = \left\|\,\frac{2m\,\|\,T(x)\,\|}{r}z\right\| \cdot \|\,z\,\| \le \frac{2m}{r}\,\|\,T(x)\,\|\,.$$

Defining $N = \frac{2m}{r}$, we have $\| x' \| \le N \| T(x) \|$.   □

Theorem 2 *Let $T : E \to F$ be a bounded linear operator. The subspace $T(E)$ is closed if and only if there is a finite constant $M > 0$, such that $y = T(x)$ and $\| x \|_E \le M \| y \|_F$.*

Proof ($\Rightarrow$) Consider the sequence $\{u_n\}_{n \in \mathbb{N}}$ constructed as follows: given $\epsilon_1 = \frac{\|y\|}{2}$, by Theorem 2, there is $u_1 \in E$ such that

$$\| y - T(u_1) \| < \frac{\| y \|}{2} \quad \text{and} \quad \| u_1 \| \le N \| y \| .$$

Given $\epsilon_2 = \frac{\|y\|}{2^2}$, by the same argument we have $u_2 \in E$ such that

$$\big\| [y - T(u_1)] - T(u_2) \big\| < \frac{\| y \|}{2^2} \quad \text{and} \quad \| u_2 \| \le N \| y - T(u_1) \| \le N . \frac{\| y \|}{2}.$$

Recursively, we have $u_n \in E$ such that

$$\| y - \sum_{i=1}^{n} T(u_i) \| < \frac{\| y \|}{2^n} \quad \text{and} \quad \| u_n \| \le N \frac{\| y \|}{2^{n-1}}.$$

Defining $x_n = \sum_{i=1}^{n} u_i$, we have $\| y - T(x_n) \| < \frac{\|y\|}{2^n}$ and

$$\| x_n \| = \| \sum_{i=1}^{n} u_k \| \le \sum_{i+1}^{n} \| u_k \| \le \sum_{i=1}^{n} N \frac{\| y \|}{2^{i-1}} \le 2N \| y \| .$$

The sequence $\{x_n\}_{n \in \mathbb{N}}$ is a Cauchy sequence since

$$\| x_n - x_m \| = \sum_{i=m+1}^{n} \| u_i \| \le \sum_{i=m+1}^{n} N \frac{\| y \|}{2^i} \to 0.$$

Let $x = \lim x_n$, so $y = T(x)$ and $\| x \| \le 2N \| y \|$.
($\Leftarrow$) Let $\{y_n\}_{n \in \mathbb{N}} \subset T(E)$ be a convergent sequence and $y = \lim y_n$. Therefore given $\epsilon > 0$, we have $n_0 \in \mathbb{N}$, such that if $n, m > n_0$, then

$$\| y_n - y_m \|_F < \epsilon.$$

Consider $y_n = T(x_n)$. So the sequence $\{x_n\}_{n \in \mathbb{N}} \subset E$ is also Cauchy, since by the hypothesis,

$$\| x_n - x_m \|_E \le | T | . \| y_n - y_m \|_F < \epsilon.$$

If $x = \lim x_n$, then by continuity, we have $y = T(x)$. Hence $y \in T(E)$.   □

Corollary 1 *If $T : E \to F$ is a closed range operator, then the left-hand side inverse T^{-1} is such that $T^{-1}T = I_E$ exists and is continuous.*

Proof ($\Rightarrow$) Since T is a closed range operator, we have $M > 0$ such that $\| x \| \leq M \| T(x) \|$. So T is injective since $T(x) = T(y)$ implies $\| x - y \| = 0$. Continuity is straightforward, noticing that

$$\| T^{-1}(T(x)) \| \leq \frac{1}{M} \| T(x) \| .$$

On the other hand, if T^{-1} is continuous, then

$$\| x \| = \| T^{-1}(T(x)) \| \leq \| T^{-1} \| . \| T(x) \| .$$

The reverse claim follows upon taking $M = \frac{1}{\|T^{-1}\|}$. $\qquad\qquad\square$

Proposition 3 *Let $T : E \to F$ be a bounded linear operator. If $T(E)$ has a topological complement, then $T(E)$ is closed.*

Proof Let $W \subset F$ be the topological complement of $T(E)$, so W is closed and $F = T(E) \oplus W$. The product space

$$E \times W = \{(x, w) \mid x \in E, w \in W\}$$

is a Banach space endowed with the norm $\| (x, w) \|_p = \| x \|_E + \| w \|_F$ since W is closed.

Define the linear operator $S : E \times W \to F$ by

$$S(x, w) = T(x) + w.$$

S is bounded since

$$\begin{aligned}
\| S(x, w) \|_p &= \| T(x) + w \|_F \leq \| T(x) \|_F + \| w \|_F \leq | T | . \| x \|_E + \| w \|_F \\
&\leq | T | . \| x \|_E + | T | . \| w \|_F + (\| x \|_E + \| w \|_F) \\
&\leq (| T | + 1) . (\| x \|_E + \| w \|_F).
\end{aligned}$$

S is surjective since $F = T(X) \oplus W$. Since $E \times W$ and F are both Banach spaces and S is bounded, then $S(E \times W)$ is closed. By Theorem 2, there is $N > 0$, such that for every $y \in F$, we have an $(x, w) \in E \times W$ such that

$$S(x, w) = T(x) + w = y, \text{ and } \| (x, w) \| \leq N \| y \|_F .$$

If $y = T(x)$, then the W-component of y is $w = 0$, i.e., $S(x, 0) = y$. So,

$$\| (x, 0) \|_p = \| x \|_E \leq N \| y \|_F .$$

Hence $T(E)$ is closed.   □

A subspace $V \subset E$ is said to be finite co-dimensional if $\dim(E/V) < \infty$.

Corollary 2 *Let E and F be Banach spaces and let T : E → F be a bounded linear operator. If T(E) is finite codimensional, then T(E) is closed.*

Proof Let $W \subset F$ be a finite dimensional algebraic complement of $T(E)$. So W is a topological complement of $T(E)$ in F. By the last proposition, $T(E)$ is closed. □

It is important to keep in mind that:

(1) bounded operators are closed operators.
(2) bounded operators may not have a closed range. Consider the embedding $C^0([0, 1]) \hookrightarrow L^1([0, 1])$.
(3) there are closed unbounded operators.
(4) there are unbounded operators with closed range.
    We could say that the bounded linear operators having closed range are tame; otherwise, they can be wild objects. Examples will be studied in the next sections. The class of differential operators is a source of examples of unbounded linear operators.

3 Dual Spaces

A linear functional defined on E is a linear operator $f : E \to \mathbb{K}$ and $\mathbb{K} = \mathbb{R}$ or $\mathbb{K} = \mathbb{C}$. The dual space E^* is the space of bounded linear functionals defined on E. Consider the following example: an integral defines the linear functional $I :$ $C^0([a, b]) \to \mathbb{R}$,

$$I(f) = \int_a^b f(x)dx.$$

In infinite dimensions, the dual space plays a relevant role. In the finite dimension, E^* is isomorphic to E. The full description of all linear functionals in E^* becomes an important question. When E is a Hilbert space, Riesz's Representation Theorem answers this question by stating that for any $f \in E^*$, we have a vector $v_f \in E$ such that $f(u) =< v_f, u >$. An operator $T \in \mathcal{L}(E, F)$ induces the linear operator $T^* : F^* \to E^*, (T^*f)(u) = f(T(u))$, for all $u \in E$. Let E and F be Hilbert spaces; given $T \in \mathcal{L}(E, F)$, we define the adjoint operator $T^* \in L(F, E)$ as

$$< T(u), v >=< u, T^*(v) > . \tag{2}$$

Both definitions of the dual operator T^* coincide on a Hilbert space.

For some purposes which we will not delve into, some concepts of convergence in infinite-dimensional spaces are necessary. Let $(E, || \cdot ||)$ be a Banach space and $\{x_n\}_{n \in \mathbb{N}} \subset E$ a sequence;

(1) the convergence $x_n \to x$ is strong if $|| x_n - x ||_E \to 0$ when $n \to \infty$.
(2) the convergence $x_n \overset{w}{\to} x$ is weak if, for all $f \in E^*$, $| f(x_n) - f(x) | \to 0$ when $n \to \infty$.

Strong convergence implies weak convergence; the reverse is false, except if $\dim(E) < \infty$. For all Banach spaces E, and for every $u \in E$, consider that the linear functional

$$u^*(v) = \begin{cases} 0, & v \text{ and } u \text{ are L.I.} \\ a, & \text{if } v = a.u, a \in \mathbb{K} \end{cases}$$

induces the sequence $E \overset{*}{\to} E^* \to (E^*)^*$. The linear operator $J : E \to (E^*)^*$, $J(x)(f) = f(x)$, $f \in E^*$, extends linearly and $|| J(x) ||=|| x ||$. We say that E is a reflexive Banach space if $J(E) = (E^*)^*$, which is trivial if $\dim(E) < \infty$. Reflexive spaces carry the following property;

Theorem 3 *If E is a reflective Banach space, then all bounded sets are compact in the weak convergence.*

Proof See [27]. □

Exercises

(1) Prove that $| T |= \sup_{||x||=1} || T(x) ||_F = \sup_{|x|\leq 1} || T(x) ||_F$.
(2) Prove that $| S.T |\leq| S | . | T |$, for all $S, T \in \mathcal{L}(E, F)$.
(3) Consider $c > 0$. Let E and F be Banach spaces and let $T \in \mathcal{L}(E, F)$ be an operator such that $|| T(x) ||\geq c. || x ||$, for all $x \in E$. Prove that $T^n(E)$ is a closed subset of F, for all $n \in \mathbb{N}$.
(4) Let $k : [0, 1] \times [0, 1] \to \mathbb{R}$, $k(x, y) = y$ and $E = (C^0([a, b]), || . ||_0)$. Find the sup norm of the integral operator $K : E \to E$ given by

$$\left(Kf\right)(x) = \int_0^1 k(x, y) f(y) dy.$$

(5) Consider $1 \leq p < \infty$ and let $E(p) = (C^0([a, b]) || . ||_{L^p})$. Assume $k : [0, 1] \times [0, 1] \to \mathbb{R}$ is continuous and prove that the operator $K : E(p) \to E(p)$, $(Kf)(x) = \int_a^b k(x, y) f(y) dy$ is bounded. (hint: use Hölder's inequality $| \int_a^b (f.g)(x) dx |\leq|| f ||_{L^p} . || g ||_{L^q}$, where $\frac{1}{p} + \frac{1}{q} = 1$.)
(6) Let E be a reflexive Banach space. Prove that a weakly convergent sequence $\{x_n\}_{n\in\mathbb{N}} \subset E$ is bounded.

4 The Spectrum of a Bounded Linear Operator

In this section, we consider $\mathbb{K} = \mathbb{C}$ and E is a Banach space. When $E = \mathbb{C}^n$, the standard procedure to study an operator $T : E \to E$ is by decomposing E into T-

invariant subspaces. For this purpose, the spectral set $\sigma(T) = \{\lambda \in \mathbb{C} \mid T_\lambda = T - \lambda I$ is non-invertible$\}$ is the main ingredient. To describe $\sigma(T)$, we decompose the characteristic polynomial $p_T(x)$ of T as the product of irreducible factors:

$$p_T(x) = \prod_{i=1}^{r}(x - \lambda_i)^{e_i}, \quad \sum_{i=1}^{r} e_i = n.$$

T-invariant subspaces are $E_{\lambda_i} = \text{Ker}(T - \lambda_i I)^{e_i}$, where $\dim(E_{\lambda_i}) = e_i$. The Primary Decomposition Theorem guarantees the decomposition $E = \bigoplus_{i=1}^{r} E_{\lambda_i}$. If T is a normal operator, i.e., $T^*.T = T.T^*$, by the Spectral Theorem, we can take $E_{\lambda_i} = \text{Ker}(T - \lambda_i I)$, and so T is diagonalized. When T is not diagonalized, T can be represented by a block matrix with blocks given by the Jordan form. When E is an infinite dimensional vector space, the spectrum $\sigma(T)$ is rather complex as a set since new phenomena arise. For every $\lambda \in \mathbb{C}$, we associate the operator $T_\lambda = T - \lambda I$. The spectrum $\sigma(T)$ of an operator is the set of elements $\lambda \in \mathbb{C}$ such that the operator T_λ is not invertible or is not lower bounded; that is, we have a sequence $\{x_n\}_{n \in \mathbb{N}} \subset E$ such that $\| T(x_n) \| \to 0$. This last condition is avoided if we have $c \in (0, \infty)$ such that $\| T(x) \| \geq c \| x \|$, which implies that the image-set $T(E)$ is closed. An operator $T \in \mathcal{L}(E)$, not lower bounded, may not be injective and its image may be an open set.

Examples

(1) Let $E = l^2(\mathbb{N})$ and consider the operator $T : l^2(\mathbb{N}) \to l^2(\mathbb{N})$ given by

$$T(x_1, x_2, x_3, \dots) = (0, x_1, x_2, x_3, \dots).$$

Clearly, T is not surjective and therefore cannot be invertible, so $0 \in \sigma(T)$. Since T has no eigenvalues, 0 is not an eigenvalue.

(2) Let $E = l^2(\mathbb{Z})$ and consider $T : l^2(\mathbb{Z}) \to l^2(\mathbb{Z})$ given by

$$T(\dots, x_{-3}, x_{-2}, x_{-1}, x_0, x_1, x_2, x_3, \dots) = (\dots, y_{-3}, y_{-2}, y_{-1}, y_0, y_1, y_2, y_3, \dots), \ y_i = x_{i-1}.$$

If $Tx = \lambda x$, $\lambda \in \mathbb{C}$, then $x \notin l^2(\mathbb{Z})$, otherwise we would have $\| x \| = \infty$. So T has no eigenvalues. However, let us see that for any $\lambda \in U_1$ ($|\lambda| = 1$), there is a sequence $\{x_n\}_{n \in \mathbb{N}}$ such that $\| Tx_n - \lambda x_n \| \to 0$ when $n \to \infty$. Fixing $\lambda \in U_1$, consider when

$$x_n = \frac{1}{\sqrt{n}}(\dots, 0, 1, \lambda^{-1}, \lambda^{-2}, \lambda^{-3}, \dots, \lambda^{2-n}, \lambda^{1-n}, 0, \dots). \tag{3}$$

The $(i - n)^{th}$ coordinate is $(x_n)_{i-n} = \lambda^{i-n}$ if $1 \leq i \leq n$ and $(x_n)_i = 0$ for $n < i$. Then $\| x_n \| = 1$, for all $x \in \mathbb{N}$, and $T(x_n)$ has coordinates

$$\left(T(x_n)\right)_{i+1-n} = \lambda^{i-n}.$$

In this way, $|| T(x_n) - \lambda x_n || = \frac{2}{\sqrt{n}}$, so,

$$\lim_{n \to \infty} || T(x_n) - \lambda x_n || = 0.$$

Therefore any $\lambda \in U_1$ belongs to $\sigma(T)$.

(3) Let H be a Hilbert space and $P : H \to H$ an orthogonal projection; so $P^2 = P$ and $P^* = P$. Therefore $| P | = 1$ and $P^n = P$ for all $n \geq 2$. Moreover, $0 \in \sigma(P)$ since Ker(P) $\neq 0$. Besides, if there is $v \neq 0 \in H$ such that $Pv = \lambda v$, then $\lambda(\lambda - 1)v = 0$. Therefore $\sigma(P) = \{0, 1\}$. If $\lambda \in \mathbb{C}$ and $| \lambda | > 1$, then the inverse of $(P - \lambda I)$ is

$$(P - \lambda.I)^{-1} = -\lambda^{-1}[I - \lambda^{-1}P]^{-1} = -\lambda^{-1}[I + \lambda^{-1}P + \lambda^{-2}P^2 + \cdots + \lambda^{-n}P^n + \ldots]$$

$$= -\lambda^{-1}[I + (\lambda^{-1} + \lambda^{-2} + \cdots + \lambda^{-n} + \ldots)P] = -\frac{1}{\lambda}\left[I + \frac{1}{\lambda - 1}P\right].$$

For any $T \in \mathcal{L}(E)$, when $T_\lambda = T - \lambda I$ is invertible, the operator $R_\lambda(T) = T_\lambda^{-1}$ is the resolvent of T. A complex number λ is a regular value of T if one of the following conditions is satisfied:

(i) $R_\lambda(T)$ exists,
(ii) $R_\lambda(T)$ is bounded,
(iii) $R_\lambda(T)$ is well-defined on a dense subset of E.

Definition 3 The resolvent set of T is $\rho(T) = \{\lambda \in \mathbb{C} \mid \lambda \text{ regular value of } T\}$. The spectrum of T is the complementary set $\sigma(T) = \mathbb{C} \setminus \rho(T)$.

The spectrum $\sigma(T)$ decomposes into $\sigma(T) = \sigma_p(T) \cup \sigma_c(T) \cup \sigma_r(T)$;

(i) $\sigma_p(T)$ is the pontual spectrum with elements λ that are such that $R_\lambda(T)$ is not well-defined. The complex numbers $\lambda \in \sigma_p(T)$ are the eigenvalues of T.
(ii) $\sigma_c(T)$ is the continuous spectrum with elements λ that are those for which $R_\lambda(T)$ is unbounded.
(iii) $\sigma_r(T)$ is the residual spectrum with elements λ that are those for which the domain of $R_\lambda(T)$ is not dense in E.

Therefore $\mathbb{C} = \rho(T) \cup \sigma(T)$. If E is a finite dimensional vector space, then $\sigma_c(T) = \sigma_r(T) = \varnothing$, that is, in the finite dimension the spectral values of T are the eigenvalues of T.

Example 3

(1) $E = C^0([a, b])$; fix $\theta \in E$ and consider $T : E \to E$ given by $T(f) = \theta.f$. For all $\lambda \in \text{Im}(\theta)$, the operator $R_\lambda(T)$ is unbounded, so $\sigma(T) = \text{Im}(\theta)$.
(2) Consider $H = L^2([0, 1])$ and let $T : H \to H$ be the bounded operator $T(f)(x) = xf(x)$ ($| T | = 1$). In this example, there are no eigenvalues since for any $\lambda \in \mathbb{C}$, the only solution in H of equation $T(f) = \lambda f$ is the distribution $f(x) = 0$ almost everywhere in $[0, 1]$. Let's analyze the following cases to describe the spectrum:

(i) $\lambda \notin [0, 1]$. The operator T_λ is invertible, since $R_\lambda g = \frac{g(x)}{x-\lambda} \in H$ and

$$\| R_\lambda g \|_{L^2}^2 = \int_0^1 \frac{| g(x) |^2}{(x - \lambda)^2} dx \leq \frac{1}{(1 - | \lambda |)^2} \| f \|_{L^2}^2 .$$

Hence $\rho(T) \subseteq \mathbb{C}\backslash[0, 1]$.

(ii) $\lambda \in [0, 1]$. The operator $T_\lambda = T - \lambda I$ is not surjective, since the function $g(x) = (x - \lambda) f(x)$ satisfies $g(\lambda) = 0$; this rules out the constant functions $g(x) = c \neq 0$ of being in the image of T_λ. Let's check that the image of R_λ is dense in H; take $f \in H$ and consider the sequence $\{f_n\}_{n \in \mathbb{N}} \subset H$ given by

$$f_n(x) = \begin{cases} f(x), & \text{if } | x - \lambda | \geq 1/n, \\ 0, & \text{if } | x - \lambda | \leq 1/n. \end{cases}$$

Therefore $f = \lim f_n$ in H and $f_n \in \text{Im}(T_\lambda)$. Hence $\sigma(T) = \sigma_c(T) = [0, 1]$ and $\rho(T) = \mathbb{C}\backslash[0, 1]$.

(3) Let $A \in U_n$ and $E_A = \{f \in C^0([a, b]; \mathbb{C}^n) \mid f(t + 1) = A.f(t)\}$. Consider the operator $T : E_A \to E_A$, $T(u) = -i\frac{du}{dt}$. Then $\text{Ker}(T_\lambda) = \{e^{i\lambda t}u_0 \mid u_0 \in \mathbb{C}^n, t \in \mathbb{R}\}$. By imposing the condition $f(t + 1) = A.f(t)$, we get $A.u_0 = e^{i\lambda t}u_0$, and so u_0 is an eigenvector associated to the eigenvalue $e^{i\lambda}$. Considering $A = \exp(i\xi)$, such that $\xi \in u_n$ (Lie algebra of U_n), it follows that $\sigma(T) = \sigma(\xi) + 2\pi \mathbb{Z}$.

The following topological properties will be proved next;

(i) the set $\rho(T) \subset \mathbb{C}$ is an open subset.
(ii) $\sigma(T) \subset \mathbb{C}$ is a compact non-empty subset.

Proposition 4 *Let $T \in \mathcal{L}(E)$. If $| T | < 1$, then $(I - T) \in \mathcal{L}(E)$ is invertible and*

$$(I - T)^{-1} = \sum_{i+0}^{\infty} T^n.$$

Proof Assume for a while that the series $\sum_{i=0}^{\infty} T^n$ converges uniformly. So

$$(I - T).\left(\sum_{i=0}^{\infty} T^n\right) = \sum_{i=0}^{\infty}(T^n - T^{n+1}) = I.$$

Convergence follows by the Weierstrass criterion since if $| T | \leq c < 1$, then

$$\left|\sum_{i=1}^{\infty} T^n\right| \leq \sum_{i=0}^{\infty} | T |^n \leq \frac{1}{1 - | T |} \leq \frac{1}{1 - c}.$$

□

Corollary 3 *If $\lambda \in \rho(T)$ and $| T | < | \lambda |$, then*

$$R_\lambda(T) = -\frac{1}{\lambda}\sum_{i=0}^{\infty}\left(\frac{1}{\lambda}T\right)^i.$$

Proof Since $| T | < | \lambda |$, the inverse $(T - \lambda I)^{-1} = \lambda^{-1}(I - \frac{1}{\lambda}T)^{-1}$ is well-defined. Indeed,

$$R_\lambda(T) = -\frac{1}{\lambda}(I - \frac{1}{\lambda}T)^{-1} = -\frac{1}{\lambda}\sum_{i=0}^{\infty}\left(\frac{1}{\lambda}T\right)^i.$$

$\square$

The set $\mathcal{GL}(E) = \{T \in \mathcal{L}(E) \mid T \text{ invertible}\}$ is a group with the operation induced by the composition $\cdot : \mathcal{GL}(E) \times \mathcal{GL}(E) \to \mathcal{GL}(E)$, $(T, S) \to T.S$. The group axioms are satisfied, i.e., we have

(i) Associativity: $S.(R.T) = (S.R).T$ for all $R, S, T \in \mathcal{GL}(E)$.
(ii) Identity element: there is an element $I \in \mathcal{GL}(E)$ such that $T.I = I.T = T$ for all $T \in \mathcal{GL}(E)$. I is the identity element in $\mathcal{GL}(E)$ and $| I | = 1$.
(iii) Inverse element: for any $T \in \mathcal{GL}(E)$, we have $T^{-1} \in \mathcal{GL}(E)$ such that $T.T^{-1} = T^{-1}.T = I$.

Next we prove that $\mathcal{GL}(E)$ is a topological Lie group, which means that the product $\mathcal{GL}(E) \times \mathcal{GL}(E) \to \mathcal{GL}(E)$, $(S, T) \to S.T^{-1}$ is continuous with respect to the topology induced by the sup norm on $\mathcal{L}(E)$.

Proposition 5 *The following items are true with respect to $\mathcal{GL}(E)$:*

(i) $\mathcal{GL}(E)$ is open in $\mathcal{L}(E)$.
(ii) The product $(T, S) \to T.S$ is continuous in $\mathcal{GL}(E)$.
(iii) $\mathcal{I} : \mathcal{GL}(E) \to \mathcal{GL}(E)$, $\mathcal{I}(T) = T^{-1}$ is continuous.
  So $(T, S) \to T.S^{-1}$ is continuous in $\mathcal{GL}(E)$.

Proof (i) Take any $T \in \mathcal{GL}(E)$; we construct an open ball in $\mathcal{GL}(E)$ centered in T. Given $S \in \mathcal{GL}(E)$, the identity

$$S = T - (T - S) = T.\left[I - T^{-1}(T - S)\right]$$

implies that S is invertible if and only if $[I - T^{-1}(T - S)]$ is invertible. Assuming its existence, the inverse of S is $S^{-1} = [I - T^{-1}(T - S)]^{-1}.T^{-1}$. Restricting to the open ball

$$B_T = \left\{S \in \mathcal{L}(E); | T - S | < \frac{1}{| T^{-1} |}\right\}, \tag{4}$$

the operator $[I - T^{-1}(T - S)]$ is invertible for all $S \in B_T$ since

$$| T^{-1}(T - S) | \leq | T^{-1} | . | T - S | < 1.$$

Therefore $B_T \subset \mathcal{GL}(E)$.

(ii) Consider the two sequences $\{T_n\}_{n \in \mathbb{N}}$ and $\{S_n\}_{n \in \mathbb{N}}$ in $\mathcal{GL}(E)$ such that $\lim T_n = T$ and $\lim S_n = S$. Therefore from

$$| T_n.S_n - T.S | < | T_n - T | . | S_n | + | T | . | S_n - S |$$

it follows that $\lim T_n.S_n = T.S$.

(iii) To prove the continuity of the inverse map, we first check that $\mathcal{I}(S) = S^{-1}$ is bounded for all $S \in B_T$. Applying the triangle inequality, we have $\| T(x) \| \leq \| S(x) \| + \| (T - S)(x) \|$. We also have

$$\| x \| = \| T^{-1}.T(x) \| \leq | T^{-1} | . \| T(x) \| \quad \Rightarrow \quad \frac{\| x \|}{| T^{-1} |} \leq \| T(x) \|,$$

Therefore

$$\frac{\| x \|}{| T^{-1} |} \leq \| T(x) \| \leq \| S(x) \| + | T - S | . \| x \| .$$

Consequently,

$$0 \leq \left(\frac{1}{| T^{-1} |} - | T - S | \right) . \| x \| \leq \| S(x) \|, \tag{5}$$

since $S \in B_T$. Taking $y \in E, x = S^{-1}(y)$, we have

$$\left(\frac{1}{| T^{-1} |} - | T - S | \right) \| S^{-1}(y) \| \leq \| y \| \quad \Rightarrow \quad \| S^{-1}(y) \| \leq \frac{| T^{-1} |}{1 - | T^{-1} | . | T - S |} . \| y \| .$$

Consequently, for all $S \in B_T$ we have $| S^{-1} | \leq \frac{|T^{-1}|}{1 - |T^{-1}|.|T - S|}$. So $S^{-1} \in \mathcal{L}(E)$. Now we are ready to prove continuity. Let $\{T_n\}_{n \in \mathbb{N}} \subset B_T$ be a sequence such that $\lim_{n \to \infty} T_n = T$. Therefore we can take $\delta > 0$ such that if $n > n_0$, then $| T_n - T | < \frac{\delta}{|T^{-1}|}$. Both inequalities

$$| T_n^{-1} - T^{-1} | = | -T_n^{-1}(T_n - T)T^{-1} | \leq | T_n^{-1} | . | T_n - T | . | T^{-1} |$$

and $| T_n^{-1} | \leq \frac{|T^{-1}|}{1 - |T^{-1}|.|T - T_n|}$ imply that

$$| T_n^{-1} - T^{-1} | \leq \frac{| T^{-1} |^2}{1 - | T^{-1} | . | T - T_n |} . | T_n - T | \leq | T^{-1} |^2 . | T_n - T | .$$

Therefore $\lim T_n^{-1} = T^{-1}$. $\qquad\square$

Corollary 4 *If $T \in \mathcal{L}(E)$, then $\rho(T) \subset \mathbb{C}$ is an open subset and $\sigma(T) = \mathbb{C}\backslash\rho(T)$ is closed.*

Proof Let $\lambda \in \rho(T)$ and consider $\epsilon > 0$ such that$| R_\lambda(T) | < \frac{1}{\epsilon}$. Let's check that if μ is such that $| \mu - \lambda | < \epsilon$, then $\mu \in \rho(T)$. Using the same trick as in the last proposition, we have

$$R_\mu(T) = [I - (\mu - \lambda).R_\lambda(T)]^{-1}.R_\lambda(T).$$

So $R_\mu(T)$ is invertible. Moreover, $R_\mu(T)$ is bounded since $| T_\mu - T_\lambda | < \epsilon$ and

$$| R_\mu(T) | < \frac{| R_\lambda(T) |}{1 - \epsilon. | R_\lambda(T) |}.$$

$\qquad\square$

Lemma 1 *Let $\phi : E \to \mathbb{C}$ be a linear functional in E^*. If $\lambda_0 \in \rho(T)$, then the function $f : \rho(T) \to \mathbb{C}$, $f(\lambda) = \phi(R_\lambda(T)x)$ is holomorphic for all $x \in E$.*

Proof From Proposition 5, we have $R_\lambda(T) = [I - (\lambda - \lambda_0)R_{\lambda_0}(T)]^{-1}.R_{\lambda_0}(T)$. So for any λ satisfying $| \lambda - \lambda_0 | < \frac{1}{|R_{\lambda_0}|}$, the resolvent R_λ is given by the absolutely convergent series

$$R_\lambda(T) = \sum_{i=0}^{\infty}(\lambda - \lambda_0)^i R_{\lambda_0}^{i+1}(T).$$

Consequently, the function $\phi : \rho(T) \to \mathbb{C}$,

$$f(\lambda) = f(R_\lambda(T)x) = \sum_{i=0}^{\infty} \phi(R_{\lambda_0}^{i+1}(T)x)(\lambda - \lambda_0)^i$$

is holomorphic in $\rho(T)$. $\qquad\square$

Theorem 4 *If $T \in \mathcal{L}(E)$, then $\sigma(T) \subset \mathbb{C}$ is a non-empty compact subset contained in the ball B_r and centered at the origin with radius $r < | T |$.*

Proof Let's start proving $\sigma(T) \neq \varnothing$. If $T = 0$, then $\sigma(T) = 0$. Let $T \neq 0$, so the series $R_\lambda(T) = -\frac{1}{\lambda} \sum_{i=0}^{\infty} \frac{1}{\lambda^i} T^i$ is absolutely convergent for all λ such that $| T | < | \lambda |$ and we have $| R_\lambda(T) | < \frac{1}{|\lambda|-|T|}$. Moreover, $\sigma(T)$ is a limited closed set contained in the ball with radius $| T |$, so it is compact. Suppose that the set $\sigma(T) = \mathbb{C}\backslash\rho(T)$ is empty. According to Corollary 1, the function $f(\lambda) = \phi(R_\lambda x)$ is holomorphic and is well-defined in the entire complex plane $\mathbb{C} = \rho(T)$ for any $\phi \in E^*$ and $x \in E$; however f is bounded. It follows from Liouville's Theorem that f has to be constant, which is absurd since the Hahn-Banach Theorem (see 14 in Appendix A) guarantees the existence of non-constant functionals in E^*. $\qquad\square$

One could ask about the size of the largest possible radius of a ball centered at the origin and containing the set $\sigma(T)$;

Definition 4 The spectral ray of a linear operator $T \in \mathcal{L}(E, F)$ is

$$r_\sigma(T) = \sup_{\lambda \in \sigma(T)} |\lambda|.$$

Exercises

Let E be a Banach space and $T \in \mathcal{L}(E)$.

(1) Prove that $r_\sigma(T) = \lim_{n \to \infty} \sqrt[n]{|T|^n}$ (see in [27]).
(2) Find an example showing that the condition $r_\sigma(T) = 0$ does not imply $T = 0$.
(3) Let $\{\lambda_1, \ldots, \lambda_n\}$ be the eigenvalues set of T and $\{u_1, \ldots, u_n\}$ an eigenvector set such that $Tu_i = \lambda_i u_i$. Prove that the set $\{u_1, \ldots, u_n\}$ is linearly independent.
(4) Let $E = C^\infty(\mathbb{R}; \mathbb{C}^n), T : E \to E, T(u) = -i\frac{du}{dt}$. Describe the spectrum $\sigma(T)$.

5 Compact Linear Operators

In this section we define a class of operators that behave similarly to finite-rank operators. We will prove that the spectrum of these operators is an enumerable set.

Definition 5 Let E and F be Banach spaces. An operator $T \in \mathcal{L}(E, F)$ is compact, if for all bounded sequences $\{x_n\}_{n \in \mathbb{N}} \subset E$, the image $\{T(x_n)\}_{n \in \mathbb{N}}$ admits a convergent subsequence.

Let $\mathcal{K}(E, F)$ be the space of compact linear operators $T : E \to F$. It follows from the definition that $\mathcal{K}(E, F) \subset \mathcal{L}(E, F)$ is a closed subset. Every finite rank operator $T \in \mathcal{L}(E, F)$ is compact, that is, if $\dim(E) < \infty$ or $\dim(T(E)) < \infty$, then T is compact. Compact operators are natural generalizations of finite rank operators. We note that if the subspace $T(E)$ is closed in F and T is compact, then it follows that $\dim(T(E)) < \infty$ as a consequence of Theorem 15 in Appendix A. The study of compact operators becomes relevant when $\dim(E) = \infty$, in which case a compact operator cannot be inverted since the image of any closed ball is compact; e.g., the identity is not a compact operator. Consequently, when $\dim(E) = \infty$, it is immediate that $0 \in \sigma(T)$ for every operator $T \in \mathcal{K}(E, F)$. Although the definition is restrictive, compact operators are fundamentally important in various theoretical and applied problems. They arose originally in the theory of integral equations.

Proposition 6 *The space $\mathcal{K}(E, F)$ satisfies the following properties:*

(i) *Is a closed subspace of $\mathcal{L}(E, F)$.*
(ii) *If $E = F$, then $\mathcal{K}(E)$ is a bilateral ideal of $\mathcal{L}(E)$, or equivalently, if $K \in \mathcal{K}(E)$ and $S \in \mathcal{L}(E)$, then $KS, SK \in \mathcal{K}(E)$.*

(iii) $T^* : F^* \to E^* \in \mathcal{K}(F^*, E^*)$.

Proof (i) Let $\{K_n\}_{n\in\mathbb{N}} \subset \mathcal{K}(E, F)$ be a sequence of compact operators such that $\lim K_n = K$ in $\mathcal{L}(E, F)$, where $\mathcal{L}(E, F)$ is endowed with the sup norm. Taking a bounded sequence $\{x_k\}_{k\in\mathbb{N}} \subset E$, we have a subsequence $\{x_{k,1}\} \subset \{x_k\}$ such that $\{K_1(x_{k,1})\} \subset F$ is convergent. In the same way, we take a sequence $\{x_{k,2}\} \subset \{x_{k,2}\}$ such that $\{K_2(x_{k,2})\} \subset F$ converges. Applying the same construction successively, we obtain a bounded sequence $\{x_{k,n}\} \subset E$ such that $\{K_p(x_{k,n})\} \subset F$ converges for all $1 \le p \le n$. Now we take the diagonal sequence $\{x_{n,n}\}$ below:

$$
\begin{array}{ccccc}
x_{1,1} & x_{1,2} & x_{1,3} & \cdots \ x_{1,n} & \cdots \\
x_{2,1} & x_{2,2} & x_{2,3} & \cdots \ x_{2,n} & \cdots \\
\vdots & \vdots & \vdots & \cdots \ \vdots & \cdots \\
x_{n,1} & x_{n,2} & x_{n,3} & \cdots \ x_{n,n} & \cdots
\end{array}
$$

The sequence $\{K(x_{n,n})\}$ converges since it is a Cauchy sequence: given $\epsilon > 0$, if n, m are large enough, then from the estimate

$$
| K(x_n^n) - K(x_m^m) | \le | K(x_n^n) - K_n(x_n^n) | + | K_n(x_n^n) - K_m(x_m^m) | + | K_m(x_m^m) - K(x_m^m) |,
$$

we have $| K(x_n^n) - K(x_m^m) | \le \epsilon$. Hence $K \in \mathcal{K}(E, F)$.

(ii) Let $\Omega \in E$ be a bounded set.
$S(\Omega)$ is bounded and $K(S(\Omega))$ is relatively compact; consequently KS is compact. Let $\{x_n\}_{n\in\mathbb{N}} \subset \Omega$ be a sequence; then we have a subsequence $\{x_{n_k}\}_{n_k\in\mathbb{N}} \subset \Omega$ such that $\{K(x_{n_k})\}_{n_k\in\mathbb{N}}$ converges. Therefore $\{SK(x_{n_k})\}$ converges. Hence SK is compact.

(iii) Let $B = \{x \in E; | x | \le 1\}$ be the unity ball. Let $\{f_n\}_{n\in\mathbb{N}} \subset F^*$ be a bounded sequence and set $R = \sup_n | f_n |_{F^*}$. Since K is compact, the closure $\overline{K(B)}$ is a compact subset in F. The sequence of functions $\{\theta_n\}_{n\in\mathbb{N}} \subset F^*, \theta_n = f_n |_{K(B)}$ (restriction) is uniformly bounded since

$$
| \theta_n(y) | \le R \, \| y \|_F \le R \, | K | .
$$

Since $| \theta_n(y') - \theta_n(y) | \le R \, \| y' - y \|_F$, the sequence $\{\theta_n\}_{n\in\mathbb{N}}$ is equicontinuous. So by Ascoli-Arzelà, there is a subsequence of $\{\theta_n\}_{n\in\mathbb{N}}$ that converges uniformly on $K(B)$. Assume $\{\theta_n\}_{n\in\mathbb{N}}$ is convergent. Therefore $\{K^*f_n\}_{n\in\mathbb{N}}$ is a Cauchy sequence in F^* since

$$
\begin{aligned}
| K^*f_n - K^*f_m |_{F^*} &= \sup_{x\in B} \left| K^*f_n(x) - K^*f_m(x) \right| = \sup_{x\in B} \left| f_n(K(x)) - f_m(K(x)) \right| \\
&= \sup_{x\in B} \left| \theta_n(K(x)) - \theta_m(K(x)) \right| = \sup_{y\in K(B)} \left| \theta_n(y) - \theta_m(y) \right|.
\end{aligned}
$$

$\square$

Example 4 Examples of compact linear operators.

(1) Let $E = L^2(S^1, \mathbb{R}) = \{f : \mathbb{R} \to \mathbb{R} \mid f(t + 2\pi) = f(t), \int_0^{2\pi} \mid f(t) \mid^2 dt < \infty\}$ be the set of square-integrable periodic functions. Taking a finite set $\mathcal{F}(n) = \{f_1, \ldots, f_n\}$ in $L^2(S^1, \mathbb{R})$, let V_n be a subspace spanned by $\mathcal{F}(n)$. The function

$$P_k(x, y) = \sum_{k=1}^{n} f_k(x)e^{iky}$$

is in $L^2(S^1 \times S^1)$. The operator $K : E \to E$, given by

$$(Kf)(x) = \int_0^{2\pi} P_k(x, y)f(y)dy$$

has finite rank. Hence K is compact.

(2) Consider $\Omega \subset \mathbb{R}^n$ a compact subset and $k : \Omega \times \Omega \to \mathbb{R}$ a continuous function. Let $E = (C^0(\Omega), \| . \|_0)$ and define the operator $K : E \to E$,

$$(Kf)(x) = \int_\Omega k(x, y)f(y)dy. \tag{6}$$

Since $k : \Omega \times \Omega \to \mathbb{R}$ is uniformly continuous, given a bounded sequence $\{f_n\}_{n \in \mathbb{N}}$, the sequence $\{Kf_n\}_{n \in \mathbb{N}} \subset C^0(\Omega)$ is also bounded and equicontinuous. Therefore by the Ascoli-Arzelà Theorem, $\{Kf_n\}$ admits a convergent subsequence. Consequently, the operator K is compact.

(3) Taking a function $k \in L^2(\Omega)$ and considering $E = (L^2(\Omega), \| . \|_{L^2})$, the operator $K : E \to E$, as defined in (6), is also compact.

(4) Let H be a separable Hilbert space and let $\Omega \subset H$ be a bounded subset. An operator $T \in L(H)$ is a Hilbert-Schmidt operator if we have an orthonormal basis $\beta = \{e_n \mid n \in \mathbb{N}\}$ such that

$$\sum_{n=1}^{\infty} \| T(e_n) \|^2 < \infty.$$

The Hilbert-Schmidt norm of T is $\| T \| = \sqrt{\sum_{n=1}^{\infty} \| T(e_n) \|^2}$. We assert that every Hilbert-Schmidt operator is compact. Any element $x \in \Omega$ is written as a linear combination of the basis β as $x = \sum_{n=1}^{\infty} < x, e_n > e_n$. Therefore $T(x) = \sum_{n=1}^{\infty} < x, e_n > T(e_n)$, and so

$$\| T(x) \|^2 \leq \| x \|^2 \sum_{n=1}^{\infty} \| T(e_n) \|^2 < \infty.$$

We also have $T(x) = \sum_n <T(x), e_n> e_n$. So given $\epsilon > 0$, we then have $n_0 \in \mathbb{N}$ such that

$$\sum_{n_0+1}^{\infty} |<T(x), e_n>|^2 < \epsilon, \quad \forall x \in \Omega, \tag{7}$$

for all $n > n_0$. For $n > n_0$, let V_n be the finite dimensional subspace generated by $\{e_1, \ldots, e_n\}$ and $P_n : H \to V_n$ the projection

$$P_n(x) = \sum_{i=1}^{n} <x, e_i> e_i, \quad (P_n^2 = P_n).$$

As for any projection, $|P| = 1$. For any Cauchy sequence $\{x_k\}_{k \in \mathbb{N}} \subset \Omega$, we have

$$\| T(x_k) - T(x_l) \| \leq \| P_n(T(x_k) - T(x_l)) \| + \| (I - P_n)(T(x_k) - T(x_l)) \| .$$

From the estimate obtained in Eq. (7), we have $\| (I - P_n)(T(x_k) - T(x_l)) \| < \epsilon$ for all $k, l \geq n_0$. Since V_n is a finite dimensional subspace and Ω is bounded, there is a subsequence $\{x_{k_i}\}$ in $\{x_k\}$ such that $\{P_n.T(x_{k_i})\}$ is a Cauchy sequence in V_n. Therefore $\{T(x_{k_i})\}$ is also a Cauchy sequence converging in E. Indeed, this proves that T can be approximated by the sequence $\{P_n.T\}_{n \in \mathbb{N}}$ of finite rank operators. Consequently, finite rank operators are dense in the Hilbert-Schmidt category.

(5) Consider the Hilbert space $E = \{(x_1, \ldots, x_n, \ldots) \mid x_i \in \mathbb{C}, \sum_i |x_i|^2 < \infty\}$. Let $\{\lambda_n\}_{n \in \mathbb{N}}$ be a sequence in E. It follows from the last example that the diagonal operator $D : E \to E$,

$$D = \begin{pmatrix} \lambda_1 & 0 & 0 & \ldots & 0 & \ldots \\ 0 & \lambda_2 & 0 & \ldots & 0 & \ldots \\ \vdots & \vdots & \vdots & \ldots & 0 & \ldots \\ 0 & 0 & 0 & \ldots & \lambda_n & \ldots \\ \vdots & \vdots & \vdots & \ldots & \vdots & \ldots \end{pmatrix},$$

is compact.

The compact operators are also characterized by transforming weakly convergent sequences into strongly convergent ones.

Theorem 5 *A bounded linear operator is compact if and only if it applies weakly convergent sequences in strongly convergent sequences.*

Proof Let E, F be Banach spaces and $K \in \mathcal{K}(E, F)$.

(i) ($\Rightarrow$); take $\{x_n\}_{n\in\mathbb{N}} \subset E$ and $x \in E$ such that $x_n \overset{w}{\to} x$; set $y_n = K(x_n)$ and $y = K(x)$. The first step is to show that $y_n \overset{w}{\to} y$, and for this consider $f \in E^*$ and $g = K^* f$. Since f is arbitrary, it follows from the inequality that

$$| g(y_n) - g(y) | = \left|g(K(x_n)) - g(K(x))\right| = | K^* f(x_n) - K^* f(x) | \to 0$$

and that $y_n \overset{w}{\to} y$. Now suppose that y_n does not converge strongly to y. Then there is a subsequence $\{y_{n_k}\}$ such that $|| y_{n_k} - y ||_F \geq \epsilon$ for some $\epsilon > 0$. The weak convergence $x_n \overset{w}{\to} x$ implies $\{x_n\}_{n\in\mathbb{N}}$ is bounded. Due to the compactness of T, the sequence $\{y_{n_k} = K(x_{n_k})\}_{n\in\mathbb{N}}$ has a strongly convergent subsequence $\{y_{n'}\}$. Taking $y' = \lim y_{n'}$, it follows that $y_{n'} \overset{w}{\to} y'$, and consequently $y' = y$; this contradicts the hypothesis that y_n does not converge strongly for y.

(ii) ($\Leftarrow$); Assume that for all sequences $\{x_n\}_{n\in\mathbb{N}} \subset E$, such that $x_n \overset{w}{\to} x$, we have $\lim K(x_n) = K(x)$ (strongly). It is immediate that K is compact since every sequence is bounded if and only if it is weakly convergent. $\qquad\square$

When H is a Hilbert space admitting an orthonormal basis $\{e_n\}_{n\in\mathbb{N}} \subset H$, we have $\sum_{i=1}^{\infty} |< x, e_n >| < \infty$ for all $x \in H$. Consequently, $\lim_{n\to\infty} < x, e_n >= 0$. Hence the sequence $\{e_n\}_{n\in\mathbb{N}}$ is weakly convergent to 0.

Corollary 5 *Let H be a Hilbert space, let $\{e_n\}_{n\in\mathbb{N}} \subset H$ be an orthonormal basis of H, and let K be a compact operator. So $\{K(e_n)\}_{n\in\mathbb{N}}$ converges strongly to 0.*

Proof Since $e_n \overset{w}{\to} 0$, Theorem 5 implies the sequence $\{K(e_n)\}_{n\in\mathbb{N}}$ strongly converges, say, $\lim K(e_n) = w \neq 0$. Then

$$|| w ||^2 = \lim_{n\to\infty} < K(e_n), w > = \lim_{n\to\infty} < e_n, K^*(w) > = 0.$$

Therefore $\{K(e_n)\}_{n\in\mathbb{N}}$ strongly converges to 0. $\qquad\square$

The spectral set of a compact operator is discrete, as we shall prove next.

Theorem 6 *The set of eigenvalues of an operator $T \in \mathcal{K}(E, F)$ is enumerable and can be finite or empty. Moreover, the set of accumulation points $\mathcal{A}(\sigma(T))$ of $\sigma(T)$ is either $\mathcal{A}(\sigma(T)) = \varnothing$ or $\mathcal{A}(\sigma(T)) = \{0\}$.*

Proof It is sufficient to prove that the set $\Lambda(c) = \{\lambda \in \sigma(T); | \lambda |> c\}$ is finite for every real number $c > 0$. Suppose that $\Lambda(c)$ is not finite; then there is a sequence $\{\lambda_n\}_{n\in\mathbb{N}}$, such that $\lambda_n \neq \lambda_m$, for all $n \neq m$ and $| \lambda_n |\geq c$. Let $E_T = \{x_1, \ldots, x_n, \ldots\}$ be the set of eigenvectors associated to the sequence $\{\lambda_n\}$, that is, $T x_n = \lambda_n x_n$. Once E_T is a linearly independent set of vectors, consider $V(n) = \{\sum_{i=1}^{n} a_i x_i \mid a_i \in \mathbb{C}\}$. Then for all $x = \sum_{i=1}^{n} a_i x_i \in V(n)$, we have

$$(T - \lambda_n I)x = a_1(\lambda_1 - \lambda_n)x_1 + \cdots + a_{n-1}(\lambda_{n-1} - \lambda_n)x_{n-1} \in V(n - 1).$$

Let's assume E is infinite dimensional; otherwise, the claim is known. From Corollary 2 in Appendix A, it follows that we have a sequence $\{y_n\}_{n\in\mathbb{N}}$ with the following properties: (i) $y_n \in V(n)$, (ii) $\| y_n \| = 1$ and (iii) $\| y_n - x \| \geq 1/2$, for all $x \in V(n-1)$. The existence of such a sequence will lead us to a contradiction. Assuming $m < n$, consider the point

$$\tilde{x} = \lambda_n y_n - (T y_n - T y_m).$$

Therefore we have $\tilde{x} \in V(n-1)$ since $(\lambda_n I - T)y_n \in V(n-1)$ and $T y_m \in V(n-1)$ $(m < n)$. Letting $x \in V(n-1)$ and taking the vector $\tilde{x} = \lambda_n x$, we get

$$\| T y_n - T y_m \| = \| \lambda_n y_n - \tilde{x} \| = | \lambda_n | \cdot \| y_n - x \| \geq \frac{\lambda_n}{2} \geq \frac{c}{2}.$$

Consequently, the sequence $\{T y_n\}_{n\in\mathbb{N}}$ does not converge, contradicting the hypothesis $T \in \mathcal{K}(E, F)$. The set $\Lambda(c) = \{\lambda \in \sigma(T); | \lambda | > c\}$ being finite, yields that either (i) $\mathcal{A}(\sigma(T)) = \varnothing$ when $\lambda_n = \lambda_0$, for all $n \geq n_0$ or (ii) $\mathcal{A}(\sigma(T)) = \{0\}$ when $\lambda_n \to 0$. □

The theory of compact operators arose from the study of integral equations

$$x(s) - \lambda \int_a^b k(s, t)x(t)dt = y(s) \tag{8}$$

which can be written in the form $(I - \lambda T)(x) = y$, and $T(x) = \int_a^b k(s, t)x(t)dt$. Fredholm proved that either the integral equation (8) has a single solution or the homogeneous equation

$$x(s) - \lambda \int_a^b k(s, t)x(t)dt = 0$$

has a finite number of linearly independent solutions. Therefore either $(I - \lambda.T)^{-1}$ exists or the kernel of $(I - \lambda.T)$ has finite dimension. The example above motivated the following definition.

Definition 6 The operator $T \in \mathcal{L}(E)$ satisfies the Fredholm alternative if one of the following conditions are satisfied;

C1 The non-homogenous equations

$$T(x) = y, \quad T^*f = g$$

have unique solutions x and f, respectively, for all $y \in E$ and $g \in E^*$.

C2 The homogenous equations

$$T(x) = 0, \quad T^*f = 0$$

have the same number of linearly independent solutions, say $x_1, \ldots, x_n$ and $f_1, \ldots, f_n$, respectively. The non-homogeneous equations

$$T(x) = y, \quad T^*f = g$$

have a solution if and only if y and g satisfy $f_k(y) = 0$, $g(x_k) = 0$, for all $1 \le k \le n$.

Theorem 7 *Let $T \in \mathcal{K}(E)$ and $\lambda \ne 0$. So T_λ satisfies the Fredholm alternative.*

Proof See in [27].   □

Exercises

(1) Let $T \in \mathcal{K}(E, F)$ and assume $\Omega \subset E$ is a bounded subset. Prove that the image $T(\Omega)$ is relatively compact, i.e., $\overline{T(\Omega)}$ is compact.
(2) Prove that $\mathcal{K}(E, F)$ is a closed subset of $\mathcal{L}(E, F)$.
(3) Assume that $T : E \to F$ is compact. Prove that $T^* : F^* \to E^*$ is also compact.
(4) Consider $\mathcal{HS}(E, F)$ the space of Hilbert-Schmidt operators defined in Example 4 with the inner product

$$< K_1, K_2 >_{HS} = \sum_{i=1}^{\infty} < K_1(e_i), K_2(e_i) > .$$

   Prove that $\mathcal{HS}(E, F)$ is a Hilbert space.
(5) Let $\Omega \in \mathbb{R}^n$ be a bounded set and $r < n$. Prove that the operator $K : L^2(\Omega) \to L^2(\Omega)$, given by

$$(Kf)(x) = \int_\Omega \frac{1}{|x - y|^r} f(y)dy$$

   is compact.

6 Fredholm Operators

A generalization of the Primary Decomposition Theorem for compact operators will be demonstrated in this section.

Definition 7 Let E, F be Banach spaces. A linear operator $T \in \mathcal{L}(E, F)$ is Fredholm if it satisfies the following conditions:

(i) Ker(T) is closed and $\dim(\text{Ker}(T)) < \infty$.
(ii) $T(E) \subset F$ is closed and $\dim(F/T(E))$ is finite $(\text{codim}(T(E)) < \infty)$.

Assuming $\dim(F/T(E)) < \infty$, then $T(E) \subset F$ is a closed subset. The space $\mathcal{F}(E, F)$ of Fredholm operators is not a vector space, since $0 \notin \mathcal{F}(E, F)$. However, if $T \in \mathcal{F}(E, F)$, then $tT \in \mathcal{F}(E, F)$ for all $t \in \mathbb{C}$.

Definition 8 Let $T \in \mathcal{F}(E, F)$. The index of T is

$$\text{ind}(T) = \dim(\text{Ker}(T)) - \dim\left(\frac{F}{T(E)}\right). \tag{9}$$

Example 5 Examples of Fredholm operators.

(1) Let $E = \{(a_1, a_2, \ldots, a_n, \ldots) \mid a_i \in \mathbb{C}, \sum_i |a_i|^p < \infty\}$. The operators $T, S : E \to E$,

$$T(a_1, a_2, \ldots, a_n, \ldots) = (a_2, a_3, \ldots, a_n, \ldots),$$
$$S(a_1, a_2, \ldots, a_n, \ldots) = (0, a_1, a_2, \ldots, a_n, \ldots)$$

are Fredholm and $\text{ind}(T) = 1$ and $\text{ind}(S) = -1$.

(2) Let $T : E \to E$ be a compact operator. So $\Phi = I - T$ is a Fredholm operator and $\text{ind}(\Phi) = 0$. This example is fundamental in the theory of integral equations and in the spectral theory of compact operators.

(a) $\dim(\text{Ker}(\Phi)) < \infty$.
Taking $x \in \text{Ker}(\Phi)$, we have that $Tx = x$, that is $T \mid_{\text{Ker}(\Phi)} = I \mid_{\text{Ker}(\Phi)}$. It follows from the compactness of T that the identity operator restricted to $\text{Ker}(\Phi)$ must be finite dimensional and so $\dim(\text{Ker}(\Phi)) < \infty$.

(b) $\Phi(E)$ is closed in E.
Consider V a closed complement of $\text{Ker}(\Phi)$ so that $E = \text{Ker}(\Phi) \oplus V$. Consider the continuous operators $\Phi_V = \Phi \mid_V : V \to E$ and $T_V : T \mid_V : V \to E$, so $\text{Ker}(\Phi_V) = \{0\}$. We will prove that $\Phi_V^{-1} : T(V) \to V$ is continuous at the origin. Suppose that Φ_V^{-1} is not continuous at the origin; then there is a sequence $\{x_n\} \subset V$ such that $\lim \Phi_V(x_n) = 0$ and $\lim x_n \neq 0$. Let $r > 0$ be such that $\mid x_n \mid \geq r$ for all $n \in \mathbb{N}$; therefore $\frac{1}{|x_n|} < \frac{1}{r}$ and, consequently, $\lim \Phi_V(\frac{x_n}{|x_n|}) = 0$. The sequence $\{\hat{x}_n = \frac{x_n}{|x_n|}\}_{n \in \mathbb{N}}$ is bounded, so we have a subsequence $\{T(\hat{x}_{n_k})\}_{n_k} \subset \{T(\hat{x}_n)\}_{n \in \mathbb{N}}$ converging in E. Since $\Phi(\hat{x}_{n_k}) = \hat{x}_{n_k} - T(\hat{x}_{n_k})$, the subsequence $\{\hat{x}_{n_k}\}$ also converges. The limit $z = \lim \hat{x}_{n_k}$ belongs to $\text{Ker}(\Phi)$ since $(z - T(z)) = 0$. This contradicts the construction of the decomposition $E = \text{Ker}(\Phi) \oplus V$. Hence Φ_V^{-1} is continuous and $\Phi(V)$ is closed.

(c) $\text{codim}(\Phi(E)) < \infty$.
Suppose that $\text{codim}(\Phi(E)) = \infty$. Let $\Phi(E) = W_0$ and consider a sequence of closed subspaces:

$$\Phi(E) = W_0 \subset W_1 \subset W_2 \subset \cdots \subset W_n \subset \cdots$$

such that $\dim(W_n) = \dim(W_{n-1}) + 1$. Given $\epsilon > 0$, Corollary 2 in Appendix A guarantees that we have a sequence $\{x_n\}_{n \in \mathbb{N}} \subset E$ such that $x_n \in W_n$, $|x_n| = 1$ and $|x_n - y| > 1 - \epsilon$ for all $y \in W_{n-1}$ and for all n. Taking $k < n$, we get

$$| T(x_n) - T(x_k) | = \left| x_n - \underbrace{(x_k + \Phi(x_n) - \Phi(x_k))}_{\in W_{n-1}} \right| \geq 1 - \epsilon.$$

So we have the non-existence of a convergent subsequence of $\{T(x_n)\}_{n \in \mathbb{N}}$. That is the assumption $\operatorname{codim}(\Phi(E)) = \infty$ leads to a contradiction since T is a compact operator.

(d) Since $\Phi : E \to E$ is Fredholm, we have the decompositions $E = \operatorname{Ker}(\Phi) \oplus V$ and $E = \Phi(V) \oplus \operatorname{Coker}(\Phi)$. The map $\Phi : V \to \Phi(V)$ is an isomorphism, so $\dim(\operatorname{Ker}(\Phi)) = \dim(\operatorname{Coker}(\Phi))$. Therefore $\operatorname{ind}(\Phi) = 0$.

(3) Let $E = (C^0([0, 1], \mathbb{R}^n), \| \cdot \|_0)$ and let $T : E \to E$ be the operator $T(f)(x) = \int_0^x f(t)dt$. Since T is linear and compact, the operator $\Phi : E \to E$, defined by

$$\Phi(f)(x) = f(x) - \int_0^x f(t)dt$$

is Fredholm with $\operatorname{ind}(\Phi) = 0$.

(4) Let $K \subset \mathbb{R}^n$ be a compact subset, $V \subset \mathbb{R}$ an open subset, and $E_j = (C^j(K, V), \| \cdot \|_j)$. Consider the differential operator $D : E_j \to E_{j-1}$ given by $\mathcal{D}(f) = f'$.

(a) The operator $\mathcal{D} : E_1 \to E_0$ is Fredholm. The kernel is $\operatorname{Ker}(\mathcal{D}) = \mathbb{R}$ and $\mathcal{D}$ is surjective ($\mathcal{D}(E_1) = E_0$), so $\operatorname{ind}(\mathcal{D}) = 1$. The surjectiveness follows from the Fundamental Theorem of Calculus; given $g \in E_0$, the curve $f(t) = \int_a^t g(s)ds$ satisfies the equation $\mathcal{D}(f) = g$ and $f \in E_1$. However, $\mathcal{D}$ is not compact since a bounded sequence $\{f_n\}_{n \in \mathbb{N}} \subset E_1$ is not enough to guarantee the equicontinuity of the sequence $\{f'_n\}_{n \in \mathbb{N}}$. Thus, the sequence $\{f_n\}_{n \in \mathbb{N}} \subset E_1$ may not converge in E_1.

(b) The operator $\mathcal{D} : E_2 \to E_0$, $\mathcal{D}(f) = f'$ is compact but is not Fredholm. The kernel is $\operatorname{Ker}(\mathcal{D}) = \mathbb{R}$. The operator is not surjective, since given a curve $g \in E_0$, the curve $f(t) = \int_a^b g(s)ds$ satisfies $\mathcal{D}(f) = g$ but it may not be C^2. However, in this case, the operator $\mathcal{D}$ is compact. Taking a bounded sequence $\{f_n\}_{n \in \mathbb{N}} \subset E_2$, such that $\| f_n \|_2 < C$, we have $\| f''_n \|_0 < C$. By the Mean Value Theorem,

$$| f'_n(y) - f'_n(x) | \leq \| f''_n \|_0 \cdot | y - x | .$$

Consequently, $\{f'_n\}_{n \in \mathbb{N}}$ is equicontinuous. This means that the sequence $\{f'_n\}_{n \in \mathbb{N}} \subset C^0(K, V)$ admits a convergent subsequence. The image of $\mathcal{D}$ is not closed since we can take a sequence $\{f'_n\}_{n \in \mathbb{N}}$ with the limit $\lim f'_n = g \notin C^1$.

(5) Let $\phi \in E = C^0(K, \mathbb{R})$ and define the operator $T : C^k \to C^k$, $T_\phi(f) = \phi.f$. So $T_\phi \in \mathcal{L}(E)$ and $| T_\phi | = \| \phi \|_0$. Let $\mathcal{D}_\phi : C^3(K, V) \to C^0(K, V)$ be the operator $\mathcal{D}_\phi = T_\phi \circ \mathcal{D}$, $\mathcal{D}_\phi(f) = \phi.f'$. Hence $\mathcal{D}_\phi$ is Fredholm.

Proposition 7 *The space $\mathcal{F}(E, F)$ is an open subset of the space $\mathcal{L}(E, F)$ and the function* ind : $\mathcal{F}(E, F) \to \mathbb{Z}$, $T \to \text{ind}(T)$ *is continuous.*

Proof Take $\Phi \in \mathcal{F}(E, F)$. The decomposition $E = \text{Ker}(\Phi) \oplus V$ is taken in such a way that $\Phi_V = \Phi : V \to \Phi(V)$ is invertible. Also, consider the decomposition $F = \Phi(V) \oplus W$, with $\dim(W) < \infty$. From Proposition 5, any operator $T : V \to T(V)$ sitting in the ball $B_r(\Phi) = \{T \in \mathcal{L}(E, F); | T - \Phi | < r\}$ is invertible by taking $r < \frac{1}{|\Phi^{-1}|}$. In this case we have a subspace $V' \subset \text{Ker}(\Phi)$ such that $\text{Ker}(\Phi) = \text{Ker}(T) \oplus V'$. Let $m = \dim(V')$ and let $W' \subset W$ be the complement of $T(V')$, that is, $F = \Phi(V) \oplus T(V') \oplus W'$, and $\dim\big((T(V'))\big) = m$ since $T : V' \to T(V')$ is an isomorphism. Therefore,

$$\text{ind}(T) = \dim(Ker(T)) - \dim(W') = [\dim(\text{Ker}(T)) + m] - [\dim(W') + m] =$$
$$= \dim(\text{Ker}(\Phi)) - \dim(W) = \text{ind}(\Phi).$$
$\square$

An important property of the Fredholm operators is their stability under perturbation, as shown in the following corollary;

Corollary 6 *If $T, S \in \mathcal{F}(E, F)$, then $T + K \in \mathcal{F}(E, F)$.*

Proof Since $T |_{\text{Ker}(T+K)} = -K |_{\text{Ker}(T+K)}$ is a compact operator with closed range, we have $\dim(\text{Ker}(T + K)) < \infty$. The operator tK is also compact for all $t \in \mathbb{C}$, and the map $t \to \text{ind}(T + tK)$ is continuous, so it is constant. Consequently, $\dim(F/(T + K)) < \infty$. Hence $(T + K)(E) \subset F$ is closed in F and $T + K$ is Fredholm. $\square$

Example 6 Examples

(1) On $E = (C^0([a, b]; \mathbb{R}), \| . \|_0)$, define the operator $T : E \to E$, $(Tf)(x) = \int_a^x f(t)dt$. The equation

$$(Tf) - \alpha f = g$$

has a solution for all $g \in E$ and $\alpha \neq 0 \in \mathbb{R}$,. It is equivalent to $-\alpha[I - \alpha^{-1}T](f) = g$. The operator $\Phi = -\alpha[I - \alpha^{-1}T]$ is Fredholm. We assert that its kernel is trivial, i.e., $\text{Ker}(\Phi) = \{0\}$. Note that if $f \in \text{Ker}(\Phi)$, then $(Tf)(x) = \alpha f(x)$. As Tf is differentiable, it follows that $f \in \text{Ker}(\Phi)$ is also differentiable and $f' = \frac{1}{\alpha} f$, that is, $f(x) = f(0).e^{-\frac{x}{\alpha}}$. However $(Tf)(a) = 0 = f(a)$ implies $f(x) = 0$, for all $x \in [a, b]$. Therefore Φ is injective. Since $\text{ind}(\Phi) = 0$, it follows that Φ is surjective.
(2) Let $K \in \mathcal{K}(E)$ and assume $T = I - K$ is injective, so T is surjective. For all $t \in \mathbb{R}$, tK is compact and the map $t \to \text{ind}(I - tK)$ is continuous, and therefore constant. Taking $t = 1$, clearly $\text{ind}(I - K) = 0$. Therefore

$$\dim\big(E/T(E)\big) = \dim(\text{Ker}(T)) = 0.$$

(3) The Laplacian operator defined in a compact region is one of the most important examples. Let $\Omega \subset \mathbb{R}^n$ be a compact subset and $E = L^2(\Omega; \mathbb{R})$. The Laplacian operator $\triangle : L^2(\Omega; \mathbb{R}) \to L^2(\Omega; \mathbb{R})$ is defined as

$$\triangle f = \sum_{i=1}^{n} \frac{\partial^2 f}{\partial x_i^2}.$$

The assertion that $\triangle$ is a Fredholm operator is proved by verifying the Inequality (14) and applying the fact that the inclusion $L^{2,k} \hookrightarrow L^2$ of the Sobolev space is a compact embedding for $k > 2$; both proofs are beyond the scope of this text (see in [18]).

Proposition 8 *Let E, F, G be Banach spaces and consider $S \in \mathcal{F}(E, F)$ and $T \in \mathcal{F}(F, G)$. So*

$$\mathrm{ind}(TS) = \mathrm{ind}(T) + \mathrm{ind}(S).$$

Proof The composition $T \circ S$ is represented by

$$E \xrightarrow{\ S\ } F \xrightarrow{\ T\ } G.$$

Due to the identity

$$\mathrm{ind}(TS) = \dim(\mathrm{Ker}(TS)) - \dim\left(G/TS(E)\right), \tag{10}$$

we will calculate each portion of $\mathrm{ind}(TS)$ as a function of $\mathrm{ind}(T)$ and $\mathrm{ind}(S)$. The inclusions $\mathrm{Ker}(S) \subset \mathrm{Ker}(TS)$ and $TS(E) \subset T(F)$ imply

$$\dim\left(\mathrm{Ker}(TS)\right) = \dim\left(\mathrm{Ker}(S)\right) + \dim\left(\frac{\mathrm{Ker}(TS)}{\mathrm{Ker}(S)}\right)$$
$$\dim\left(\frac{G}{TS(E)}\right) = \dim\left(\frac{T(F)}{TS(E)}\right) + \dim\left(\frac{G}{T(F)}\right). \tag{11}$$

All spaces above have finite dimension. Applying the Rank-Nullity Theorem to the linear operator $T : F/S(E) \to T(F)$, we get

$$\dim\left(\frac{F}{S(E)}\right) = \dim\left(\frac{\mathrm{Ker}(T)}{S(E) \cap \mathrm{Ker}(T)}\right) + \dim\left(\frac{T(F)}{TS(E)}\right). \tag{12}$$

Therefore Eqs. (11) and (12) imply that

$$\text{ind}(TS) - \text{ind}(T) - \text{ind}(S) =$$

$$= \dim\left(\frac{\text{Ker}(TS)}{\text{Ker}(S)}\right) - \dim\left(\frac{T(F)}{TS(E)}\right) - \dim(\text{Ker}(T)) + \dim\left(\frac{F}{S(E)}\right) =$$

$$= \dim\left(\frac{\text{Ker}(TS)}{\text{Ker}(S)}\right) - \dim(\text{Ker}(T)) + \dim\left(\frac{\text{Ker}(T)}{S(E) \cap \text{Ker}(T)}\right) = \qquad (13)$$

$$= \dim\left(\frac{\text{Ker}(TS)}{\text{Ker}(S)}\right) - \dim\left(\frac{\text{Ker}(T)}{S(E) \cap \text{Ker}(T)}\right) - \dim(S(E) \cap \text{Ker}(T)) +$$

$$+ \dim\left(\frac{\text{Ker}(T)}{S(E) \cap \text{Ker}(T)}\right) = \dim\left(\frac{\text{Ker}(TS)}{\text{Ker}(S)}\right) - \dim(S(E) \cap \text{Ker}(T)).$$

Considering the decompositions $\text{Ker}(TS) = \text{Ker}(S) \oplus W$ and $E = \text{Ker}(TS) \oplus U = \text{Ker}(S) \oplus W \oplus U$, we have $S(E) = S(W) \oplus S(U)$. We claim $S(E) \cap \text{Ker}(T) = S(W)$; it is clear that $S(W) \subset S(E) \cap \text{Ker}(T)$. To prove the reverse, let $y = S(x) \in S(E)$ such that $TS(x) = 0$; by taking $x = x_1 + x_2 \in \text{Ker}(S) \oplus W$, then $S(x) = S(x_2)$, and so, $y \in S(W)$. Consequently $S(E) \cap \text{Ker}(T) \subset S(W)$. Hence $\text{ind}(TS) = \text{ind}(T) + \text{ind}(S)$. □

A sufficient condition often used to check if an operator is Fredholm is the following;

Proposition 9 *Let E, F, G be Banach spaces. Let $T : E \to F$ be a bounded operator and let $K : E \to G$ be a compact operator. If there is a constant $c > 0$ such that*

$$\| x \|_E \leq c\left(\| T(x) \|_F + \| K(x) \|_G \right), \qquad (14)$$

for all $x \in E$, then the following conditions are satisfied:

(i) $\dim(\text{Ker}(T)) < \infty$.
(ii) $T(E)$ *is closed.*

Proof (i) Let $\{x_n\}_{n \in \mathbb{N}} \subset \text{Ker}(T)$ be a sequence such that $\| x_n \| \leq 1$ for all n. Since $\{x_n\}_{n \in \mathbb{N}}$ is bounded, we have a subsequence $\{x_{n_k}\}$ so that $\{K(x_{n_k})\}$ converges. From the Inequality (14), $\{x_{n_k}\}$ is a Cauchy sequence, and so it converges in E. Therefore $\text{Ker}(T)$ is finite dimensional.
(ii) Let $V \subset E$ be the complementary subspace of $\text{Ker}(T)$ and $y \in \overline{T(V)}$. Now we have a sequence $\{x_n\} \subset V$ such that $y = \lim T(x_n)$. Let's check that $\{x_n\}$ is bounded; by contradiction, suppose $\{x_{n_k}\} \subset V$ is unbounded and $\| x_{n_k} \| \to \infty$. Considering the sequence $\hat{x}_{n_k} = \frac{x_{n_k}}{\|x_{n_k}\|}$, we have $\lim T(\hat{x}_{n_k}) = 0$ and $\{K(\hat{x}_{n_k})\}$ converges. Therefore $\{x_{n_k}\} \subset V$ is a Cauchy sequence. If $\lim x_{n_k} = \hat{x}$, then $\| \hat{x} \| = 1$ and $T(\hat{x}) = 0$. However this contradicts the decomposition that $E = \text{Ker}(T) \oplus V$. □

The operators $T, P \in \mathcal{L}(E, F)$ are considered congruent $T \equiv P \mod \mathcal{K}(E, F)$ whenever $T - P \in \mathcal{K}(E, F)$. This is an equivalent relation defined in $\mathcal{L}(E, F)$. When $F = E$, if $T_1 \equiv P_1 \mod \mathcal{K}(E)$ and $T_2 \equiv P_2 \mod \mathcal{K}(E)$, then $T_1.T_2 \equiv P_1.P_2 \mod \mathcal{K}(E)$. An operator $T \in \mathcal{L}(E, F)$ is invertible module $\mathcal{K}(E, F)$ if there is $T' \in \mathcal{L}(E, F)$ such that

$$T'.T \equiv I_E \bmod \mathcal{K}(E, F) \quad \text{and} \quad T.T' \equiv I_F \bmod \mathcal{K}(E, F).$$

Therefore T' is the inverse mod $\mathcal{K}(E, F)$ of T.

Proposition 10 *Let $T \in \mathcal{L}(E, F)$. T is a Fredholm operator if and only if T is invertible   mod $\mathcal{K}(E, F)$. The inverse T' of T can be chosen so that* $\dim(F/T'(E)) < \infty$.

Proof $(\Rightarrow)$ Consider $E = \mathrm{Ker}(T) \oplus V$ and $F = T(V) \oplus W$, and now V and W are closed subspaces. Let $P : F \to E$ be the continuous linear operator $P = i \circ T^{-1} \circ pr_1$ given by

$$F = T(V) \oplus W \xrightarrow{pr_1} T(V) \xrightarrow{T^{-1}} V \xrightarrow{i} E = \mathrm{Ker}(T) \oplus V.$$

Since $TP(x) = 0$, for all $x \in W$, we get either cases:

(i) if $x \in W$, then $(I_F - TP)(x) = x$ or
(ii) if $x \notin W$, then $(I_F - TP)(x) = 0$.

Therefore $pr_W = I_F - TP$ is the projection over W and a compact operator since $\dim(W) < \infty$. With the same argument, $pr_{N(T)} = I_E - PT$ is the projection over $\mathrm{Ker}(T)$, so it is also a compact operator.
$(\Leftarrow)$ Let $K_F = I_F - TP \in \mathcal{K}(F, E)$ and $K_E = I_E - PT \in \mathcal{K}(E, F)$. Since $\mathrm{Ker}(T) \subset \mathrm{Ker}(PT)$ and $(I_E - PT) \mid_{\mathrm{Ker}(PT)} = I_E$, we have $\dim(\mathrm{Ker}(PT)) < \infty$. So $\dim(N(T)) < \infty$. Once the operators $PT = I_E - K_E$ and $TP = I_F - K_F$ are Fredholm, the inclusion $TP(E) \subset T(E)$ implies $T(E)$ has finite codimension; therefore $T(E) \subset F$ is closed. $\square$

The proof above gives a geometric interpretation showing that the operator $I - TP$ is a projection over the finite-dimensional space W.

The linear Fredholm operators may not be invertible, however, they are pseudo-invertible. Let $T : E \to F$ be a bounded linear operator, the pseudo-inverse of T is the operator $Q : F \to E$ such that:

(i) $QTQ = Q$,
(ii) $TQT = T$.

Proposition 11 *A linear operator $T : E \to F$ has a pseudo-inverse if and only if the following conditions are satisfied:*

(i) *$T(E)$ is closed.*
(ii) *We have a subspace $E_0 \subset E$ such that $E = \mathrm{Ker}(T) \oplus E_1$.*
(iii) *We have a subspace $F_1 \subset F$ such that $F = F_0 \oplus T(E)$.*
      In particular, every Fredholm operator has a pseudo-inverse.

Proof $(\Leftarrow)$ Let's assume T satisfies items (i)–(iii). Taking $E_0 = \mathrm{Ker}(T)$ and $F_1 = T(E)$, let $E_1 \subset E$ and $F_0 \subset F$ be subspaces such that

$$E = E_0 \oplus E_1, \quad F = F_0 \oplus F_1.$$

Therefore $T_1 = T \mid_{E_1}: E_1 \to F_1$ is a bounded isomorphism. For any $v = v_0 + v_1 \in F$, with $v_0 \in F_0$ and $v_1 \in F_1$, define

$$Q : F \to E, \quad Q(v_0 + v_1) = T_1^{-1}(v_1).$$

Therefore we have (i) $QTQ = Q$ and (ii) $TQT = T$.

($\Rightarrow$) Now we assume T has a pseudo-inverse $Q : F \to E$. Define the subspaces $E_1 = Q(F)$ and $F_0 = \mathrm{Ker}(Q)$. The operator $P = QT : E \to E$ is a projection, since $P^2 = (QT)(QT) = QT = P$, $\mathrm{Ker}(P) = \mathrm{Ker}(T)$ and $P(E) = Q(F)$. Then P projects E over $Q(F)$. Analogously, $P' = TQ : F \to F$ is a projection such that $\mathrm{Ker}(P') = \mathrm{Ker}(Q)$ and $P'(F) = T(E)$ since P' projects F over $T(E)$. Now the following three items are satisfied:

(i) the subspace $T(E)$ is closed since it is the image of the projection P'.
(ii) the complement of $\mathrm{Ker}(T)$ is $Q(F)$.
(iii) the complement of $T(E)$ is $\mathrm{Ker}(Q)$.   □

Exercises

(1) Let E, F be finite-dimensional Banach spaces and $T \in \mathcal{L}(E, F)$. Prove that T is Fredholm and
$$\mathrm{ind}(T) = \dim(E) - \dim(F).$$

(2) Let $T : E \to F$ be a Fredholm compact operator. Prove that $\dim(E)$ and $\dim(F)$ are finite.

(3) Let $K \subset \mathbb{R}^n$ be a compact subset, $V \subset \mathbb{R}^m$ an open subset and $E_j = (C^j(K, V), \| \cdot \|_j)$. Is the derivative operator $\mathcal{D} : E_k \to E_{k-1}, \mathcal{D}(f) = f'$, a Fredholm operator?

(4) Let $T_1, T_2 \in \mathcal{F}(E)$. Prove that $T_2.T_1 \in \mathcal{F}(E)$.

(5) Let $T \in \mathcal{F}(E, F)$. Prove that T satisfies the Fredholm alternative, defined in (6).

(6) Let H be a Hilbert space and $T \in \mathcal{F}(H)$. Prove that

$$\mathrm{ind}(T) = \dim\big(\mathrm{Ker}(T)\big) - \dim\big(\mathrm{Ker}(T^*)\big).$$

(7) Let P_1, P_2 be the pseudo-inverses of $T \in \mathcal{F}(E, F)$. Prove that $P_1 \equiv P_2 \bmod \mathcal{F}(E, F)$.

(8) Assume $T_1, T_2 \in \mathcal{F}(E, F)$. Prove that $T_1.T_2 \in \mathcal{F}(E, F)$.

(9) Let E, F be Banach spaces and $T \in \mathcal{F}(E, F)$. Prove that there is a Banach space G and a bounded linear operator $S : G \to F$, such that the operator $T \oplus G : E \oplus G \to F$, $T \oplus S(u \oplus v) = T(u) + S(v)$, is surjective. Moreover, the projection $P : \mathrm{Ker}(T \oplus S) \to G$ is Fredholm and $\mathrm{ind}(T) = \mathrm{ind}(P)$.

6.1 The Spectral Theory of Compact Operators

When E is a finite dimensional vector space, we associate to every linear operator $T \in \mathcal{L}(E)$ the characteristic polynomial

$$p_T(x) = (x - \lambda_1)^{n_1} . (x - \lambda_2)^{n_2} \ldots (x - \lambda_k)^{n_k}, \tag{15}$$

and λ_i, $1 \le i \le k$, are the eigenvalues of T and $\sum_{i=1}^{k} n_k = \dim(E)$. The Cayley-Hamilton Theorem asserts that $p_T(T) = 0$. For every eigenvalue λ_i, we associated the operator $T_i = T - \lambda_i I$ and the generalized eigenspace

$$E_i(\lambda_i, n_i) = \mathrm{Ker}(T^{n_i}). \tag{16}$$

The Primary Decomposition Theorem asserts that the space E decomposes as

$$E = E_1(\lambda_1, n_1) \oplus E_2(\lambda_2, n_2) \oplus \ldots E_k(\lambda_k, n_k). \tag{17}$$

Consider E a Banach space. We will extend the Primary Decomposition Theorem for compact operators $K \in \mathcal{K}(E)$. For any $\lambda \ne 0 \in \mathbb{C}$, the linear operator $T_\lambda = K - \lambda I = -\lambda(I - \lambda^{-1}K)$ is Fredholm. The same is true for the powers

$$T_\lambda^p = (K - \lambda I)^p = (-\lambda)^p (I - K_1),$$

and for the operator $K_1 = \sum_{j=1}^{p} \binom{p}{j}(\lambda^{-1}K)^j$ is compact. Since $\mathrm{ind}(T_\lambda) = 0$, we get $\dim(\mathrm{Ker}(T_\lambda^p)) = \dim(\frac{E}{T_\lambda^p(E)})$, for all $p \in \mathbb{N}$. Therefore either T_λ is invertible or $\lambda \in \sigma(T)$. By Theorem 6, the spectrum $\sigma(K)$ is enumerable and either its only accumulation point is $\mathcal{A}(\sigma(K)) = \{0\}$ or $\mathcal{A}(\sigma(K)) = \varnothing$. We fix an order $\sigma(K) = \{\lambda_1, \lambda_2, \ldots, \lambda_n, \ldots\}$ with $|\lambda_i| > |\lambda_{i+1}|$ and $\lim \lambda_n = 0$. For $1 \le i \le n$, define the operator $T_i = K - \lambda_i I$.

Proposition 12 *Let $K \in \mathcal{K}(E)$ and $\lambda_i \in \sigma(K)$, $\lambda_i \ne 0$. So for $i \in \{1, \ldots, n, \ldots\}$ there is an integer $n_i > 0$ such that $\mathrm{Ker}\big((K - \lambda_i I)^n\big) = \mathrm{Ker}\big((K - \lambda I)^{n_i}\big)$ for all $n \ge n_i$. The integer n_i is the exponent of the eigenvalue λ_i in the characteristic polynomial defined in (15).*

Proof By contradiction, assume that the assertion is false. Therefore there is an ascending chain of subspaces

$$\mathrm{Ker}\big([K - \lambda I]\big) \subsetneqq \mathrm{Ker}\big([K - \lambda I]^2\big) \subsetneqq \cdots \subsetneqq \mathrm{Ker}\big([K - \lambda I]^n\big) \subsetneqq \ldots .$$

However this is not possible since the operator $(K - \lambda I)^n$ is Fredholm. Therefore we get $\dim(\mathrm{Ker}\big([K - \lambda I]^n\big)) < \infty$, for all $n \in \mathbb{N}$, and we also have $n_\lambda \in \mathbb{N}$ so that now the chain stops. $\square$

Theorem 8 *Consider $K \in \mathcal{K}(E)$ and let $\lambda_i \in \sigma(K)$ be a non-null eigenvalue with exponent n_i. Then the space E admits a decomposition*

$$E = \operatorname{Ker}(T_i^{n_i}) \oplus \operatorname{Im}(T_i^{n_i}). \tag{18}$$

The subspaces of the decomposition are K-invariants. Moreover, if $\lambda_j \in \sigma(K)$ is an eigenvalue distinct of λ_i with exponent n_j, then

$$\operatorname{Ker}(T_j^{n_j}) \subset \operatorname{Im}(T_i^{n_i}). \tag{19}$$

Proof The spaces $\operatorname{Ker}(T_i^{n_i})$ and $\operatorname{Im}(T_i^{n_i})$ are K-invariants closed subspaces. Suppose the closed subspace $F = \operatorname{Ker}(T_i^{n_i}) \oplus \operatorname{Im}(T_i^{n_i}) \subsetneq E$. Since $\operatorname{ind}(T_i^{n_i}) = 0$, we have $\dim(E/\operatorname{Im}(T_i^{n_i})) = \dim(\operatorname{Ker}(T_i^{n_i}))$. Once $E = \operatorname{Im}(T_i^{n_i}) \oplus (E/\operatorname{Im}(T_i^{n_i}))$, it follows that $F = E$. So we have $E = \operatorname{Ker}(T_i^{n_i}) + \operatorname{Im}(T_i^{n_i})$. To prove that $\operatorname{Ker}(T_i^{n_i}) \cap \operatorname{Im}(T_i^{n_i}) = \{0\}$, consider $x \in \operatorname{Ker}(T_i^{n_i}) \cap \operatorname{Im}(T_i^{n_i})$ and assume that $x = T_i^{n_i}(y)$, $y \in E$. Since $T_i^{n_i}(x) = 0$, we get $(T_i^{n_i})^2(y) = 0$. Since $\operatorname{Ker}((T_i^{n_i})^2) = \operatorname{Ker}(T_i^{n_i})$, we have $y \in \operatorname{Ker}(T_i^{n_i})$ and $x = 0$. Therefore $E = \operatorname{Ker}(T_i^{n_i}) \oplus \operatorname{Im}(T_i^{n_i})$. Now, let $\lambda_j \neq \lambda_i$ be another eigenvalue in $\sigma(K)$. The commutativity $T_i^{n_i}.T_j^{n_j} = T_j^{n_j}.T_i^{n_i}$ implies that the subspaces in the decomposition (19) are $T_j^{n_j}$-invariants. Let $z \in \operatorname{Ker}(T_j^{n_j})$ and consider $z = x + y$, where $x \in \operatorname{Ker}(T_i^{n_i})$ and $y \in \operatorname{Im}(T_i^{n_i})$. The identity $T_j^{n_j}(z) = T_j^{n_j}(x) + T_j^{n_j}(y) = 0$ now implies that $T_j^{n_j}(x) = 0$, i.e., $x \in \operatorname{Ker}(T_j^{n_j})$. Using the fact that the polynomials $p(x) = (x - \lambda_i)^{n_i}$ and $q(x) = (x - \lambda_j)^{n_j}$ are primes, we now have polynomials $r(x), s(x)$ such that

$$r(x).p(x) + s(x).q(x) = 1. \tag{20}$$

Evaluating the identity (20) on K, we get

$$r(K).p(K) + s(K).q(K) = I. \tag{21}$$

Then,

$$x = I(x) = r(K).T_i^{n_i}(x) + s(K).T_j^{n_j}(x) = 0. \tag{22}$$

Therefore $z \in \operatorname{Im}(T_i^{n_i})$. □

For every $k \in \mathbb{N}$, let $\mathcal{K}_k = \bigoplus_{i=1}^{k} \operatorname{Ker}(T_i^{n_i})$ and $\mathcal{I}_k = \bigcap_{i=1}^{k} \operatorname{Im}(T_i^{n_i})$. Applying Theorem 8, we have the decomposition $E = \mathcal{K}_k \oplus \mathcal{I}_k$ corresponding to the Primary Decomposition Theorem for compact operators on Banach spaces. If the exponents of all eigenvalues $\lambda_i \in \sigma(K)$ satisfy $n_i = 1$, then K is diagonalizable. The subspaces satisfy the inclusion chain

$$\mathcal{N}_1 \subsetneq \mathcal{N}_2 \subsetneq \cdots \subsetneq \mathcal{N}_n \subsetneq \cdots,$$
$$\mathcal{I}_1 \supsetneq \mathcal{I}_2 \supsetneq \cdots \supsetneq \mathcal{I}_n \supsetneq \cdots.$$

7   Linear Operators on Hilbert Spaces

A sesquilinear form $< .,. >: H \times H \rightarrow \mathbb{C}$ defined on a Hilbert space imposes a more efficient structure to address the linear algebra questions. As a result of the Riesz Representation Theorem 12, in Appendix A, the Hilbert spaces H and H^* are isomorphic.

Definition 9 Let H be a Hilbert space and $T \in L(H)$. The adjoint operator T^* : $H \rightarrow H$ is the linear operator such that

$$< v, T(u) >=< T^*(v), u >,$$

for any $u, v \in H$. In particular, if f is a linear functional and $f(g) =< v_f, g >$ for all $g \in H$, then $T^*(v_f) = v_{T^*f}$.

Considering $R : H^* \rightarrow H$, the Riesz isomorphism $f \rightarrow v_f$, then the following diagram commutes:

$$
\begin{array}{ccc}
H & \xrightarrow{\ T^*\ } & H \\
{\scriptstyle R}\uparrow & & \uparrow{\scriptstyle R} \\
H^* & \xrightarrow{\ T^*\ } & H^*.
\end{array}
$$

Proposition 13 If $T, S \in L(H)$, then $T^*, S^* \in L(H)$. The adjoint operator $* : H \rightarrow H^*, T \rightarrow T^*$ satisfies the following items:

(i) $(T + S)^* = T^* + S^*$.
(ii) $(T^*)^* = T$.
(iii) $(a.T)^* = \bar{a}.T^*$.
(iv) $(T.S)^* = S^*.T^*$.
(v) $\mid T^* \mid = \mid T \mid$ and $\mid T^*.T \mid = \mid T \mid^2$.

Proof The items (i)–(iv) are straightforward to prove. We now prove item (v). From the inequality

$$\big| < T^*(u).v > \big| \ = \ \big| < u, T(v) > \big| \ \leq \mid T \mid . \mid u \mid . \mid v \mid,$$

we have $\mid T^* \mid \leq \mid T \mid$. Since $(T^*)^* = T$, the same argument applied twice implies that $\mid T \mid \leq \mid T^* \mid$. Therefore $\mid T^* \mid = \mid T \mid$ and

$$\mid T^*.T \mid \leq \mid T^* \mid . \mid T \mid = \mid T \mid^2 .$$

The reverse inequality follows easily from

$$\mid T(u) \mid^2 = < T(u), T(u) >=< u, T^*.T(u) > \leq \mid T^*.T \mid . \mid u \mid^2 \Rightarrow \mid T \mid^2 \leq \mid T^*.T \mid.$$

Hence $| T |^2 = | T^*.T |$ ☐

Definition 10 Let H be a Hilbert space. An operator $T \in L(H)$ is self-adjoint if

$$< T(u), v > = < u, T(v) > \iff T^* = T,$$

for any $u, v \in H$.

Proposition 14 *Let H be a Hilbert space and let $T \in L(H)$ be a self-adjoint operator. So*

$$| T | = \sup_{||x|| \leq 1} | < T(x), x > |. \tag{23}$$

Proof Taking $c = \sup_{||x|| \leq 1} | < T(x), x > |$, we have $| < T(x), x > | \leq c. || x ||^2$ for all $x \in H$. The Cauchy-Schwartz inequality implies that $c \leq | T |$. From the inequality

$$| T | = \sup_{|x| \leq 1} || T(x) || \leq \sup_{|x| \leq 1} \sup_{|y| \leq 1} | < T(x), y > |, \tag{24}$$

we learn that it is sufficient to prove $| < T(x), y > | \leq c$ whenever $|| x || \leq 1$ and $|| y || \leq 1$. Applying the polarization identity, we get

$$4 < T(x), y > = \{ < T(x + y), x + y > - < T(x - y), x - y > \} + $$
$$+ i \{ < T(x + iy), x + iy > - < T(x - iy), x - iy > \}.$$

The parallelogram identity, as described in Appendix A (11), gives the following results:

$$16 | < T(x), y > |^2 = [< T(x + y), x + y > - < T(x - y), x - y >]^2 + $$
$$+ [< T(x + iy), x + iy > - < T(x - iy), x - iy >]^2. \tag{25}$$

The polar representations $< T(x), y) >= e^{i\theta} | < T(x), y > |$ and $x' = e^{-i\theta} x$ applied into the sesquilinear product $< T(x'), y >= | < T(x), y > |$ show the product is real. Therefore we can assume that $< T(x), y) > \in \mathbb{R}$. As a consequence,

$$< T(x + iy), x + iy > - < T(x - iy), x - iy >= 0$$
$$< T(x + iy), x + iy > - < T(x - iy), x - iy >= 2i[- < T(x), y > + < T(y), x >] =$$
$$= 2i[- < T(x), y > + < y, T(x) >] = 2i[- < T(x), y > + \overline{< T(x), y >}] = 0.$$

Identity (25) implies that

$$4 | < T(x), y > | \leq | < T(x + y), x + y > | + | < T(x - y), x - y > | \leq$$
$$\leq | T | . | x + y |^2 + | T | . | x - y |^2 \leq | T | . (| x + y |^2 + | x - y |^2) \leq$$
$$\leq 2 | T | (| x |^2 + | y |^2).$$

Since $\mid x \mid\leq 1$ and $\mid y \mid\leq 1$, we have

$$4 \mid< T(x), y >\mid \leq 2c\left(\parallel x \parallel^2 + \parallel y \parallel^2 \right) \leq 4c \quad \Rightarrow \quad \mid T \mid\leq c.$$

Hence $\mid T \mid\leq c$.

$\square$

Taking into account that $\sup_{\parallel x\parallel\leq 1} \mid< T(x), x >\mid= \sup_{\parallel x\parallel=1} \mid< T(x), x >\mid$, define

$$m(T) = \inf_{\parallel x\parallel=1} < T(x), x >, \quad M(T) = \sup_{\parallel x\parallel=1} < T(x), x > . \tag{26}$$

Theorem 9 *If $T \in L(H)$ is a self-adjoint operator, then the following conditions are satisfied:*

(i) $\sigma(T) \subset [m(T), M(T)] \subset \mathbb{R}$.
(ii) Both values $m(T)$ and $M(T)$ belong to $\sigma(T)$.

Proof Let $\lambda \in \sigma(T)$ be a non-null eigenvalue and $x \neq 0 \in N(T - \lambda I)$. From

$$\lambda \parallel x \parallel^2 = < T(x), x > = < x, T(x) > = \bar{\lambda} \parallel x \parallel^2 \quad \Rightarrow \quad (\lambda - \bar{\lambda}) \parallel x \parallel^2 = 0, \tag{27}$$

we have $\bar{\lambda} = \lambda$, and so $\sigma(T) \subset \mathbb{R}$. To prove Item (i), we verify the operator $T - \lambda I \in \mathcal{L}(E)$ is bijective for all $\lambda \notin [m(T), M(T)]$. Consider $\lambda \notin \sigma(T)$;
(ia) $T - \lambda I$ is injective.
Consider $x \in H$ such that $\alpha =< T(x), x >\in [m(T), M(T)]$. Therefore $< (T - \alpha I)x, x >= 0$ and

$$\begin{aligned}
\parallel (T - \lambda I)x \parallel^2 &= \parallel (T - \alpha I)x + (\alpha - \lambda)x \parallel^2 \\
&= \mid \lambda - \alpha \mid^2 \parallel x \parallel^2 + \parallel (T - \alpha I)x \parallel^2 + 2(\alpha - \lambda) < (T - \alpha I)x, x > \\
&\geq \mid \lambda - \alpha \mid^2 \parallel x \parallel^2 .
\end{aligned}$$

Since $\mid \lambda - \alpha \mid^2> 0$, then $T - \lambda I$ is injective. Moreover, $T(H)$ is a closed subset in H.

(ib) $T - \lambda I$ is surjective.
For all $v \in (\text{Im}(T - \lambda I))^\perp$ we have $< (T - \lambda I)x, v >= 0$ for all $x \in H$. Then $< x, (T - \lambda I)v >= 0$, and so $v = 0$. Therefore $(T - \lambda I)$ is surjective.
(ic) The resolvent $R_\lambda = (T - \lambda I)^{-1}$ is bounded since the norm of $y = (T - \lambda I)x$ satisfies $\parallel y \parallel^2\geq d \parallel x \parallel^2$, where $d = \text{dist}(\lambda, [m(T), M(T)])$. Therefore $\parallel R_\lambda y \parallel\leq \frac{1}{d} \parallel y \parallel$, i.e., $\mid R_\lambda \mid\leq \frac{1}{d}$.
(ii) We observe that either $\mid T \mid= -m(T)$ or $\mid T \mid= M(T)$. Assuming $\mid T \mid= M(T) = M$, we have a sequence $\{x_n\}_{n\in\mathbb{N}} \subset H$ of unitary vectors such that $\lim < T(x_n), x_n >= M$. However, we have $\lim \parallel (T - MI)x_n \parallel\to 0$ since

$$\parallel (T - M.I)x_n \parallel^2 = \parallel T(x_n) \parallel^2 +M^2 - 2M < T(x_n), x_n > \leq 2M^2 - 2M < T(x_n), x_n > \to 0.$$

This means that $(T - M.I)$ is not invertible in $L(H)$. Hence $M \in \sigma(T)$. The proof is similar to the case $| T | = -m(T)$. □

The spectrum of a self-adjoint linear operator is a non-empty subset in $\mathbb{R}$. Moreover, its norm is equal to its spectral ray;

$$r_\sigma(T) = \sup_{\lambda \in \sigma} \{| \lambda |\}. \tag{28}$$

Exercises

(a) Prove identity (24).

(b) Prove that $| T^*T | = | T |^2$ for all $T \in L(H)$.

(c) Let $T \in L(H)$ be a self-adjoint operator. Taking $M(T) = M$ and $m(T) = m$, prove that

$$| M.I - T | = \sup_{||x||=1} < (T - M.I)x, x >= M - m.$$

Apply this identity to prove that $M - m \in \sigma(T)$. Equivalently, we have that the operator $(M - m)I - (M.I - T) = T - m.I$ is not invertible.

(d) Assuming T is a self-adjoint operator, prove the identity $| T | = r_\sigma$ (norm is equal to the spectral ray).

(e) Let H be a Hilbert space. Prove that a linear operator $F : H \to H$ is Fredholm if and only if there exist orthogonal decompositions $H = H_1 \oplus H_2$ and $H = H_3 \oplus H_4$, such that

(a) H_1 and H_3 are both closed subspaces.

(b) H_2 and H_4 are finite dimensional subspaces.

(c) F can be represented in blocks as

$$F = \begin{pmatrix} F_{11} & F_{12} \\ F_{21} & F_{22} \end{pmatrix} : H_1 \oplus H_2 \to H_3 \oplus H_4.$$

where $F_{11} : H_1 \to H_3$ is invertible.

(d) The index of F is $\mathrm{ind}(F) = \dim(H_2) - \dim(H_4)$.

7.1 Characterization of Compact Operators on Hilbert Spaces

Compact operators behave in a similar way with finite-dimensional operators. We will give the criterion for deciding whether an operator on a Hilbert space is compact.

Proposition 15 *Let H be a Hilbert space and $K \in L(H)$. So K is compact if and only if we have a sequence of finite rank operators $\{K_n\} \in L(H)$ such that $K_n \to K$ in $L(H)$.*

Proof Since $\overline{K(H)}$ is compact, it has an enumerable and dense subset, so $\overline{K(H)}$ is a separable subset of H. Let $\{\phi_n\}_{n\in\mathbb{N}}$ be an orthonormal basis of $\overline{K(H)} \subset H$ and $P_N(y) = \sum_{i=1}^{N} < y, \phi_i > \phi_i$ the orthogonal projection over the subspace generated by $\{\phi_1, \ldots, \phi_N\}$. Now we have $\lim_{N\to\infty} \| P_N(y) - y \| = 0$ for all $y \in K(H)$. The operator $K_N = P_N K \in \mathcal{K}(H)$ has finite rank. Since K is compact, taking a bounded sequence $\{x_n\} \subset H$, let $\lim K(x_n) = y \in H$. In this way,

$$\| (K - K_N)(x_n) \| = \| (I - P_N)K(x_n) \| \leq \| (I - P_N)(K(x_n) - y) \| + \| (I - P_N)(y) \|$$
$$\leq \| K(x_n) - y \| + \| (I - P_N)(y) \| \to 0, \text{ when } N \to \infty.$$

The reverse is straightforward. Consider $\{T_n\} \in L(H)$ a sequence of finite rank linear operators such that $\lim_{n\to\infty} | T - T_n | = 0$. Let $\{x_k\} \subset H$ be a bounded sequence. Taking $y_n = \lim_{k\to\infty} T_n(x_k)$, we get

$$| T(x_k) - T(x_l) | \leq | (T(x_k) - T_n(x_k) | + | (T_n(x_k) - T_n(x_l) | + | (T_n(x_l) - T(x_l) | .$$

Considering $k, l \in N$ large enough, the sequence $\{T(x_k)\}$ is a Cauchy sequence. Hence T is compact.   □

The above theorem is false in the category of Banach spaces. Per Enflo gave a counterexample in [15].

7.2   Self-adjoint Compact Operators on Hilbert Spaces

Let H be a Hilbert space. We will use the acronym $SACO(H)$ to represent the class of self-adjoint compact operators defined on H. The main properties of a self-adjoint operator T are (i) $T^* = T$, (ii) the eigenvalues of T are real numbers, (iii) the generalized eigenspaces associated with distinct eigenvalues are orthogonal and invariant by T. We also learned from Theorem 9 that $| T | \in \sigma(T)$. The proof of the Spectral Theorem for compact self-adjoint operators defined on an infinite dimensional vector space goes along the same reasoning to prove when we have a finite dimensional vector space.

Theorem 10 *Let H be a Hilbert space and $T \in SACO(H)$. So:*

(i) *the spectrum $\sigma(T) = \{0\} \cup \{\lambda_1, \ldots, \lambda_n, \ldots\}$ is an ordered enumerable set such that λ_n is an eigenvalue of T, $| \lambda_k | \geq | \lambda_{k-1} | > 0$ for all $k \in \mathbb{N}$ and $\lim \lambda_n = 0$.*
(ii) *there is an orthonormal basis of eigenvalues $\beta = \{\phi_1, \ldots, \phi_n \ldots\}$ spanning a closed subspace V such that $H = \text{Ker}(T) \oplus V$.*
(iii) *For any $v \in H$,*

$$T(f) = \sum_{n=1}^{\infty} \lambda_n < v, \phi_n > \phi_n.$$

Proof Let $\lambda_1 \in \{\pm \mid T \mid\}$ and let $\phi_1 \in H$ be a unitary vector such that $T\phi_1 = \lambda_1\phi_1$. Taking the subspace $V_1 = <\phi_1>$ generated by ϕ_1, we get an orthogonal decomposition $H = V_1 \oplus V_1^{\perp}$, where $T(V_1) \subset V_1$ and $T(V_1^{\perp}) \subset V_1^{\perp}$. The restriction $T_1 = T\mid_{V_1^{\perp}}$ is also compact and self-adjoint. Suppose $T_1 \neq 0$. Let $\lambda_2 \in \{\pm \mid T_1 \mid\}$ and let $\phi_2 \in V_1^{\perp}$ be a unitary vector such that $T_1\phi_2 = \lambda_2\phi_2$. Considering $V_2 = V_1 \oplus <\phi_2>$ with basis $\beta_H^2 = \{\phi_1, \phi_2\}$, we have the decomposition $H = V_2 \oplus V_2^{\perp}$. Observing that $\mid T_i \mid \geq \mid T_{i+1} \mid$ for all i, the same construction is performed to obtain a decomposition for all $i \in \mathbb{N}$. According to Theorem 6, the set $\sigma(T) = \{\lambda_1, \ldots, \lambda_n, \ldots\}$ is enumerable, $\mid \lambda_i \mid \geq \mid \lambda_{i-1} \mid$ and $\lim \lambda_n = 0$. So for any $v \in H$, represented as the linear combination of the basis $\beta_H = \{\phi_n\}$ as $v = \sum_{n=1}^{\infty} < f, \phi_n > \phi_n$, we get the uniformly convergent series $T(v) = \sum_{n=1}^{\infty} \lambda_n < v, \phi_n > \phi_n$. $\square$

The theorem above extends to compact operators.

Theorem 11 *Let E, F be Hilbert spaces and $K \in \mathcal{K}(E, F)$. So we have orthonormal bases $\beta_E = \{\phi_n\}_{n\in\mathbb{N}} \subset E$ and $\beta_F = \{\psi_n\}_{n\in\mathbb{N}} \subset F$, and a sequence $\{\lambda_n\}_{n\in\mathbb{N}} \subset \mathbb{C}$, $\lim \lambda_n = 0$, such that*

$$K(f) = \sum_{n=1}^{\infty} \lambda_n < f, \phi_n > \psi_n,$$

for all $f \in E$.

Proof Since the operator K^*K is compact and positive (self-adjoint), the set of eigenvectors $\{\phi_n\}$ forms an orthonormal basis of H. Let $\sigma(K^*K) = \{\mu_1, \ldots, \mu_n, \ldots\} \subset [0, \infty)$ be the set of eigenvalues. For every n, define $\lambda_n = \sqrt{\mu_n}$ and let $(K^*K)^{1/2} \in \mathcal{K}(E, F)$ be the operator given by

$$(K^*K)^{1/2}(f) = \sum_{n=1}^{\infty} \lambda_n < f, \phi_n > \phi_n. \tag{29}$$

The inverse operator $(K^*K)^{-1/2} \in \mathcal{K}(E, F)$ is

$$(K^*K)^{-1/2}(f) = \sum_{n=1}^{\infty} \lambda_n^{-1} < f, \phi_n > \phi_n. \tag{30}$$

Let $U = K(K^*K)^{-1/2}$ and $U(f) = \sum_{n=1}^{\infty} < f, \phi_n > U(\phi_n)$. From Eq. (30), we have

$$U(f) = \sum_{n=1}^{\infty} \lambda_n^{-1} < f, \phi_n > K(\phi_n). \tag{31}$$

Therefore we get $U(\phi_n) = \lambda_n^{-1} K(\phi_n)$. Since $\| U(f) \| = \| f \|$ for all $f \in E$, U is a unitary operator and the basis $\beta = \{\lambda_n^{-1} K(\phi_n)\}$ is orthonormal. So taking $\psi_n = \lambda_n^{-1} K(\phi_n)$, we have

$$K(f) = \sum_{n=1}^{\infty} \lambda_n < f, \phi_n > \psi_n. \tag{32}$$

□

It is interesting to remark that a Hilbert space provided with a compact, self-adjoint operator is necessarily separable. This indicates that the properties of operators can reveal the intrinsic properties of the space.

Exercises

(1) The operator $T : H \rightarrow H$ is positive if it is self-adjoint and $< T(u), u > \geq 0$ for all $u \in H$. Given a bounded linear operator $T : H \rightarrow H$, prove that T^*T and TT^* are positives.

(2) The square root of a positive operator T is the self-adjoint operator S such that $S^2 = T$. Prove that any positive operator T has a unique square root S.

(3) Find the eigenvalues and eigenvectors of $T : L^2([0, 2\pi]) \rightarrow L^2([0, 2\pi])$ given by

$$T(u)(x) = \int_0^{2\pi} \sin(x - t)u(t)dt.$$

(4) Let $k : [0, 2\pi] \rightarrow \mathbb{R}$ be a periodic integrable function. Find the eigenvalues and eigenvectors of $T : L^2([0, 2\pi]) \rightarrow L^2([0, 2\pi])$ given by (hint: try $u_n(x) = e^{inx}$)

$$T(u)(x) = \int_0^{2\pi} k(x - t)u(t)dt.$$

7.3   Fredholm Alternative

Let H be a Hilbert space and let $K \in \mathcal{K}(H)$ be a compact self-adjoint operator. The Spectral Theorem 10 guarantees the existence of a sequence of eigenvalues $\sigma(K) = \{\lambda_n \mid n \in \mathbb{N}\}$ satisfying the conditions (i) $\mid \lambda_i \mid > \mid \lambda_{i+1} \mid$, (ii) if $\#\sigma(K) = \infty$, then $\lim \lambda_n = 0$ and (iii) we have an orthonormal basis $\beta = \{\phi_n\}$ of H with vectors that are eigenvectors of K. Let's consider the following two cases;
Case 1: If $\lambda \neq 0$ and $\lambda \notin \sigma(K)$, then the equation

$$\lambda.x - K(x) = y \tag{33}$$

admits a unique solution x for each $y \in H$.

Proof Let's check that the operator $T_\lambda = \lambda.I - K$ has a continuous inverse $T_\lambda^{-1} \in L(H)$. Any vector $x \in H$ can be written as a linear combination $x = \sum_n x_n \phi_n$ of the basis β, where $x_n = < x, \phi_n >$. Then from Eq. (33), we have

$$\sum_n (\lambda - \lambda_n) x_n \phi_n = \sum_n y_n \phi_n \Rightarrow x_n = \frac{y_n}{\lambda - \lambda_n}.$$

The series $x = \sum_n \frac{y_n}{\lambda - \lambda_n} \phi_n$ is a candidate to solve the equation $x = (\lambda.I - K)^{-1}(y)$. To verify, first let's check the uniform convergence of the series. Let $\alpha = \sup_n | \frac{1}{\lambda - \lambda_n} |$ and apply the Cauchy-Schwartz inequality;

$$\left| \sum_n \frac{y_n}{\lambda - \lambda_n} \phi_n \right|^2 \leq \sum_n \left| \sum_n \frac{y_n}{\lambda - \lambda_n} \right|^2 . | \phi_n |^2 \leq \alpha \sum_n | y_n |^2 . | \phi_n |^2 \leq \alpha. | y |^2 .$$

Then $| T_\lambda^{-1}(y) | \leq \alpha. | y |^2$ and $| T_\lambda^{-1} | \leq \alpha$. If the equation $T_\lambda^{-1}(y) = x$ has a solution, then it must be unique; otherwise, having a solution x' that is different from x would imply $\lambda \in \sigma(K)$. This is not allowed since it contradicts the hypothesis. Hence the solution is unique. □

Case 2: If $\lambda_p \neq 0$ and $\lambda_p \in \sigma(K)$, then the solution set of the Eq. is a finite dimensional vector space.

Proof The necessary and sufficient condition for the existence of a solution x to the equation $\lambda_p x - K(x) = y$ is $< y, \phi >= 0$ for all eigenvectors ϕ associated to λ_p. Let's prove the necessity first. Let x be a solution and $K(\phi) = \lambda_p \phi$, so

$$< \lambda_p x - K(x), \phi > = < x, \lambda_p \phi - K(\phi) > = 0 \quad \Rightarrow \quad < y, \phi > = 0.$$

To check sufficiency, suppose $< y, \phi >= 0$ for all ϕ so that $K(\phi) = \lambda_p \phi$. Expanding in the basis β, we get

$$\lambda_p x - K(x) = y \Rightarrow \sum_{n \neq p} (\lambda_p - \lambda_n) x_n \phi_n = \sum_{n \neq p} y_n \phi_n.$$

It follows that $x_p = \sum_{n \neq p} \frac{y_n}{\lambda - \lambda_n} \phi_n$ is a solution. Now, for all ϕ solving $(\lambda_p I - K)\phi = 0$, the general solution is $x = x_p + \phi$ since

$$\lambda_p x - K(x) = \left(\lambda_p x_p - K(x_p) \right) - \left(\lambda_p \phi - K(\phi) \right) = 0.$$

Since t $(\lambda_p I - K)$ is a Fredholm operator, its kernel has finite dimension. Therefore the solution set

$$\{ x_p + \phi \mid \phi \in \text{Ker}(\lambda_p I - K) \}$$

has finite dimension. □

7.4   *Hilbert-Schmidt Integral Operators*

An important category of compact operators is the Hilbert-Schmidt (HS) integral operators.

Definition 11 Let $\Omega \subset \mathbb{R}^n$ be a connected open subset. The Hilbert-Schmidt kernel is a function $k : \Omega \times \Omega \to \mathbb{C}$ such that

$$\int_\Omega \int_\Omega \mid k(x, y) \mid^2 dx dy < \infty.$$

The Hilbert-Schmidt operator associated to k is

$$T_k(f)(x) = \int_\Omega k(x, y) f(y) dy, \tag{34}$$

and we have $\parallel T_k \parallel_{HS} = \parallel k \parallel_{L^2}$. The operator T_k is linear, continuous and compact. Assuming the condition $\overline{k(x, y)} = k(x, y)$, T_k becomes a self-adjoint operator. So it is diagonalizable. The kernel k is non-negative defined when, for any finite sequence of points $\{x_1, \ldots, x_n\} \subset [a, b]$ and any choice of real numbers $c_1, \ldots, c_n$, we have

$$\sum_{i=1}^n \sum_{j=1}^n k(x_i, x_j) c_i c_j \geq 0.$$

Theorem 12 (Mercer) *Let $\Omega = [a, b]$ and let k be a continuous, self-adjoint and non-negative HS kernel. So we have an orthonormal basis $\beta = \{\phi_n\}$ of $L^2([a, b])$ consisting of eigenfunctions of the operator T_k defined in (34), with eigenvalues that are non-negative. The eigenfunctions corresponding to an eigenvalue $\lambda_n \neq 0$ are continuous in $[a, b]$, and k is given by the absolutely convergent series*

$$k(x, y) = \sum_i \lambda_i \phi_i(x) \overline{\phi_i(y)}, \tag{35}$$

Proof The operator T_k defined in (34) is non-negative, compact and self-adjoint in $L^2[a, b]$. Expanding $k(x, y)$ as a linear combination of the basis β, we get

$$k(x, y) = \sum_i \alpha_i(x) \phi_i(y),$$

where $\alpha_i(x) = \int_a^b k(x, y) \phi_i(y) dy$. However since β is a set of eigenvectors, we have

$$\int_a^b k(x, y) \phi_i(y) dy = \lambda_i . \phi_i(x).$$

Consequently, $k(x, y) = \sum_i \lambda_i \phi_i(x)\overline{\phi_i(y)}$. When $\lambda_i \neq 0$, the eigenfunction ϕ_i is continuous in $[a, b]$. From the inequality

$$\sum_i \lambda_i \mid \phi_i(x).\overline{\phi_i(y)} \mid \leq \sup_{x \in [a,b]} \mid k(x, x) \mid^2,$$

we conclude that the series in (35) uniformly converges to k. $\qquad\square$

Exercises

(1) Prove that a compact operator is bounded.
(2) Given the hypothesis of Theorem 12, prove that

$$\int_a^b k(t, t)dt = \sum_n \lambda_n.$$

In this case, the operator T_k belongs to the class of trace operators, and we define

$$tr(T_k) = \int_a^b k(t, t)dt = \sum_n \lambda_n.$$

(3) Given the hypothesis of Theorem 12, prove that (hint: use Bessel's inequality)

$$\sum_n \lambda_n^2 \leq \int_a^b \int_a^b \mid k(s, t) \mid^2 ds dt.$$

(4) Let $K : L^2([0, 1]) \to L^2([0, 1])$ be the operator given by

$$K(f)(x) = \int_0^x f(y)dy.$$

    (a) Find the adjoint K^*.
    (b) Find $\mid K \mid$.
    (c) Prove that the spectral ray is $r_\sigma(K) = 0$.
    (d) Prove that 0 belongs to the continuous spectrum of K.

(5) Consider $S : l^2(\mathbb{Z}) \to l^2(\mathbb{Z})$ the operator $S(x)_k = x_{k-1}$, where $x = (x_k)_{-\infty}^{\infty} \in l^2(\mathbb{Z})$ for all $k \in \mathbb{Z}$. Prove the following items:

    (a) $\sigma(S) = \varnothing$.
    (b) $(\lambda I - S)$ is surjective for all $\lambda \in \mathbb{C}$ such that $\mid \lambda \mid \neq 1$.
    (c) The spectrum $\sigma(S) = \{\lambda \in \mathbb{C}; \mid \lambda \mid = 1\}$ of S is only continuous.

8   Closed Unbounded Linear Operators on Hilbert Spaces

So far the theory developed deals only with bounded linear operators; however there are important classes of unbounded linear operators. Let E, F be normed spaces, a linear operator $T : E \to F$ is unbounded if it is not bounded. That is, T is unbounded if we have a sequence $\{x_n\} \subset E$ such that $\lim \frac{|T(x_n)|_F}{|x_n|_E} = \infty$.

Example 7   Among the unbounded operators, we highlight those that are differential operators.

(1) Let $E = (C^0([a, b]), \| \cdot \|_0)$ and let $\mathcal{D} = C^1([a, b]) \subset E$ be the domain of T. The differential operator $T : \mathcal{D} \to E, T(f) = f'$ is unbounded. For $n \in \mathbb{N}$, consider the function $f_n : [-\pi, \pi] \to \mathbb{R}, f_n(x) = \cos(nx)$, so $f'_n(x) = -n \sin(nx)$. Since

$$\| f_n \|_0 = \sup_{x \in [-\pi, \pi]} | f_n(x) | = 1 \quad \text{and} \quad \| f'_n \|_0 = \sup_{x \in [-\pi, \pi]} | f'_n(x) | = n,$$

we have $\lim \frac{\| T(f_n) \|_0}{\| f_n \|_0} = \infty$.

(2) Let $E = L^2(\mathbb{R})$ and $\mathcal{D} = H^2(\mathbb{R})$. Consider $T : \mathcal{D} \to L^2(\mathbb{R})$ the operator $T(f) = -f''$. The sequence $\{f_n\}_{n \in \mathbb{N}}$, $f_n(t) = e^{-n|t|}dt$, T is unbounded since

$$\| f_n \|_{L^2}^2 = \frac{1}{n}, \quad \| T(f_n) \|_{L^2}^2 = n^3 \quad \Rightarrow \quad \lim_{n \to \infty} \frac{\| T(f_n) \|_{L^2}}{\| f_n \|_{L^2}} = \lim_{n \to \infty} n = \infty.$$

(3) Let ϕ be an unbounded continuous function defined on $\mathbb{R}$ and $D = C_0^\infty(\mathbb{R})$ (set of functions with compact support). Define $T_\phi^c : D \to L^2(\mathbb{R})$ by $(T_\phi^c(f))(x) = \phi(x).f(x)$. Therefore T is unbounded.

(4) In the last example, consider $D = \{f \in L^2(\mathbb{R}) \mid \phi.f \in L^2(\mathbb{R})\}$ and $T : D \to L^2(\mathbb{R})$. Therefore T is unbounded.

(5) Let $H = l^2(\mathbb{Z})$ be the Hilbert space $\mathcal{D} = \{(x_1, \ldots, x_n, \ldots) \mid \sum_{i=0}^{\infty} | jx_j |^2 < \infty\}$ and $T : \mathcal{D} \to H$ the operator $T(x_1, \ldots, x_n, \ldots) = (x_1, 2x_2, 3x_3, \ldots, nx_n, \ldots)$. Therefore T is unbounded.

The differential operator, as defined in Example (1) above, has the following property: if $\{f_n\}_{n \in \mathbb{N}} \subset C^1([a, b])$ is a convergent sequence such that (i) $f_n \to f \in (C^0([a, b]), \| \cdot \|_0)$ and (ii) $T(f_n) \to g \in (C^0([a, b]), \| \cdot \|_0)$, then $f \in C^1([a, b])$ and $f' = g$ (Theorem 11, Appendix A). This property is explored in [20] for studying the spectral properties of T. So in this case the operator is closed, as defined in Appendix A, Definition 2.

Definition 12   Let E, F be Banach spaces and let $T : E \to F$ be a linear operator with the domain $\mathcal{D}(T) \subset E$. T is a closed operator if given any sequence $\{x_n\}_{n \in \mathbb{N}} \subset \mathcal{D}(T)$ such that $\lim x_n = x \in E$ and $\lim T(x_n) = y \in F$, then $x \in \mathcal{D}(T)$ and $T(x) = y$.

We emphasize the importance of the domain $\mathcal{D}(T)$ and the range $\mathcal{R}(T)$ to study the properties of the operator. For example, the differential operator $T : (C^1(\mathbb{R}), \|\cdot\|_1) \to (C^0(\mathbb{R}), \|\cdot\|_0), T(f) = f'$, is bounded while the operator $T : C^1(-1, 1) \to L^2(-1, 1), T(f) = f'$, is unbounded and not closed, as shown in the following case: consider the sequence of functions $\{f_n\}_{n \in \mathbb{N}}$ given by $f_n(x) = \sqrt{x^2 + \frac{1}{n^2}}$. So,

$$f_n \xrightarrow{L^2} f(x) =| x |, \quad f_n'(x) = \frac{nx}{\sqrt{n^2x^2 + 1}} \xrightarrow{L^2} g(x) = \begin{cases} \frac{x}{|x|}, & x \neq 0, \\ 0, & x = 0. \end{cases}$$

Obviously, $g \notin C^1$ and $f'(x) \neq g$. A large class of differential operators belongs to the category of unbounded and closed operators.

When we perform operations with the operators T and S, with domains $\mathcal{D}(T)$ and $\mathcal{D}(S)$ respectively, it is important to note that:

(i) $\mathcal{D}(T + S) = \mathcal{D}(T) \cap \mathcal{D}(S)$;
(ii) $\mathcal{D}(T \circ S) = \{x \in \mathcal{D}(S) \mid S(x) \in \mathcal{D}(T)\}$.

The operations $T + S$ or $T \circ S$ make sense only if the respective domain is non-empty.

From now on, consider H a Hilbert space.

Definition 13 Let $\mathcal{D}(T) \subset H$ be the domain of a linear operator $T : \mathcal{D}(T) \to H$.

(i) T is well-defined if $T(u) = f$ and $T(u) = g$ implies $f = g$. Equivalently, there is no $f \in \mathcal{D}(T)$ such that $T(0) = f$ since $T(0) = 0$.
(ii) T is densely defined in H if $\mathcal{D} \subsetneq H$ is dense in H.
(iii) The linear operator $S : \mathcal{D}(S) \to H$ is an extension of $T : \mathcal{D}(T) \to H$ if $\mathcal{D}(T) \subset \mathcal{D}(S)$ and $S(u) = T(u)$ for all $u \in \mathcal{D}(T)$. In this case, we denote $T \subset S$.

Next, we will give a sufficient condition to guarantee the existence of a closed extension $\overline{T} : \mathcal{D}(\overline{T}) \to H$.

Definition 14 Let $\mathcal{D}(T) \subset H$ and let $T : \mathcal{D}(T) \to H$ be an unbounded linear operator densely defined. Consider $\mathcal{D}(T^*) = \{v \in H \mid u \to < T(u), v >$ is continuous for all $u \in \mathcal{D}(T)\}$. The adjoint operator $T^* : \mathcal{D}(T^*) \to H$ is defined by the identity

$$< T(u), v >_H = < u, T^*(v) >_H, \forall u \in \mathcal{D}(T).$$

The operator T^* is well-defined; otherwise there are v, w such that $T^*(u) = v = w$, and so $< u, v - w >= 0$ for all $u \in \mathcal{D}(T)$. Since $\mathcal{D}(T)$ is densely defined, we take any element $\hat{u} \in H$ and let $\hat{u} = \lim u_n \in H$, where $\{u_n\}_{n \in \mathbb{N}} \subset \mathcal{D}(T)$. It follows from the continuity of the product that $< \hat{u}, v - w >= 0$ for any $\hat{u} \in H$, so $v = w$. In addition, T^* is a linear operator, since for any $a, b \in \mathbb{C}, u \in \mathcal{D}(T)$ and $v \in \mathcal{D}(T^*)$,

$$< T(u), av_1 + bv_2 > = a < T(u), v_1 > + b < T(u), v_2 >=< u, aT^*(v_1) + bT^*(v_2) >,$$
$$< T(u), av_1 + bv_2 > = < u, T^*(av_1 + bv_2) >$$
$$\Rightarrow < u, T^*(av_1 + bv_2) - aT^*(v_1) - bT^*(v_2) >= 0.$$

Hence $T^*(av_1 + bv_2) = aT^*(v_1) + bT^*(v_2)$.

Example 8 Examples of adjoint operators.

(1) $H = L^2([0, 1])$, $\quad \mathcal{D}(T) = \{u \in C^1([0, 1]) \mid u(0) = u(1) = 0\} \subset H$ $\quad$ and $T(u) = u'$.

$$< T(u), v >= \int_0^1 u'(x)v(x)dx = \int_0^1 u(x).v'(x)dx = < u, -v >, \forall v \in C^1([0, 1]).$$

So we have $T^*(v) = -v$ and $\mathcal{D}(T) \subsetneq \mathcal{D}(T^*) = C^1([0, 1])$.
(2) $H = L^2([0, 1])$, $\mathcal{D}(T) = \{u \in C^1([0, 1]) \mid u(0) = 0\} \subset H$ and $T(u) = u'$.
$$< T(u), v >= \int_0^1 u'(x)v(x)dx = u(1)v(1) + \int_0^1 u(x).(-v'(x))dx =< u, -v > .$$

So we have $D(T^*) = \{v \in C^1([0, 1]) \mid v(1) = 0\}$. Therefore, $\mathcal{D}(T) \not\subset \mathcal{D}(T^*)$.

Proposition 16 Let $T : \mathcal{D}(T) \subset H \to H$ and $S : \mathcal{D}(S) \subset H \to H$ be densely defined operators. So:

(i) If $T \subset S$, then $S^* \subset T^*$.
(ii) If $\mathcal{D}(T^*)$ is dense in H, then $T \subset T^{**}$.
(iii) If T is injective and T^{-1} is densely defined, then T^* is injective and $(T^*)^{-1} = (T^{-1})^*$.
(iv) $S^*T^* \subset (TS)^*$.

Definition 15 Let $\mathcal{D}(T) \subset H$. A densely defined operator $T : \mathcal{D}(T) \to H$ is self-adjoint if $\mathcal{D}(T^*) = \mathcal{D}(T)$ and $T = T^*$.

Example 9 Let's consider the following classical examples. Consider $H = L^2([0, 1])$.

(1) Let $\mathcal{D}(T) = \{u \in C^2([0, 1]) \mid u(0) = u(1) = 0\}$ and let $T : \mathcal{D}(T) \to H$ be the operator $T(u) = u''$.

$$< T(u), v) >= u'(1)v'(1) - u'(0)v'(0)+ < u, T(v) > .$$

Then $\mathcal{D}(T^*) = \{v \in C^2([0, 1]) \mid v'(0) = v'(1) = 0\} \neq \mathcal{D}(T)$. Restricting to the intersection $\mathcal{D}(T^*) \cap \mathcal{D}(T)$, we have $T^* = T$. Therefore T is not self-adjoint.
(2) As above, consider the same operator $T(u) = u''$, $T : \mathcal{D}(T) \to H$ defined on the space $\mathcal{D}(T) = \{u \in C^2([0, 1]) \mid u(0) = u(1) = 0$ and $u'(0) = u'(1) = 0\}$. Now we have the identity $< T(u), v >=< u, T(v) >$ for all $u, v \in \mathcal{D}(T)$. Consequently, $\mathcal{D}(T^*) = \mathcal{D}(T)$, $T^* = T$. Hence T is self-adjoint.

The last examples motivate the next definition.

Definition 16 Let H be a Hilbert space $\mathcal{D}(T) \subset H$ and let $T : \mathcal{D}(T) \to H$ be a densely defined linear operator. T is symmetric if $< T(u), v >=< u, T(v) >$ for all $u, v \in \mathcal{D}(T)$.

The classic example of a symmetric operator is the operator $T(u) = i\frac{d}{dt}$ defined on

$$\mathcal{D}(T) = \left\{ f \in L^2([a, b]) \mid \frac{df}{dt} \in C^0 \text{ and } f(a) = f(b) = 0 \right\} \subset H = L^2([a, b]).$$

In the example above we can easily compute the adjoint as shown next;

$$< T(f), g >= \int_a^b i\frac{df}{dt}\overline{g}dt = i\big[f(b)g(b) - f(a)g(a)\big] + < f, i\frac{dg}{dt} >=< f, T^*(g) > .$$

So we have $\mathcal{D}(T^*) = H$ and $T^* \mid_{\mathcal{D}(T)}= T$. Clearly, T^* is an extension of T.

Proposition 17 *Let H be a Hilbert operator and $\mathcal{D}(T) \subset H$. If $T : \mathcal{D}(T) \to H$ is a densely defined linear operator, then T^* is closed.*

Proof Take a sequence $\{v_n\}_{n \in \mathbb{N}} \subset D(T^*)$ such that $\lim v_n = v$ and $\lim T^*(v_n) = w$. Therefore

$$< T(u), v >= \lim < T(u), v_n >= \lim < u, w > \Rightarrow < u, T^*(v) - w >= 0$$

for any $u \in \mathcal{D}(T)$. Therefore $T^*(v) = w$ and T^* is closed. □

The operator $T : \mathcal{D}(T) \to H$ is symmetric if and only if T^* is an extension of T. The next two theorems give us enough conditions to find a closed operator S extending T.

Theorem 13 *Let H be a Hilbert space. Let $\mathcal{D}(T)$ and $\mathcal{D}(T^*)$ be dense subsets in H and consider the linear operators $T : \mathcal{D}(T) \to H$ and $T^* : D(T^*) \to H$. So the operator $(T^*)^* = T^{**}$ is a closed extension of T.*

Proof Since T^* is densely defined, the operator T^{**} is closed. To prove $\mathcal{D}(T) \subset D(T^{**})$, consider $u \in D(T^{**})$ and $w = T^{**}(u)$; therefore we have

$$< w, v >=< T^{**}(u), v >=< u, T^*(v) > .$$

By definition, $< u, T^*(v) >=< T(u), v >$ for all $v \in D(T^*)$ and $u \in \mathcal{D}(T)$. Therefore if $u \in \mathcal{D}(T)$, then $T^{**}(u) = T(u)$. Hence $\mathcal{D}(T) \subset D(T^{**})$. □

Theorem 14 *Let H be a Hilbert space and $\mathcal{D}(T) \subset H$. Consider $T : \mathcal{D}(T) \to H$ a symmetric densely defined linear operator. So we have a closed symmetric operator S such that S is an extension of T.*

Proof Let's start by defining the domain of S. Let $\mathcal{D}(S)$ be the set of elements $u \in H$ for which there is a sequence $\{u_n\}_{n \in \mathbb{N}} \subset \mathcal{D}(T)$ and an element $v \in H$ such that $u = \lim u_n$ and $\lim T(u_n) = v$. $\mathcal{D}(S)$ is a vector space and $\mathcal{D}(T) \subset \mathcal{D}(S)$. Let $S : D(S) \to H$ be the operator given by $S(u) = \lim T(u_n)$, where $\{x_n\}_{n \in \mathbb{N}} \subset \mathcal{D}$ is a sequence so that $\lim u_n = u$. By the definition of $\mathcal{D}(S)$, the limit $\lim T(u_n)$ exists.

(i) S is well-defined.

Suppose there are sequences $\{v_n\}$ and $\{w_n\}$ in $\mathcal{D}(T)$ such that $\lim v_n = v, \lim w_n = w$ and $\lim T(u_n) = \lim T(w_n)$. So,

$$< u, T(u_n) - T(w_n) >=< u, T(u_n - w_n) >=< T(u), u_n - w_n >=< T(u), v - w >= 0.$$

Therefore $(v - w) \perp \mathcal{D}$. The density condition on $\mathcal{D}(T) \subset H$ yields $v = w$.

(ii) $S(u) = T(u)$, for any $u \in \mathcal{D}$ and $\mathcal{D}(S) \neq \varnothing$

Let $u \in \mathcal{D}(T)$ and consider the constant sequence $u_n = u$ for all $n \in \mathbb{N}$. So we have $S(u) = T(u)$ for all $u \in \mathcal{D}(T)$. Hence S is an extension of T.

(iii) S is symmetric.

Let $\{u_n\}$ and $\{v_n\}_{n \in \mathbb{N}}$ be sequences in $\mathcal{D}(T)$ such that $\lim u_n = u$, $\lim v_n = v$. In this way, $S(u) = \lim T(u_n)$ and $S(v) = \lim T(v_n)$. Consequently,

$$\lim < T(u_n), v_n >= \lim < u_n, T(v_n) > \quad \Rightarrow \quad < S(u), v >=< u, S(v) > .$$

(iv) S is closed.

Let $\{u_n\}_{n \in \mathbb{N}} \subset \mathcal{D}(S)$ be a sequence such that $\lim u_n = u$ and $\lim S(u_n) = w$, where $u, w \in H$. For every $n \in \mathbb{N}$, consider the sequence $\{u_{n,k}\}_{k \in \mathbb{N}} \subset \mathcal{D}(T)$ such that $u_n = \lim_k u_{n,k}$. Therefore the subsequence $\{u_{n,n}\}_{n \in \mathbb{N}} \subset \mathcal{D}(T)$ converges to $\lim u_{n,n} = u$. Consequently we have $u \in \mathcal{D}(S)$ and $S(u) = \lim T(u_{n,n})$. $\square$

An example of an operator that does not allow a closed extension to be obtained is as follows: let H be a Hilbert space, let $\mathcal{D}(U) \subset H$ be a subspace and let $U : \mathcal{D}(U) \to H$ be an unbounded linear functional. We fix $w \neq 0$ belonging to $H \backslash \mathcal{D}(U)$ and define the linear operator $T : \mathcal{D}(U) \to H$ by $T(u) = U(u)w$. Since U is not continuous at 0, there is a sequence $\{u_n\}_{n \in \mathbb{N}} \subset \mathcal{D}(U)$ such that $u_n \to 0$ and $U(u_n) \nrightarrow 0$, that is, there is $c > 0$ such that $| U(u_n) |\geq c > 0 \ \forall n \in \mathbb{N}$. The sequence $v_n = \frac{u_n}{U(u_n)}$ has the following properties:

(i) $v_n \to 0$, (ii) $T(v_n) = w \neq 0$ for all $n \in \mathbb{N}$.

Therefore if there were an extension of S to the operator T, then we would have $S(0) = \lim T(v_n) = T(0) = 0$. However, $S(0) = w$.

Exercises

(1) Prove Proposition 16.
(2) Let H be a Hilbert space, $\mathcal{D}(T) \subset H$ a dense subset in H and $T : \mathcal{D}(T) \to H$ a linear operator. Consider the sets

$$\mathcal{R}(T) = \{v \in H \mid v = T(u) \text{ for some } u \in \mathcal{D}(T)\}, \ \mathrm{Ker}(T) = \{u \in \mathcal{D} \mid T(u) = 0\},$$
$$\mathcal{R}(T^*) = \{u \in H \mid u = T^*(v) \text{ for some } v \in \mathcal{D}(T^*)\}, \ \mathrm{Ker}(T^*) = \{v \in \mathcal{D}(T^*) \mid T(v) = 0\}.$$

Prove that (i) $\mathcal{R}(T)^{\perp} = \mathrm{Ker}(T^*)$, (ii) $\left(\mathcal{R}(T)^{\perp}\right)^{\perp} = \overline{\mathcal{R}(T)} = \mathrm{Ker}(T^*)^{\perp}$.

(3) Let H be a Hilbert space, $\mathcal{D} \subset H$ a dense subset in H and $T : \mathcal{D} \to H$ a self-adjoint linear operator. Suppose there is $C > 0$ and $\| T(u) \|_H \geq C \| u \|_H$ for all $u \in \mathcal{D}$. Prove the following claims;

(i) $T(u) = f$ has a unique solution for every $f \in H$.

(ii) If $f \in N(T^*)^{\perp}$, then there is a solution but there is no uniqueness.

(4) Consider T a closed densely defined linear operator and suppose we have a constant $C > 0$ such that $< T(u), u >\geq C. \| u \|_H^2$ for all $u \in \mathcal{D}(T)$. Prove that T has closed range.

(5) Let H be a Hilbert space and $\mathcal{D} \subset H$. Let $T : \mathcal{D} \to H$ be an unbounded linear operator. The inverse operator T^{-1} of T is defined as follows;

$$f \in D(T^{-1}) \text{ satisfies } T^{-1}(f) = u \iff \exists u \in \mathcal{D} \text{ such that } T(u) = f.$$

(a) Prove T^{-1} is well-defined if and only if T is injective.

(b) Assume T is densely defined and there is $C > 0$ such that $< T(u), u >\geq C. \| u \|_H^2$ for all $u \in \mathcal{D}(T)$. Prove that T has a bounded inverse.

(c) Suppose T is closed and injective, prove that the image subspace $\mathcal{R}(T)$ is closed if and only if T^{-1} is bounded.

(d) Assume T is closed and densely defined. Prove that:

(i) $\mathcal{R}(T) = H$ if and only if $(T^*)^{-1}$ is bounded.

(ii) $\mathcal{R}(T^*) = H$ if and only if $(T)^{-1}$ is bounded.

To conclude this section, we will make a short comment on the spectral properties of unlimited closed operators found in [20]. If $T : \mathcal{D} \to H$ is closed and densely defined, then the operator $\lambda I - T$ is closed and $D(\lambda I - T) = \mathcal{D}$ for any $\lambda \in \mathbb{C}$. If T is also symmetric, then it follows that:

(i) the eigenvalues of T belong to $\mathbb{R}$.

(ii) the eigenvectors associated with distinct eigenvalues are orthogonal, that is,

$$\lambda \neq \mu \implies \mathrm{Ker}(\lambda I - T) \perp \mathrm{Ker}(\mu I - T).$$

Chapter 3
Differentiation in Banach Spaces

In this chapter we will introduce the concept of differentiability of maps defined in Banach spaces. The Inverse Function Theorem (InFT) is the main result; some examples of optimization in Variational Calculus are given, as well as some properties of the Fredholm maps are proved along with some applications of the InFT.

1 Maps on Banach Spaces

Let E, F be Banach spaces. The purpose of this section is to give examples of differentiable maps $f : E \to F$ between Banach Spaces. In the preceding chapter, we proved that the spectrum $\sigma(T)$ of a linear operator $T \in \mathcal{L}(E)$ is a non-void compact subset of $\mathbb{C}$. Let $K \subset \mathbb{C}$ be a compact subset. From the Stone-Weierstrass Theorem, any function $f \in (C^0(K; \mathbb{C}), || \, . \, ||_0)$ can be approximated using a sequence of polynomials $\{p_n\}_{n \in \mathbb{N}}$. Therefore assuming $T \in \mathcal{L}(E)$ and $| \, T \, | \in K$, we define the operator

$$f(T) = \lim_{n \to \infty} p_n(T) : E \to E.$$

For any $f = \lim p_n$ and $T \in \mathcal{L}(E)$, let's check that $f(T) \in \mathcal{L}(E)$. Any polynomial $p : \mathbb{C} \to \mathbb{C}, p(\lambda) = \sum_{i=1}^{n} a_i \lambda^i$, can be extended to the map $p : \mathcal{L}(E) \to \mathcal{L}(E)$, $p(T) = \sum_{i=1}^{n} a_i T^i$, since

$$| \, p(T) \, | \leq \sum_{i=1}^{n} | \, a_i \, | \, . \, | \, T \, |^i < \sup_{\lambda \in K} | \, p(\lambda) \, | \, .$$

© Springer Nature Switzerland AG 2021
C. M. Doria, *Differentiability in Banach Spaces, Differential Forms and Applications*,
https://doi.org/10.1007/978-3-030-77834-7_3

Let $\{p_n\}_{n\in\mathbb{N}}$ be a sequence of polynomials, $p_n(\lambda) = \sum_{i=1}^{n} a_{n,i}\lambda^i$, so

$$| p_n(T) - p_m(T) | \leq \sum_{i=1}^{n} | a_{n,i} - a_{m,i} | . | T |^i \leq \sup_{\lambda\in\sigma(T)} | p_n(\lambda) - p_m(\lambda) | = || p_n - p_m ||_0 .$$

Assume $\{p_n\}_{n\in\mathbb{N}}$ is a Cauchy sequence converging uniformly in $(C^0([m, M]; \mathbb{R}), || . ||_0).$[1] So the sequence $\{p_n(T)\}_{n\in\mathbb{N}}$ is also a Cauchy sequence in $\mathcal{L}(E)$ since

$$| p_n(T) - p_m(T) | = \sup\{| p_n(\lambda) - p_m(\lambda) |; \lambda \in \sigma(T)\} = || p_n - p_m ||_0 .$$

Taking another polynomial sequence $\{q_n\}_{n\in\mathbb{N}}$ converging uniformly to f, we can check that the value $f(T)$ does not depend on the polynomial. Let $\lim p_n(T) = A$ and $\lim q_n(T) = B$, so

$$| A - B | \leq | A - p_n(T) | + | p_n(T) - q_n(T) | + | q_n(T) - B | \leq$$
$$\leq | A - p_n(T) | + || p_n - q_n ||_0 + | q_n(T) - B | .$$

Passing to the limit, we have $A = B = f(T)$. The following properties are satisfied with the limits:
(i) $(af + bg)(T) = af(T) + bg(T)$,
(ii) $(f.g)(T) = f(T).g(T)$,
(iii) $\bar{f}(T) = [f(T)]^*$.
The spectrum of $f(T)$ is quite simple and will be described, as we show next.

Theorem 1 (Spectral Map Theorem)
  Let $T \in \mathcal{L}(E)$. Let $p \in \mathbb{C}[X]$ be a polynomial, so

$$\sigma(p(T)) = p(\sigma(T)) = \{p(\lambda) \mid \lambda \in \sigma(T)\}.$$

Proof For any $\lambda \in \mathbb{R}$, the polynomial $\hat{p}(x) = p(x) - p(\lambda)$ can be written as the product of $\hat{p}(x) = (x - \lambda)q(x)$. Therefore we have

$$\hat{p}(T) = p(T) - p(\lambda)I = (T - \lambda I)q(T).$$

• $p(\sigma(T)) \subset \sigma(p(T))$.
Suppose $p(\lambda) \in \rho(p(T))$; in this case, the operator $\hat{p}(T)$ has an inverse S such that

$$S.\hat{p}(T) = \hat{p}(T).S = I \Rightarrow (S.q(T)).(T - \lambda I) = (T - \lambda I).(q(T).S) = I \Rightarrow \lambda \in \rho(T).$$

Given the above, if $p(\lambda) \in \sigma(p(T))$, then $\lambda \in \sigma(T)$.

---

[1]Recall $m = \inf \sigma(T)$ and $M = \sup \sigma(T)$.

- $\sigma(p(T)) \supset p(\sigma(T))$.

Let $k \in \sigma(p(T))$. When factored over $\mathbb{C}$, the polynomial $p(x) - k$ is written as

$$p(x) - k = (x - \lambda_1) \ldots (x - \lambda_r), \quad 1 \le r \le n.$$

The inverse of $p(T) - kI = \prod_{i=1}^{r}(T - \lambda_i I)$ is

$$(p(T) - kI)^{-1} = \prod_{i=r}^{1}(T - \lambda_i I)^{-1}.$$

However since $k \in \sigma(p(T))$, there is one λ_i for which the operator $T - \lambda_i I$ has no inverse. Hence $\lambda_i \in \sigma(T)$ and $p(\lambda_i) = k \in p(\sigma(T))$. $\qquad \square$

Corollary 1 *If* $f \in (C^0(\sigma(T); \mathbb{C}), \| \cdot \|_0)$ *and* $T \in \mathcal{L}(E)$, *then* $\sigma(f(T)) = f(\sigma(T))$.

Corollary 2 *Let* $p \in \mathbb{C}[x]$ *be a polynomial, and let* H *be a Hilbert space. If* $T \in \mathcal{L}(H)$ *is a self-adjoint operator, then*

$$| p(T) | = \sup\{| p(\lambda) |; \lambda \in \sigma(T)\}.$$

Proof Let's first assume that $p \in \mathbb{R}[x]$. The fact that T is self-adjoint implies that $p(T)$ is also self-adjoint, and so the norm is equal to the spectral radius:

$$| p(T) | = \sup\{| \mu |; \mu \in \sigma(p(T))\} = \sup\{| p(\lambda) |; \lambda \in \sigma(T)\}.$$

To address the case $p \in \mathbb{C}[x]$; we consider the polynomial $\bar{p}.p \in \mathbb{R}[x]$. Since $(\bar{p}.p)(T) = [p(T)]^*.p(T)$, then

$$| p(T) |^2 = \| [p(T)]^*.p(T) \| = \sup\{| (\bar{p}.p)(\lambda); \lambda \in \sigma(T)\} = $$
$$= \sup\{| p(\lambda) |^2; \lambda \in \sigma(T)\}.$$

$\qquad \square$

Examples of maps between Banach spaces are shown using analytical functions. Let $f : \mathbb{C} \to \mathbb{C}$ be an analytic function with the radius of convergence R. When we fix a point $x_0 \in \mathbb{C}$, there is a unique sequence $\{a_n\}_{n \in \mathbb{N}}$ so that $f(x)$ is equal to $f(x) = \sum_{i=0}^{\infty} a_n(x - x_0)^n$ for every $x \in (x_0 - R, x_0 + R)$. Therefore we define the map $f : \mathcal{L}(E) \to \mathcal{L}(E)$ as $f(A) = \sum_{i=0}^{\infty} a_n A^n$. In this way, $f(A)$ is well-defined in $\{A \in \mathcal{L}(E); | A | < R\}$ since

$$| f(A) | \le \sum_{i=0}^{\infty} a_n | A |^n < \infty.$$

Example 1 Examples of maps.

(1) Exponential of a linear operator.
   Let $f(x) = e^x = \sum_{k=0}^{\infty} \frac{x^k}{k!}$. For every $A \in \mathcal{L}(E)$, consider the series $\exp(A) = \sum_{k=0}^{\infty} \frac{A^k}{k!}$. This series is convergent for all $A \in \mathcal{L}(E)$ since

$$| (A) | \leq \sum_{k=0}^{\infty} \frac{| A |^k}{k!} = e^{|A|}. \tag{1}$$

So $\exp(0) = I$. In general, $\exp(A + B) \neq \exp(A).\exp(B)$. If we assume $AB = BA$, then we have $\exp(A + B) = \exp(A).\exp(B)$ since

$$\exp(A + B) = \sum_{k=0}^{\infty} \frac{(A + B)^k}{k!} = \sum_{k=0}^{\infty} \left[\sum_{j=0}^{k} \frac{A^{k-j}}{(k - j)!} \frac{B^j}{j!} \right] =$$

$$= \sum_{k=0}^{\infty} \left[\sum_{i+j=k} \frac{A^{k-j}}{(k - j)!} \frac{B^j}{j!} \right] = \left[\sum_{i=0}^{\infty} \frac{A^i}{i!} \right] \left[\sum_{j=0}^{\infty} \frac{B^j}{j!} \right] = \exp(A).\exp(B).$$

An immediate consequence is that $\exp(A) \in \mathcal{GL}(E)$ for all $A \in \mathcal{GL}(E)$ since $[\exp(A)]^{-1} = \exp(-A)$. We treat the case $AB \neq BA$ by introducing the commutator $[A, B] = AB - BA$ and the operator $C = \exp(A + B)$. In this way,

$$C = A + B + \frac{1}{2}[A, B] + \frac{1}{12} ([A, [A, B]] + [[A, B], B]) + \dots.$$

When $E = \mathbb{R}^n$ or $E = \mathbb{C}^n$, the exponential of $A \in \mathcal{L}(E)$ is explicitly computed using Jordan's canonical form of A.

(2) $\cos(x) = \sum_{k=0}^{\infty}(-1)^k \frac{x^{2k}}{(2k)!}$, $\sin(x) = \sum_{k=0}^{\infty}(-1)^k \frac{x^{2k+1}}{(2k+1)!}$. Both functions are analytic functions with the radius of convergence being $R = \infty$. This allows us to extend the real functions $\cos, \sin : \mathbb{R} \to \mathbb{R}$ to the maps $\cos, \sin : \mathcal{L}(E) \to \mathcal{L}(E)$.

(3) The function $\tan^{-1}(x) = \sum_{i=1}^{\infty} \frac{(-1)^{i+1}}{i} x^i$ is analytic with the radius of convergence $R = 1$. Its extension $\tan^{-1} : B_1 \to \mathcal{L}(E)$ is given in the series $\tan^{-1}(A) = \sum_{i=1}^{\infty} \frac{(-1)^{i+1}}{i} A^i$ defined as $B_1 = \{A \in \mathcal{L}(E); | A |< 1\}$.

Exercises

(1) Find the values of the maps $\exp(A)$, $\sin(A)$, $\cos(A)$ and $\tan^{-1}(A)$ at

$$A = \begin{pmatrix} 5 & 3 \\ -6 & -4 \end{pmatrix}.$$

1.1 Extension by Continuity

Let E be a normalized space, $F \subset E$ is a vector subspace, and G is a Banach space. We will briefly address the issue of extending the linear operator $T : F \to G$ to the closing $\overline{F}$;

Theorem 2 *Let $T : F \to G$ be a bounded linear operator, so we have a unique extension $\overline{T} : \overline{F} \to G$ such that $\overline{T}(x) = T(x)$ for all $x \in F$ and $\mid \overline{T} \mid = \mid T \mid$ in F.*

Proof To define $\overline{T}$, we take $x \in \overline{F}$ and let $\{x_n\}_{n \in \mathbb{N}} \subset F$ be a sequence such that $\lim x_n = x$. Then $\{T(x_n)\}_{n \in \mathbb{N}}$ is a Cauchy sequence since

$$\| T(x_n) - T(x_m) \| \le \mid T \mid . \| x_n - x_m \| .$$

Therefore $\{T(x_n)\}_{n \in \mathbb{N}}$ is convergent. Let $\overline{T}(x) = \lim_{n \to \infty} T(x_n)$. The limit does not depend on the sequence $\{x_n\}_{n \in \mathbb{N}}$. Let's consider two convergent sequences $\{x_n\}_{n \in \mathbb{N}}$ and $\{y_n\}_{n \in \mathbb{N}}$ with $x_n \to x$ and $y_n \to x$. Given $\epsilon > 0$, we have $n(\epsilon) \in \mathbb{N}$, such that for all $n, m > n(\epsilon)$, we have

$$\| T(x_n) - T(y_n) \| \le \mid T \mid . \| x_n - y_n \| < \epsilon.$$

Hence $\overline{T}(x) = \lim_{n \to \infty} T(x_n) = \lim_{n \to \infty} T(y_n)$. We define $\overline{T}(x) = \lim_{n \to \infty} T(x_n)$, with $\{x_n\}_{n \in \mathbb{N}}$ being any sequence converging to x. The operator $\overline{T}$ is bounded and $\mid \overline{T} \mid = \mid T \mid$ since

$$\| \overline{T}(x) \| = \lim_{n \to \infty} \| T(x_n) \| \le \mid T \mid . \lim_{n \to \infty} \| x_n \| = \mid T \mid . \| x \| .$$

$\square$

Exercises

(1) Show that if $F \subset E$ is a subspace, then $\overline{F} \subset E$ is also a subspace.
(2) Show the following assertions:

  (a) If $\dim(F) < \infty$, then $\overline{F} = F$.
  (b) If $\dim(F) = \infty$, then item (a) is false.

2 Derivation and Integration of Functions $f : [a, b] \to E$

Formally, the theory goes along the same lines as in the case when $\dim(E) < \infty$. However, additional conditions must be assumed for the techniques to have the same efficiency.

2.1 Derivation of a Single Variable Function

A curve in a Banach space E is a function $f : [a, b] \to E$.

Definition 1 A curve $f : [a, b] \to E$ is differentiable at $t \in [a, b]$, if there is $A \in \mathcal{L}(E)$ such that

$$f(t + h) - f(t) = Ah + r(h),$$

and $\lim_{h \to 0} \frac{\|r(h)\|}{|h|} = 0$.

Therefore we have $A = \lim_{h \to 0} \frac{f(t+h)-f(t)}{h}$ justifying the notation $A = f'(t)$. The linear operator $D(f) = f'$ defined on differentiable functions satisfies the following properties:
(i) $(f + g)'(t) = f'(t) + g'(t)$.
(ii) $(f.g)'(t) = f'(t)g(t) + f(t)g'(t)$.
(iii) (chain rule) Let $\alpha : [c, d] \to [a, b]$ and $h(t) = f(\alpha(t))$, so $h'(t) = f'(\alpha(t)).\alpha'(t)$.

The differentiability of f implies its continuity, since the linear operator $f'(t) :$ $\mathbb{R} \to E$, $f'(t)(h) = f'(t).h$ is bounded and $\lim_{h \to 0} r(h) = 0$. Then passing to the limit, we have

$$\| f(t + h) - f(t) \| = \| f'(t)h + r(h) \| \leq | f'(t) | \cdot \| h \| + \| r(h) \| \to 0.$$

In the theory of differentiable maps between Banach spaces, to prove the Mean Value inequality requires more care than in the case of a map between finite dimensional spaces. When $\dim(E) < \infty$, a function is described in terms of its coordinates $f(t) = (f_1(t), \ldots, f_n(t))$, which cannot be done for a function defined in E. When $E = \mathbb{R}^n$, in considering the linear functional $\pi_i : \mathbb{R}^n \to \mathbb{R}$ given by $\pi(x_1, \ldots, \pi_n) = x_i, i \in \{1, \ldots, n\}$, each coordinate of f is described as $f_i = \pi_i \circ f$. To get around the lack of a coordinate system, we use Corollary 1 in Appendix A, which we have as a consequence of the Hahn-Banach Theorem 14 in Appendix A.

Proposition 1 *Let $I \subset \mathbb{R}$ be an interval, let E be a normed space and let $f : I \to E$ be a differentiable curve. If $f'(t) = 0$ for all $t \in I$, then f is constant.*

Proof Take $t_0 \in I$ and assume that there is $t_1 \in I$ such that $f(t_1) \neq f(t_0)$; otherwise f is constant. The Hahn-Banach Theorem guarantees the existence of a functional $\lambda : E \to \mathbb{R}$, such that $\lambda(f(t_1) - f(t_0)) \neq 0$, and by linearity we get $\lambda(f(t_1)) \neq \lambda(f(t_0))$. Define the function $g = \lambda \circ f : I \to \mathbb{R}$, $g(t) = \lambda(f(t))$. Since $g'(t) = \lambda(f'(t)) = 0$, we have that g is constant, contradicting the fact that $g(t_0) \neq g(t_1)$. Hence $f(t)$ is constant. $\qquad \square$

2.2 Integration of a Single Variable Function

We will briefly present the theory of integration of functions $f : [a, b] \to E$. The reader can find the complete approach to this topic in Ref. [28].

Let $B([a, b], E) = \{f : [a, b] \to E; |\ f\ |_\infty < \infty\}$ be the set of bounded functions provided with the norm $||\ .\ ||_0$ so that $(B([a, b], E), ||\ .\ ||_0)$ is a Banach space. A function $f \in B([a, b], E)$ is a step function if there is a partition $\mathcal{P}_f = \{a = x_0, x_1, \ldots, x_n = b\}$ such that f is constant in the interval $[x_{i-1}, x_i]$, $i = 1, \ldots, n$, that is, there is a collection of values $\{k_1, \ldots, k_n\} \subset E$, such that $f\ |_{[x_{i-1}, x_i]} = k_i$ and $||\ f\ ||_0 = \sup_i\{|\ k_i\ |\}$. Let $S([a, b], E) \subset B([a, b], E)$ be the subset of step functions.

Definition 2 The integral of $f \in S([a, b], E)$ with respect to the partition $\mathcal{P}_f$ is

$$I(f, \mathcal{P}) = \sum_{i=1}^{n} k_i (x_i - x_{i-1}). \tag{2}$$

The integral of the function $f \in S([a, b], E)$ is independent of the partition used, i.e., $I(f, \mathcal{P}') = I(f, \mathcal{P})$ for any partition $\mathcal{P}$ and $\mathcal{P}'$ of $[a, b]$. Now we consider the functional $I_a^b : S([a, b], E) \to E$, $I_a^b(f) = I(f, \mathcal{P})$.

Lemma 1 *The map $I_a^b : S([a, b], E) \to E$ is linear and continuous.*

Proof Let $f, g \in S([a, b], E)$, and let $\mathcal{P}$ be a partition of $[a, b]$ such that both functions are step functions with respect to $\mathcal{P}$. Linearity follows easily, since

$$I_a^b(af + bg) = \sum_{i=1}^{n}(ak_i + bl_i)(x_i - x_{i-1}) = aI_a^b(f) + bI_a^b(g).$$

Continuity follows from the boundedness of I_a^b;

$$||\ I_a^b(f)\ ||_E = \Big|\Big| \sum_i k_i (x_i - x_{i-1}) \Big|\Big|_E \leq \sum_i |\ x_i - x_{i-1}\ |\ .\ ||\ k_i\ ||_E \leq$$
$$\leq (b - a) \sup_i ||\ k_i\ ||_E \leq (b - a)\ ||\ f\ ||_0 .$$

$\square$

If $\{f_n\} \subset S([a, b], E)$ is a sequence converging uniformly to f, then the integral of $f \in \overline{S([a, b], E)}$ is obtained by taking the limit

$$\int_a^b f(x)dx = \lim_{n\to\infty} \int_a^b f_n(x)dx = \lim_{n\to\infty} I_a^b(f_n).$$

It follows from Theorem 2 that the operator $I_a^b : S([a, b], E) \to E$ admits a unique extension $\bar{I}_a^b : \overline{S([a, b], E)} \to E$ to the closing of $S([a, b], E)$. If $f \in \overline{S([a, b], E)}$, then define $\bar{I}_a^b(f) = \int_a^b f(x)dx$.

Theorem 3 $C^0([a, b], E) \subset \overline{S([a, b], E)}$.

Proof Since f is continuous and $[a, b]$ is compact, given $\epsilon > 0$, we have $\delta > 0$ such that if $|x - y| < \delta$, then $|f(x) - f(y)| < \epsilon$ and f is uniformly continuous. Now choose $n \in \mathbb{N}$ such that $\frac{b-a}{n} < \delta$. Consider the partition $\mathcal{P}(n) = \{a = x_0, \ldots, x_n = b\}$ satisfying $x_i - x_{i-1} = \frac{b-a}{n}$ and define the step function $g_n : [a, b] \to \mathbb{R}$,

$$g_n(x) = \begin{cases} f(x_{i-1}), & x \in [x_{i-1}, x_i), \\ f(b), & x = b. \end{cases}$$

Therefore g_n depends on $\mathcal{P}(n)$, and so g_n depends on n. Taking n sufficiently large, we have $\| f - g_n \|_0 < \epsilon$. $\qquad\square$

The integral satisfies the following properties: let $f, g \in \overline{S([a, b], E)}$ and $c, d \in \mathbb{R}$;

(1) $\int_a^b (cf + dg)(x)dx = c\int_a^b f(x)dx + d\int_a^b g(x)dx$.
(2) $\int_a^b f(x)dx = \int_a^c f(x)dx + \int_c^b f(x)dx$.
(3) $|\int_a^b f(t)dt| \le \int_a^b |f(t)| \, dt \le \| f \|_0 (b - a)$.

Next, we will prove a version of the Fundamental Theorem of Calculus (FTC) for functions (curves) $f : [a, b] \to E$;

Theorem 4 (TFC) *Let $f \in C^0([a, b], E)$ and consider $F : [a, b] \to \mathbb{R}$ the function given by*

$$F(x) = k + \int_a^x f(t)dt,$$

and $k \in E$ is constant. So $F'(x) = f(x)$ and $F(a) = k$.

Proof Since

$$F'(x) = \lim_{h \to 0} \frac{F(x + h) - F(x)}{h} = \lim_{h \to 0} \frac{1}{h} \int_x^{x+h} f(t)dt,$$

we have

$$\left\| \frac{F(x + h) - F(x)}{h} - f(x) \right\|_E \le \frac{1}{h} \int_x^{x+h} \| f(t) - f(x) \|_E \, dt.$$

Then,

$$\lim_{h \to 0} \left\| \frac{F(x + h) - F(x)}{h} - f(x) \right\|_E = \lim_{h \to 0} \frac{1}{h} \int_x^{x+h} \| f(t) - f(x) \|_E \, dt.$$

Since f is uniformly continuous on $[a, b]$, given $\epsilon > 0$, there is $\delta > 0$ such that if $|t - x| < \delta$ for all $x, t \in [a, b]$, then $\| f(t) - f(x) \|_E < \epsilon$. If $|h| < \delta$, then

$$\left\| \frac{F(x + h) - F(x)}{h} - f(x) \right\|_E < \epsilon.$$

since $| t - x | \leq | h |$. Hence $F'(x) = \lim_{h \to 0} \frac{F(x+h) - F(x)}{h} = f(x)$. □

Corollary 3 Let $f \in C^0([a, b], E)$ and $F \in C^1([a, b], E)$ be such that $F'(x) = f(x)$. Then,

$$\int_a^b f(x) dx = F(b) - F(a).$$

Proof Defining $I(x) = \int_a^x f(t) dt$, we have $I(a) = 0$. Let $F : [a, b] \to E$ be another function such that $F'(x) = f(x)$. By Proposition 1, we have a constant $k \in E$ such that $F(x) - I(x) = k$. In this way, $k = F(a)$ and $I(x) = F(x) - F(a)$, that is, $I(b) = F(b) - F(a)$. □

3 Differentiable Maps II

We must be aware that the concept of distance in Banach Spaces depends on the norm, which does not occur when the dimension is finite since the norms are all equivalent.

Definition 3 Let $U \subset E$ be an open subset and $f : E \to \mathbb{R}$ a map. The Gâteaux derivative of f at $p \in U$ and in the direction of the vector v is the directional derivative

$$\frac{\partial f}{\partial v}(p) = \lim_{t \to 0} \frac{f(p + tv) - f(p)}{t}.$$

The nickname Gâteux derivative is often used in the context of Variational Calculus, it is also called the "functional derivative". Regardless of how it is called, it is the directional derivative as we defined it in Chap. 1. Of course, just as in $\mathbb{R}^n$, the directional derivative may not be linear in the argument for v, as shown in the following example: let $p = (0, 0)$ and $v = (a, b)$;

$$f(x, y) = \begin{cases} \frac{x^3}{x^2 + y^2}, & (x, y) \neq (0, 0), \\ 0, & (x, y) = (0, 0). \end{cases} \qquad \frac{\partial f}{\partial v}(p) = \begin{cases} \frac{a^3}{a^2 + b^2}, & (a, b) \neq (0, 0), \\ 0, & (a, b) = (0, 0). \end{cases}$$

Nonlinearity in the examples above is because the partial derivatives are not continuous at the origin. The concept of differentiability for maps between Banach spaces is called the Fréchet derivative.

Definition 4 A function $f : E \to \mathbb{R}$ is Fréchet differentiable at p whenever we have a continuous linear functional $df_p : E \to \mathbb{R}$ and a function $r : E \to \mathbb{R}$ such that

$$f(p + v) - f(p) = df_p(v) + r(v)$$

and $\lim_{v \to 0} \frac{|r(v)|}{||v||} = 0$. Let $C^1(E, \mathbb{R})$ be the set of Fréchet differentiable maps f : $E \to \mathbb{R}$.

The differential operator df_p is unique and satisfies the following properties:
(i) linearity: $df_p(av + bw) = a df_p(v) + b df_p(w)$,
(ii) Leibnitz's rule: $d(f.g)_p(v) = df_p(v).g(p) + f(p).dg_p(v)$,
(iii) chain rule: $\phi : E \to E, h = f \circ \phi : E \to \mathbb{R}$, so $dh_q(v) = df_{\phi(q)}.d\phi_q(v)$.
(iv) If f is differentiable at p (Fréchet differentiable), then f is also Gâteaux differentiable at p and $\frac{\partial f}{\partial v}(p) = df_p(v)$.

The reverse is false, but if the Gâteaux derivative is linear and bounded at p, then it follows that f is Fréchet differentiable at p. We emphasize the need for the linear operator df_p to be continuous in defining differentiability; otherwise, it does not imply the continuity of the application.

Proposition 2 *If $f : E \to \mathbb{R}$ is Fréchet differentiable, then f is continuous.*

Proof Since we have

$$|| f(p + v) - f(p) ||_E \leq || df_p(v) ||_E + || r(v) ||_E \leq | df_p | . || v ||_E + || r(v) ||_E,$$

taking $v \to 0$, yields $\lim_{v \to 0}(f(p + v) - f(p)) = 0$. $\square$

The Mean Value inequality is fundamental in the theory of differentiability in finite dimensional spaces. It can extend to Fréchet differentiable maps (application of the FTC).

Theorem 5 *Let $U \subset E$ be a convex open subset, $f \in C^1(U, E)$ and $x, y \in U$. Assuming the line $r : [0, 1] \to E, r(t) = x + t(y - x)$ is contained in U, then*

$$|| f(y) - f(x) || \leq \sup_{t \in [0,1]} | df_{r(t)} | . || y - x || . \tag{3}$$

Proof Let the function $g : [0, 1] \to E$ be given by $g(t) = f(r(t))$. Then we have $g'(t) = df_{r(t)}.(y - x)$. By the FTC

$$|| f(y) - f(x) || = || g(1) - g(0) || = || \int_0^1 g'(t)dt || = || \int_0^1 df_{r(t)}.(y - x)dt || \leq$$

$$\leq \sup_{t \in [0,1]} | df_{r(t)} | . || y - x || .$$

Since $f \in C^1(U, E)$, the supreme of $df_{r(t)}$ is attained in $[0, 1]$. $\square$

Proposition 3 *Let $U \subset E$ be a convex subset and let $f : U \to F$ be a Fréchet differentiable map for all points belonging to the line segment $r : [1.1] \to U, r(t) = p + tv$. Then there is $\tau \in (0, 1)$ such that*

$$|| f(p + v) - f(p) ||_F \leq | df_{r(\tau)} | . || v ||_E .$$

Proof Consider the function $h : [0, 1] \to F$, $h(t) = f(p + tv)$. Using the chain rule, we have $h'(t) = df_{p+tv}.v$. Applying the inequality from the last proposition, we get $\| h'(t) \|_F \leq M \| v \|_v$, $M = \sup_{t \in [0,1]} | df_{p+tv} |$. □

Definition 5 Let $U \subset E$ and $V \subset F$ be open subsets of Banach spaces. A map $f : U \to V$ is Fréchet differentiable at p if we have a continuous linear functional $df_p : E \to F$ and a map $r : U \to V$ such that

$$f(p + v) - f(p) = df_p(v) + r(v)$$

and $\lim_{v \to 0} \frac{|r(v)|}{\|v\|} = 0$. Let $C^1(U, V)$ be the set of Fréchet differentiable maps $f : U \to V$.

The differential df_p is unique and satisfies the following properties:
(i) linearity: $df_p(av + bw) = a df_p(v) + b df_p(w)$,
(ii) chain rule: Let E, F and G be Banach spaces. Let $\phi : E \to F$ and let $f : F \to G$ be Fréchet differentiable maps, so the composite $h = f \circ \phi : E \to G$ is Fréchet differentiable and $dh_p(v) = df_{\phi(p)}.d\phi_p(v)$.
(iii) If $f : U \to V$ is Fréchet differentiable, then f is continuous.

4 Inverse Function Theorem (InFT)

In this section, we give the statement and the proof of the more general version of the Inverse Function Theorem (InFT) for differentiable maps defined on Banach Spaces. In Chap. 2, the InFT was enunciated and some applications given in the finite dimensional context.

4.1 Prelude for the Inverse Function Theorem

We now introduce the main ideas to prove the InFT through a simple example. We will consider the case of a single real variable function so that the similarity with the Newton-Raphson Method (NR) becomes apparent. If the reader finds it appropriate, this section can be ignored. The N-R Method is a practical tool for finding the roots of an equation, while InFT is a theoretical tool with widespread applications. Now let's address the problem to find a solution to the equation $f(x) = 0$, we assume $f : (a, b) \to \mathbb{R}$ is a C^2-function. The proof of InFT will follow with the same arguments.

Method of Newton–Raphson

Let $x_1 \in (a, b)$ be a point such that $f'(x_1) \neq 0$. The tangent line $\mathcal{L}_1$ to the graph of f at $(x_1, f(x_1))$ is given by the equation

$$\mathcal{L}_1 : \ y - f(x_1) = f'(x_1)(x - x_1).$$

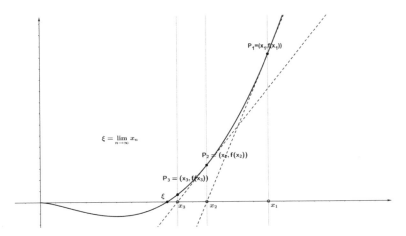

Fig. 1 Sequence $\{x_{n+1}\}_{n\in\mathbb{N}}$

The intersection of $\mathcal{L}_1$ with the x-axis is the point $x_2 = x_1 - \frac{f(x_1)}{f'(x_1)}$. If $f'(x_2) \neq 0$, then we repeat the process to obtain the point $x_3 = x_2 - \frac{f(x_2)}{f'(x_2)}$, as indicated in Fig. 1. After successive steps, if we have $f'(x_n) \neq 0$, then we can repeat the process to obtain

$$x_{n+1} = x_n - \frac{f(x_n)}{f'(x_n)}. \tag{4}$$

If $f'(x_n) = 0$, then the algorithm does not go further since the tangent line $\mathcal{L}_n$ will be parallel to the x-axis.

The N-R method to solve equation $f(x) = 0$ consists of applying the following algorithm:

(1) Choose two points $\alpha, \beta \in (a, b)$ such that $f(\alpha).f(\beta) < 0$ and $f'(x) \neq 0$ for all $x \in (\alpha, \beta)$. As a result, we can guarantee the existence of a unique solution in the interval (α, β).
(2) Now take an initial point x_1 to apply the algorithm induced by the recursive formula (4).

Geometrically, once we choose a point $x_1 \in [\alpha, \beta]$, we can intuit from the figure that $\lim x_n = \xi$ is a solution. If item (2) above fails at step n, then the algorithm will not be efficient to find a root. Given an arbitrary point y, we can generalize the N-R method to solve the equation $f(x) = y$ given that the function $f_y(x) = y - f(x)$ and defining the recursive sequence x_{n+1} to be

$$x_{n+1} = x_n + \frac{1}{f'(x_n)}(y - f(x_n)). \tag{5}$$

If the sequence $\{x_n\}_{n \in \mathbb{N}}$ converges to $\xi = \lim x_n$, then $y = f(\xi)$. Let's consider $y = 0$. To prove that the method indeed solves the equation, it is crucial to show that $\{x_n\}_{n \in \mathbb{N}}$ converges to a solution of the equation $f(x) = 0$. Consider the function $\phi : [\alpha, \beta] \to \mathbb{R}$ given by

$$\phi(x) = x - \frac{f(x)}{f'(x)}, \tag{6}$$

we note that $\xi \in [\alpha, \beta]$ is a solution of $f(x) = 0$ if and only if $\phi(\xi) = \xi$. Its derivative is

$$\phi'(x) = \frac{f''(x) . f(x)}{[f'(x)]^2}. \tag{7}$$

Taking $M = \sup_{t \in [x_0, x_1]} | \phi'(t) |$, it follows from the Mean Value Theorem that

$$| \phi(x') - \phi(x) | \le M . | x' - x |, \ \forall x, x' \in [\alpha, \beta].$$

Theorem 6 *Let $f : (a, b) \to \mathbb{R}$ be a C^2-function. Let $\alpha, \beta \in (a, b)$ be such that $f(\alpha) . f(\beta) < 0$ and $f'(x) \ne 0$ for all $x \in [\alpha, \beta]$. If*

$$| f''(x) . f(x) | < | f'(x) |^2,$$

then the sequence $\{x_n\}_{n \in \mathbb{N}}$ given by $x_{n+1} = x_n - \frac{f(x_n)}{f'(x_n)}$ converges to a solution of the equation $f(x) = 0$.

By the Contraction Lemma 1 in Appendix A, ϕ has a fixed point ξ. Hence $f(\xi) = 0$. So if $\{x_n\}_{n \in \mathbb{N}}$ converges, then $\lim x_n = \xi$.

4.2 InFT for Functions of a Single Real Variable

Consider subsets $X, Y \subset \mathbb{R}$. A function $f : X \to Y$ is well-defined once we know the subset $X = D(f)$ (domain of f), the subset Y (codomain of f) and the map $x \to f(x)$ for all $x \in X$. The image set of f is $\mathrm{Im}(f) = \{y \in Y \mid y = f(x)\}$. The composition of two functions $f : X \to Y$ and $g : Y \to Z$ is the function $g \circ f : X \to Z$ defined by $(g \circ f)(x) = g(f(x))$. The composition satisfies the associative law

$$h \circ (g \circ f) = (h \circ g) \circ f) = h \circ g \circ f.$$

The identity function on X is $\mathrm{id}_X(x) = x$, and we have $f \circ \mathrm{id}_X = \mathrm{id}_Y \circ f = f$ for all $f : X \to Y$. When $f : X \to Y$ defines an injective and surjective map $X \xrightarrow{f} \mathrm{Im}(f)$, then it admits an inverse function $f^{-1} : \mathrm{Im}(f) \to X$ with respect to the composition of functions such that $f \circ f^{-1} = \mathrm{id}_Y$ and $f^{-1} \circ f = \mathrm{id}_X$.

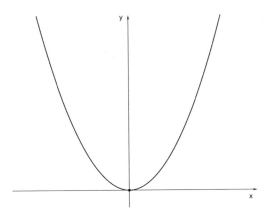

Fig. 2 $f(x) = x^2$

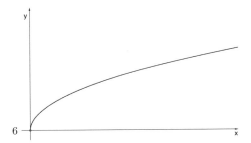

Fig. 3 $f^{-1}(y) = \sqrt{y}$

Example 2 Several functions are not invertible, as some examples are shown next. However, if we restrict the domain, these funtions become invertible.

(1) Let $f : \mathbb{R} \to [0, \infty)$ be given by $f(x) = x^2$. Since $f(-x) = f(x)$ it is not injective. If we restrict to $f : [0, \infty) \to [0, \infty)$, then for $y \in [0, \infty)$, there corresponds only one x, such that $y = x^2$ as illustrated in Fig. 2. Therefore f admits an inverse after restricting its domain. Define $x = f^{-1}(y) = \sqrt{y}$ (Fig. 3).

(2) Let $f : (-\pi, \pi) \to [-1, 1]$ be given by $f(x) = \cos(x)$. The restriction $f : [0, \pi] \to [-1, 1]$ admits an inverse since there is a unique $x \in [0, \pi]$ such that $y = \cos(x)$, for $y \in [-1, 1]$, as illustrated in Fig. 4. Define $x = f^{-1}(y) = \arccos(y)$ (Fig. 5).

The above examples show that restricting the domain $D(f)$ allows us to define an inverse for f; in this case, we say that f is locally invertible. In this section, we will address the question of finding sufficient conditions, so that the function $f :$

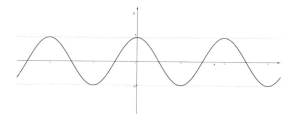

Fig. 4 $f(x) = \cos(x)$

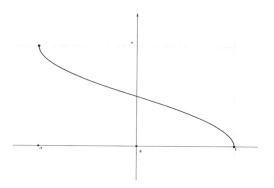

Fig. 5 $f^{-1}(y) = \arccos(y)$

$(a, b) \to \mathbb{R}$ admits a local inverse f^{-1} in the neighborhood of a point $x_0 \in D(f)$ and f^{-1} is differentiable. The strategy is to prove the existence of an open neighborhood U_0 of x_0, such that $f : U_0 \to f(U_0)$ is bijective. This means that for any $y \in f(U_0)$, we have a unique $x \in U_0$ such that $y = f(x)$. Indeed, we want to solve the equation $y = f(x)$ and, restricted to U_0, the solution is unique. As it has been enunciated, the problem may not be solvable since f^{-1} may not be differentiable. To get around this, we will always assume that $f \in C^1$. As in the N-R method, now we use the function $\phi(x)$ given by Eq. (6). Let $x_1 \in D(f)$ be such that $f'(x_1) \neq 0$, and let $y \in \mathbb{R}$ be a given point. We consider $\phi_y : D(f) \to \mathbb{R}$ to be the function

$$\phi_y(x) = x + \frac{1}{f'(x_0)}(y - f(x)). \tag{8}$$

Again, equation $y = f(x)$ has a solution if and only if x is a fixed point of ϕ_y. Also, the number of solutions is equal to the number of fixed points. In this way, we are going to study conditions so that ϕ_y admits fixed points. By restricting f, beyond proving the existence of f^{-1}, we will also show that it is differentiable.

Theorem 7 *Let $I \subset \mathbb{R}$ be an interval and let $f : I \to \mathbb{R}$ be a C^1-function. Consider $x_0 \in I$ a point such that $f'(x_0) \neq 0$; then there is a neighborhood $U_0 \subset I$ containing x_0, such that $f : U_0 \to f(U_0)$ admits a differentiable inverse.*

Proof of Theorem 6 Under the hypotheses of Theorem 6, the function ϕ is a contraction. Therefore we have a fixed point ξ and, consequently, $f(\xi) = 0$. Although the existence of ξ is guaranteed, to obtain its numerical value we need a numerical sequence converging to ξ. For this purpose, we will use the sequence $\{x_n\}_{n \in \mathbb{N}}$ defined in (4). Since we are working with functions of a real variable, some arguments are simple. From the hypotheses of Theorem 6, it follows that in the interval $[x_0, x_1]$, we have a unique solution for $f(x) = 0$ denoted by ξ. Making $\xi = x_n + h_n$, it follows that

$$0 = f(x_n + h) = f(x_n) + f'(x_n)h + \frac{f''(x_n)}{2}h^2 + o(h^2)$$

$$\Rightarrow |x_{n+1} - \xi| < \leq |\frac{f''(x_n)}{2f'(x)}| \cdot |x_n - \xi|^2 + \frac{|o(h_n^2)|}{|h_n|}.$$

Let $C = \sup \frac{f''(x_n)}{2f'(x)}$. Since $\lim_{h_n \to 0} \frac{|o(h_n^2)|}{|h_n|} = 0$, let $\delta > 0$ (as small as we wish) be such that $\frac{|o(h_n^2)|}{|h_n|} < \delta$. Then for all $n \in \mathbb{N}$,

$$|x_{n+1} - \xi| \leq C \cdot |x_n - \xi|^2 + \delta.$$

So

$$|x_{n+1} - \xi| \leq C \cdot |x_n - \xi|^2.$$

We choose $r > 0$ such that $Cr < 1$, and we take the starting point x_1 so that $|x_1 - \xi| < r$. Therefore we have

$$|x_{n+1} - \xi| \leq \frac{1}{C} \cdot (C.r)^{2^{n-1}}.$$

Hence $\lim x_n = \xi$.
Proof of Theorem 7 Let's start by analyzing an example from Ref. [6].

Example 3 To find a solution to the equation $\cos(x) + 3xe^{-x} = 0$ is equivalent to finding the fixed points of the function $f(x) = -\frac{1}{3}e^x \cos(x)$; the graph is illustrated in Fig. 6. Let z_0 be the only negative fixed point; the others are $\{z_k\}, k \geq 1$, all being positive with an approximate value of $z_k \sim (2k + 1)\frac{\pi}{2}$. The derivative of f satisfies

$$|f'(x)| < \frac{2}{3}e^x.$$

If $x < 0$, then $\mid f'(x) \mid < \frac{2}{3} < 1$. So f is a contraction in the neighborhood of z_0. In the neighborhood of z_k, $k \geq 1$, we have $\mid f'(x) \mid > 1$. The numeric behavior of the sequence $\{x_n\}_{n \in \mathbb{N}}$ defined in (4) can be understood by taking different initial values x_0 close to z_k. Only the point z_0 is an attractor, that is, choosing x_0 near z_0, the sequence converges to z_0 since f is a contraction. In the cases $k \neq 0$, so that x_0 is close to z_k, the sequence does not converge to z_k, so they are called repulsive points since f is no longer a contraction. We will introduce another strategy to approach the question; we will consider the functions $f(x) = \cos(x) + 3xe^{-x}$ and $g(x) = x - a.f(x)$. Finding a fixed point for g is equivalent to finding a zero for $f(x)$. We can choose a so that the derivative

$$g'(x) = 1 - a[-\sin(x) + 3e^{-x}(1 - x)]$$

satisfies the condition $\mid g'(x) \mid < 1$ in the neighborhood of z_k, therefore making $g(x)$ a contraction. In this example, the size of the neighborhood depends on z_k, so it is not uniform.

We now consider the following case. Let $\hat{f} : \mathbb{R} \to \mathbb{R}$ be a differentiable map. Given $c \in E$, we will study the question about the existence of solutions for the equation $\hat{f}(x) = c$. First, let's look at some of the ideas that motivate the techniques we will use. To solve the equation $\hat{f}(x) = c$ is equivalent to solving $f(x) = \hat{f}(x) - c = 0$, so we can reduce the problem to find the zeros (roots) of a function. Given the function

$$g(x) = x - af(x),$$

the equation $f(x) = 0$ has a solution if and only if $g(x) = x$. This brings us to the Contraction Lemma 1 in Appendix A to show that g has a fixed point.

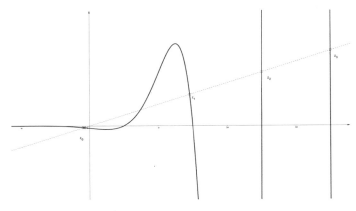

Fig. 6   fixed points of $f(x) = -\frac{1}{3}e^x \cos(x)$

Proposition 4 *Let $U \subset E$ be an open subset and let $f : U \to U$ be a differentiable map. If $\mid df_p \mid \leq \lambda < 1$, then there is a neighborhood $U_0 \subset U$ of p such that $f \mid_{U_0}$ is a contraction.*

Proof Let $x, y \in U$ and $r : [0, 1] \to E$ be the line $r(t) = x + t(y - x)$. By Theorem 5, there is $t_0 \in (0, 1)$ such that

$$\mid f(y) - f(x) \mid \leq \mid df_{t_0} \mid . \mid y - x \mid < \lambda \mid y - x \mid .$$

$\square$

The constant a in $g(x) = x - af(x)$ is chosen so that the root α of $f(x) = 0$ satisfies $\mid g'(\alpha) \mid \leq 1$. So α is an attractor and a fixed point of g since

$$\mid g'(\alpha) \mid \leq \mid 1 - a.f'(x) \mid < 1.$$

We note that the optimal choice for a is $\frac{1}{f'(\alpha)}$.

Let's proceed to prove Theorem 7. The first step is to show that under some restrictions, the function ϕ_y defined in Eq. 8 has a fixed point.

Definition 6 Let $a = f'(x_0)$. Consider the open set

$$U_0 = \left\{ x \in I; \mid f'(x) - f'(x_0) \mid < \frac{\mid a \mid}{2} \right\} \tag{9}$$

and $V_0 = f(U_0)$ its image.

The proof of Theorem 7 is achieved once the next series of five propositions is proved.

Proposition 5 *If $x_0 \in U_0$, then ϕ_y is a contraction for any $y \in \mathbb{R}$.*

Proof The derivative of $\phi_y(x) = x + \frac{1}{a}(y - f(x))$ is $\phi'_y(x) = 1 - \frac{1}{a} f'(x)$. If $x \in U_0$, then

$$\mid f'(x) - f'(x_0) \mid = \mid a \mid . \mid \phi'_y(x) \mid < \frac{\mid a \mid}{2} \Rightarrow \mid \phi'_y(x) \mid < \frac{1}{2}.$$

Applying the Mean Value Theorem, we have

$$\mid \phi_y(x') - \phi_y(x) \mid \leq \frac{1}{2} \mid x' - x \mid .$$

$\square$

Taking $r > 0$ and $\alpha = \frac{\mid a \mid}{2}$, we consider the open balls $B_r(x_0) = \{x \in U_0; \mid x - x_0 \mid < r\}$ and $B_{\alpha r}(y_0) = \{y \in V_0; \mid y - y_0 \mid < \alpha r\}$.

Proposition 6 *If $y \in B_{ar}(y_0)$, then there is $x \in U_0$ such that $y = f(x)$.*

Proof We have $\phi_y(\overline{B_r(x_0)}) \subseteq \overline{B_r(x_0)}$ since

$$| \phi_y(x) - x_0 | \leq | \phi_y(x) - \phi_y(x_0) | + | \phi_y(x_0) - x_0 | \leq \frac{1}{2} | x - x_0 | + \frac{1}{|a|} | y - y_0 | \leq r.$$

Therefore $\phi_y : \overline{B_r(x_0)} \rightarrow \overline{B_r(x_0)}$ is a contraction for all $y \in B_r(y_0)$. Consequently, ϕ_y has a fixed point $\xi \in \overline{B_r(x_0)}$. Hence $y = f(\xi)$. □

Proposition 7 *V_0 is an open set.*

Proof Given $\epsilon > 0$, let $y_0 = f(x_0) \in V_0$ and $y \in B_\epsilon(y_0)$. According to the statement of Proposition 6, we have $\phi_y(\overline{B_\epsilon(x_0)}) \subset \overline{B_r(x_0)}$ for all $y \in B_\epsilon(y_0)$. So the contraction $\phi_y : \overline{B_\epsilon(x_0)} \rightarrow \overline{B_r(x_0)}$ has a unique fixed point $\xi \in \overline{B_r(x_0)}$, and therefore $y = f(\xi) \in V_0$ and $B_\epsilon(y_0) \subset V_0$. □

Proposition 8 *The map $f : U_0 \rightarrow V_0$ is a homeomorphism.*

Proof First, let's check that $f : U_0 \rightarrow V_0$ is injective. To argue by contradiction, assume there are points x_1, x_2 such that $y = f(x_1) = f(x_2)$, then

$$\phi_y(x_1) = x_1 \quad \text{and} \quad \phi_y(x_2) = x_2.$$

Consequently, we have $x_1 = x_2$ due to the uniqueness of the fixed point. Since $f^{-1}(V_0) = U_0$ is open, then $f^{-1} : V_0 \rightarrow U_0$ is continuous. □

Proposition 9 *The map $f^{-1} : V_0 \rightarrow U_0$ is differentiable.*

Proof For any $x \in U_0$, we have $f'(x) \neq 0$. Now let's consider $y = f(x)$, $y + k = f(x + h)$;

$$f^{-1}(y + k) - f^{-1}(y) - \frac{k}{f'(x)} = f^{-1}(f(x + h)) - f^{-1}(f(x)) - \frac{k}{f'(x)} =$$

$$= h - \frac{k}{f'(x)} = -\frac{1}{f'(x)}(k - f'(x)h) = \frac{f(x + h) - f(x) - f'(x)h}{f'(x)}.$$

Passing to the limit $k \rightarrow 0$ and taking into account the inequality $| h | < \frac{2}{|f'(x_0)|} | k |$, we have $h \rightarrow 0$ and

$$\lim_{k \to 0} \left(f^{-1}(y + k) - f^{-1}(y) - \frac{k}{f'(x)} \right) < \lim_{h \to 0} \frac{f(x + h) - f(x) - f'(x)h}{f'(x)} = 0.$$

Therefore at $y = f(x)$, the derivative of f^{-1} is $(f^{-1})'(y) = \frac{1}{f'(x)}$. To prove that $|h| < \frac{2}{|f'(x_0)|} |k|$, we note that

$$| \phi_y(x+h) - \phi_y(x) | = \left| h - \frac{k}{f'(x_0)} \right| \leq \frac{|h|}{2}.$$

Using the triangular inequality $\big| |a| - |b| \big| < |a - b|$, we get

$$\left| |h| - \left| \frac{k}{f'(x_0)} \right| \right| \leq \frac{|h|}{2} \quad \Rightarrow \quad \frac{|h|}{2} < \frac{|k|}{|f'(x_0)|}.$$

$\square$

Exercises

Considering $f \in C^0([a, b])$, answer the following;

(1) (Secant's Method) Given the points x_0 and x_1, show that the sequence $\{x_n\}_{n \in \mathbb{N}}$,

$$x_{n+1} = \frac{x_{n-1} f(x_n) - x_n f(x_{n-1})}{f(x_n) - f(x_{n-1})}$$

converges to a solution of $f(x) = 0$.

(2) Let $U \subset \mathbb{R}^n$ be an open subset, let $f \in C^2(U, \mathbb{R}^n)$ and let $p \in U$ be a root of f such that df_p is invertible. In U, consider the sequence $\{x_n\}_{n \in \mathbb{N}}$ defined by the Eq. (4). Show that there is a neighborhood $V \subset U$ of p and a constant $C > 0$ such that if $x_0 \in V$, then $\lim x_n = p$.

4.3   Proof of the Inverse Function Theorem (InFT)

We will prove the general version of InFT for differentiable maps on a Banach space. In Chapter I, the Implicit Function Theorem (ImFT) used to solve equation $f(x) = c$ is a consequence of the InFT.

Theorem 8 (InFT) *Let E be a Banach space, and let $U \subset E$ be an open subset, $f \in C^1(U, E)$ and $p \in U$. If the differential operator df_p is invertible, then there are open subsets U_0, V_0 in E such that:*
(i) $U_0 \subset U$, $p \in U_0$ and $f(p) \in V_0$.
(ii) $V_0 = f(U_0)$ and $f : U_0 \to V_0$ is a diffeomorphism.

Proof The proof is divided into four steps;

(i) The choice of U_0;

Let $A = df_p$ and $\lambda = \frac{1}{2|A^{-1}|}$. Define $U_0 = \{x \in U; | df_x - A | < \lambda\}$. Given the continuity of df_p, the subset U_0 is well-defined. The existence of df_x^{-1} for all $x \in U_0$ is guaranteed due to Proposition 5 in Chap. 2.

(ii) The map $f \mid_{U_0}$ is injective.

Let $y \in E$ and define the map $\phi_y : U \to E$,

$$\phi_y(x) = x + A^{-1}(y - f(x)).$$

There are two important remarks to be stressed;

(1^a) If we have $x \in U$ such that $y = f(x)$, then $\phi_y(x) = x$,

(2^a) $\phi_y \mid_{U_0}$ is a contraction.

Indeed, for all $x, \bar{x} \in U_0$, we claim the inequality

$$\| \phi_y(\bar{x}) - \phi_y(x) \| \leq \frac{1}{2} \| \bar{x} - x \| .$$

Since

$$| d(\phi_y)_x | = | I - A^{-1} df_x | \leq | A^{-1} | . | A - df_x | \leq \frac{1}{2},$$

for all $x \in U_0$, we can apply the Mean Value Theorem 5 to obtain a contraction.

Suppose there are $x_0, \ x_1 \in U_0$ such that $y = f(x_0) = f(x_1)$; then ϕ_y has two fixed points contracting the Contraction Lemma. Therefore $x_0 = x_1$ and $f \mid_{U_0}$ is injective.

(iii) The set $V_0 = f(U_0)$ is open.

Before proving the statement, we note it implies $f^{-1} \mid_{U_0}$ is continuous. Set $y_0 \in V_0$ and let $x_0 \in U_0$ be such that $y_0 = f(x_0)$. Let's show that there is an open ball $B_\epsilon(y_0)$ contained in V_0. We choose $r > 0$ so that $\overline{B_r(x_0)} \subset U_0$ and define the ball $B_{\lambda r}(y_0) = \{y \in E; | y - y_0 | < \lambda r\}$. We will prove that $B_{\lambda r}(y_0) \subset V_0$, i.e., for any $y \in B_{\lambda r}(y_0)$, then we have $\xi \in U_0$ such that $y = f(\xi)$. Given $y \in B_{\lambda r}(y_0)$, we consider the map $\phi_y : U_0 \to U_0$, $\phi_y(x) = x + A^{-1}(y - f(x))$, and we note that $\phi_y(\overline{B_r(x_0)}) \subseteq \overline{B_r(x_0)}$ for any $x \in \overline{B_r(x_0)}$ given that

$$| \phi_y(x) - x_0 | \leq | \phi_y(x) - \phi_y(x_0) | + | \phi_y(x_0) - x_0 | \leq \frac{1}{2} | x - x_0 | + | \phi_y(x_0) - x_0 | \leq$$

$$\leq \frac{r}{2} + | A^{-1}(y - f(x_0)) | \leq \frac{r}{2} + | A^{-1} | . | y - y_0 | \leq \frac{r}{2} + \frac{r}{2} = r.$$

Since $\overline{B_r(x_0)}$ is a complete metric space, it follows that the map ϕ_y when restricted to $\overline{B_r(x_0)}$ is a contraction having a unique fixed point $\xi \in \overline{B_r(x_0)}$. Therefore $y = f(\xi)$, and so $B_{\lambda r}(Y_0) \subset V_0$.

(iv) The map $g = f^{-1} : V_0 \to U_0$ is differentiable and $dg_y = [df_{g(y)}]^{-1}$.
     Since the map $f : U_0 \to V_0$ is bijective, we can define $g = f^{-1} : V_0 \to U_0$. Let
     $y, y + k \in V_0$ and $x, x + h \in U_0$ be such that $y = f(x)$ and $y + k = f(x + h)$. For all $x \in U_0$, the operator $[df_x]^{-1}$ is well-defined. Let $B = [df_x]^{-1}$;

$$g(y + k) - g(y) - B(k) = g(f(x + h)) - g(f(x)) - B(k)$$
$$= x + h - x - B(k) = h - B(k) = B(B^{-1}(h) - k)$$
$$= -B(f(x + h) - f(x) - B^{-1}(h)).$$

In this way, we have

$$\frac{g(y + k) - g(y) - B(k)}{|k|} = \frac{|B(f(x + h) - f(x) - B^{-1}(h))|}{|k|} \leq$$
$$\leq |B| \cdot \frac{|f(x + h) - f(x) - B^{-1}(h)|}{|k|}.$$

Passing to the limit $k \to 0$, we use the inequality $|h| \leq 2|A^{-1}| \cdot |k| = \lambda^{-1}|k|$ with a proof to be given at the end. Now, taking the limit $h \to 0$ on the right-hand side of

$$\frac{g(y + k) - g(y) - B(k)}{|k|} \leq \frac{|B|}{\lambda} \cdot \frac{|f(x + h) - f(x) - df_x(h)|}{|h|},$$

we get that g is differentiable. Hence $B = dg_y = [df_{g(y)}]^{-1}$. The continuity of dg_y in relation to the argument y is verified by observing that the map $\mathcal{I} : GL(E) \to GL(E)$, $\mathcal{I}(S) = S^{-1}$ is continuous. The maps $df : x \to df_x$ and g are continuous, so the map $dg_y = [df_{g(y)}]^{-1} = \mathcal{I}(df_{g(y)})$ is a composite of continuous maps, so it is continuous. The inequality $|h| \leq 2|A^{-1}| \cdot |k| = \lambda^{-1}|k|$ follows from

$$|\phi_y(x + h) - \phi_y(x)| = |h + A^{-1}k| \leq \frac{1}{2}|h|$$

and $|h| - |A^{-1}k| \leq |h + A^{-1}k| \leq \frac{1}{2}|h|$.   □

4.4 Applications of InFT

(1) Implicit Function Theorem in finite dimension.
    Consider $\mathbb{R}^{k+m} = \{(x, y) \mid x \in \mathbb{R}^k, y \in \mathbb{R}^m\}$ and let $f : \mathbb{R}^{k+m} \to \mathbb{R}^m$ be a differentiable map given by $f(x, y) = (f_1(x, y), \ldots, f_m(x, y))$. Let $c = (c_1, \ldots, c_m) \in \mathbb{R}^m$ be fixed and let $p \in \mathbb{R}^{k+m}$ be a point satisfying $f(p) = c$. We will

assume that the differential $df_p = \left(\frac{\partial f}{\partial x} \cdots \frac{\partial f}{\partial y} \right) : \mathbb{R}^{k+m} \to \mathbb{R}^m$ is surjective and $\frac{\partial f}{\partial y}(p)$ is an isomorphism. The differential of

$$F(x, y) = (f_1(x, y), \ldots, f_m(x, y), x)$$

at p is

$$dF_p = \begin{pmatrix} A & B \\ I & 0 \end{pmatrix},$$

and $A = \frac{\partial f}{\partial x}_{(p)}$ is an $(m \times k)$ matrix, $B = \frac{\partial f}{\partial y}_{(p)}$ is an $(m \times m)$, I is the $(k \times k)$ identity matrix, and 0 is the null matrix $(k \times m)$. So dF_p is an isomorphism. Due to the InFT, there are open sets $U \ni p$ and $V \ni F(p)$ such that $F : U \to V$ is a diffeomorphism. Letting $\Phi = F^{-1}$, $u = f(x, y)$ and $v = x$, we have $\Phi(u, v) = (x, y)$. Therefore $f \circ \Phi(u, v) = f(x, y) = u$. We see that up to a local diffeomorphism, the map f is a projection. Letting $\Phi(u, v) = (\phi_1(u, v), \phi_2(u, v))$, if $f(x, y) = c$, then $x = v$ and $y = \phi_2(x, c)$. The solution set of the equation $f(x, y) = c$ is locally described in $\mathbb{R}^{k+m}$ by the map $x \to (x, \phi_2(x, c))$.

(2) Let $f : M_n(\mathbb{R}) \times M_n(\mathbb{R}) \to M_n(\mathbb{R})$ be the map $f(X, Y) = X.Y - I$. The entries of $f(X, Y)$ are polynomials, so f is differentiable. The differential of f at (X, Y) is the linear operator $df_{(X,Y)} : \mathbb{R}^{n^2} \times \mathbb{R}^{n^2} \to \mathbb{R}^{n^2}$, $df_{(X,Y)}.(U, V) = UY + XV$. Then $df_{(X,Y)}$ is surjective if X or Y belong to $GL_n(\mathbb{R})$. The point (I, I) is a solution to the equation $f(X, Y) = 0$, so the solution set is locally diffeomorphic to $\mathbb{R}^{n^2}$.

(3) Let $U \subset \mathbb{R}^n$ be an open subset and $f : U \to \mathbb{R}^n$ a C^1-map. The InFT can be applied to show that the dependence on initial conditions $x_0 = x(0)$ for solutions of ODE $x'(t) = f(x(t))$ is differentiable. Consider the Banach spaces

$$E = \left(C^0([a, b]), \| \cdot \|_0 \right), \quad F = \{g \in E \mid g(x_0) = 0\} \subset E,$$

and let f be a function such that $\| f \|_0 < M$. Define the map $\mathcal{F} : E \to F$ by

$$\mathcal{F}(x)(t) = x(t) - x_0 - \int_0^t f(x(s))ds. \tag{10}$$

The zeros of $\mathcal{F}$ correspond to solutions of the initial value problem $x'(t) = f(x(t))$, $x(0) = x_0$, which has a unique solution due to the Existence and Uniqueness Theorem.[2] The operator $d\mathcal{F}_x : F \to F$, $d\mathcal{F}_x.h = h(t) - \int_a^t f'(x(s))h(s)ds$ is written as $d\mathcal{F}_x = I - T$, $T : F \to F$ is given by $T(h) = \int_a^t f(x(s))h(s)ds$ and $|T| \leq M |t|$. Let's choose a neighborhood $(-\epsilon, \epsilon)$ so that $M |t| < 1$, and so $|T| < 1$. Therefore the operator $d\mathcal{F}_x = I - T$ is an isomorphism. So there are open subsets $U, V \subset E$ and a diffeomorphism

---

[2] Proved in Chap. 4.

$\Phi : U \to V$. We conclude that the equation $\mathcal{F}(x) = y$ has a solution for all y belonging to the neighborhood $V \subset E$ of 0 and it is defined for all t sufficiently small. Moreover, its dependence on the initial condition $x(0)$ is differentiable.

(4) Let $E_1 = \left(C^1([0, 1]), \| f \|_1\right)$ and $E_2 = \left(C^0([0, 1]), \| f \|_0\right)$. Let $\mathcal{F} : E_1 \to E_2$ be a differentiable map, and let $\mathcal{F}(f + h) = \mathcal{F}(f) + d\mathcal{F}_f.h + r(h)$ be the Taylor formula of degree 1. We want to show that for all $g \in E_2$ satisfying $\| g \|_0 < \epsilon$, the differential equation

$$\frac{df}{dx} + \mathcal{F}(f) = g, \ f(0) = k$$

admits a unique solution. The derivative of $\Phi : E_1 \to E_2$ given as

$$\Phi(f) = \frac{df}{dx} + \mathcal{F}(f)$$

defines the linear operator $d\Phi_f.h = \frac{dh}{dx} + \mathcal{F}'(f).h : E_1 \to E_2$. For $f = 0$, we have $\Phi(0) = 0$ and $d\Phi_0.h = \frac{dh}{dx}$. The kernel of the $d\Phi_0$ is the set of constant functions in E_1, so it is isomorphic to $\mathbb{R}$. Consider the decomposition $E_1 = \text{Ker}(d\Phi_0) \oplus W$, with $W \subset E_1$ as a closed complement. By the Fundamental Theorem of Calculus, the operator $d\Phi_0 : W \to E_2$ is bijective. We therefore have open subsets $U \subset W$ and $V \subset E_2$ such that $\hat{\Phi} : U \to V$ is a diffeomorphism. Therefore given $\epsilon > 0$ sufficiently small, the equation $\hat{\Phi}(f) = g$ has a unique solution in $U \subset E$.

Exercises

(1) Let L be an invertible linear operator and f a map such that $\| f(x) \| < M \| x \|^2$. Show that the map $g(x) = L(x) + f(x)$ is a diffeomorphism.

(2) Show that there are open neighborhoods U, V of the identity matrix in which every matrix A admits the square root, so that we have $X \in U$ such that $X^2 = A$. Extend the result to the neighborhood of an arbitrary matrix.

(3) Define the exponential map $\exp : M_n(\mathbb{R}) \to GL_n(\mathbb{R})$ by $\exp(A) = \sum_{i=0}^{\infty} \frac{A^n}{n}$. Show that:

   (a) The map $\exp : M_n(\mathbb{R}) \to GL_n(\mathbb{R})$ is well-defined and $\exp \in C^1$.
   (b) The identity has a neighborhood $U \subset M_n(\mathbb{R})$ in which $\exp |_U$ is a local diffeomorphism.
   (c) If $| A | < \ln(2)$, then $| \exp(A) - I | < 1$.

(4) Define $\ln(A) = \sum_1^{\infty} (-1)^{n+1} \frac{(A-I)^n}{n}$. Show that;

   (a) If $| I - A | < 1$, then the series $\ln(A)$ converges absolutely and $| \ln(A) | < 1$.
   (b) If $| A | < \ln(2)$, then $| \exp(A) - I | < 1$ and $\ln(\exp(A)) = A$ (hint: first consider A diagonalizable. Next, apply the Jordan Canonical form to the general case).

(5) Let $f : \mathbb{R}^n \subset \mathbb{R}^n$ be differentiable and $f(0) = 0$. If 1 is not in the spectrum $\sigma(df_0)$, prove there is a neighborhood $V \subset \mathbb{R}^n$ of 0 such that $f(x) \neq x$ for all $x \in V - 0$.

(6) Let $A = \{A \in M_n(\mathbb{R}) \mid A^t = -A\}$ be the space of skew-symmetric matrices. Define the exponential map $\exp : A \to SO_n$ by $\exp(A) = \sum_{i=0}^{\infty} \frac{A^n}{n}$. Show that:

(a) The map exp is well-defined.
(b) There is a neighborhood U of identity $I \in A$ such that $\exp |_U$ is a diffeomorphism.

(7) Let $k : [a, b] \times [a, b] \to \mathbb{R}$ be a continuous function such that $\| k \|_0 < C, C > 0$ and consider $r \in \mathbb{R}$ a real number satisfying $| r | \leq \frac{1}{C(b-a)}$. For $n \in \mathbb{N}$, consider the map

$$\Phi_n(f(x)) = f(x) + r \int_a^b k(x, t) f^n(t) dt$$

and solve the following items:

(a) Show that $\Phi_n : C^0([a, b]) \subset C^0([a, b])$.
(b) Assuming $n = 1$, prove that given an arbitrary function $g \in C^0([a, b])$ there is always a unique function $f \in C^0([a, b])$ such that $g = \Phi_1(f)$.
(c) Assuming $n = 2$ and $\| f \|_0 \leq 1$, prove there are neighborhoods $U(f) \ni f$ and $V(f) \ni \Phi_2(f)$ such that, given an arbitrary $h \in V(f)$, we have a unique function $g \in U(f)$ such that $\Phi_2(g) = h$.

(8) For which values of (r, θ) is the map $P : \mathbb{R}^2 \to \mathbb{R}^2$, $P(r, \theta) = (r. \cos(\theta), r. \sin(\theta))$, a local diffeomorphism?

(9) Given the functions $f, g \in C^1(\mathbb{R}^2)$, let $\phi : \mathbb{R}^2 \to \mathbb{R}^2$ be the map

$$\phi(x, y) = (x + x^2 f(x, y), y^2 + y^2 g(x, y)).$$

Show that there is a neighborhood V of the origin and a solution for the equation $p = \phi(x, y)$ for all $p \in V$.

(10) (Implicit Function Theorem) Let E_1, E_2 and F be Banach spaces, let $U_1 \subset E_1$, $U_2 \subset E_2$ be open subsets and $E = E_1 \times E_2$. Let $U = U_1 \times U_2 \subset E$ and $f \in C^1(E, F)$. For a point $p \in U$, the differential operator df_p admits a decomposition into $d_1 f_p \oplus d_2 f_p : E_1 \oplus E_2 \to F$ with $d_i f = df |_{E_i}$. Assume the differential $d_2 f_p : E_2 \to F$ is in $GL(E_2, F)$ at $p = (p_1, p_2) \in E$, $p_i \in E_i$. Let $q = f(p) \in F$. So there are open sets $U_1' \subset E_1 \ni p_1$ and $V \subset E_2 \ni p_2$ satisfying the following condition: there is a C^1-map $\xi : U_2' \to V$ such that $f(x, \xi(x)) = c$. Moreover, the differential of $\xi : V \to W$ at $x \in U$ is $d\xi_x = -[d_2 f(x, \xi(x))]^{-1}.d_1 f(x, \xi(x))$.

5   Classical Examples in Variational Calculus

The theory of differentiability in Banach spaces is a powerful tool that is often used in several contexts to address questions in Pure and Applied Mathematics. There are a variety of problems where solutions require optimizing a function $F : \Omega \to \mathbb{R}$, defined on a Banach space Ω, and given by

$$F(\gamma) = \int_a^b L\big(t, \gamma(t), \dot{\gamma}(t)\big)dt, \quad \left(\dot{\gamma} = \frac{d\gamma}{dt}\right), \tag{11}$$

with $\gamma \in \Omega$, $\gamma : [a, b] \to \mathbb{R}^n$, and

$$\begin{aligned} L : &[a, b] \times \mathbb{R}^n \times \mathbb{R}^n \to \mathbb{R}, \\ &(t, x, y) \mapsto L(t, x, y) \end{aligned} \tag{12}$$

is a C^2-function. L is the Lagrangian. The Calculus of Variations deals with this kind of optimization problem. To optimize F, using the differential techniques developed so far, it means finding the critical points of F which may be local minimums, local maximums or saddle points. The Calculus of Variations has its origins in the famous Brachistochrone problem formulated by Johann Bernoulli in 1696. After the Lagrangian and Hamiltonian formulation of Classical Newtonian Mechanics, the topic had a tremendous advance and technical growth culminating in a variety of applications; e.g., in the formulation of Quantum Mechanics, in addressing global problems in Differential Geometry, as well as in Engineering and many other areas. Nowadays, the Calculus of Variations is one of the most important subjects in mathematics. Often the topic uses its own jargon such as the terms;
– the function $F : \Omega(p, q) \to \mathbb{R}$ is called the functional.
– the Banach space E is the set of admissible functions.
– the directional derivative is called the functional derivative.

   The initial formulation of a variational problem is settled when we know the functional F and the space of admissible functions. The examples and the approach we will present are classics, so in most of them the space Ω of admissible functions is a subset of the Banach space $C^1([a, b], \mathbb{R}^n)$ with the norm $\| f \|_1 = \| f \|_0 + \| f' \|_0$. In general, the closing of a bounded subset of the space of admissible functions is not compact, so there is no *apriori* guarantee that the optimization problem will have a solution. The critical points of a function f are roots of the equation $f'(x) = 0$. The equivalent version of this equation requires finding a critical point of a functional F. Since the critical points are minimal in our examples, a variational problem in our context is well posed when it is given as follows:

$$\begin{cases} \text{minimize: } F(\gamma) = \int_a^b L(t, \gamma(t), \dot{\gamma}(t))dt, \\ \text{constrain to: } \gamma \in \Omega. \end{cases} \tag{13}$$

5.1 Euler-Lagrange Equations

Let's fix the points $p, q \in \mathbb{R}^n$ and consider the space of admissible functions as the affine space

$$\Omega(p, q) = \{\gamma : [a, b] \to \mathbb{R}^n \mid \gamma \in C^1, \gamma(a) = p, \ \gamma(b) = q\}.$$

Let $\mathbb{R}^n = \{(x_1, \ldots, x_n) \mid x_i \in \mathbb{R}, 1 \le i \le n\}$ and let $\beta = \{e_1, \ldots, e_n\}$ be the canonical basis of $\mathbb{R}^n$. For all $\gamma \in \Omega(p, q)$, $\gamma(t) = (x_1(t), \ldots, x_n(t))$, we consider the map $\Gamma : [a, b] \to \mathbb{R} \times \mathbb{R}^n \times \mathbb{R}^n$, $\Gamma(t) = (t, \gamma(t), \dot{\gamma}(t))$. Let $T_\gamma \Omega(p, q)$ be the tangent space to $\Omega(p, q)$ at γ. So the functional is written as $F(\Gamma) = \int_a^b (L \circ \Gamma)(t) dt$.
Fixing the origin of $\Omega(p, q)$ at γ, we get the vector space

$$V = \{v : [a, b] \to \mathbb{R}^n \mid v \in C^1, \ v(a) = v(b) = 0\}.$$

Then $T_\gamma \Omega(p, q) = V$ for all $\gamma \in \Omega(p, q)$. The vector space $T_\gamma \Omega(p, q)$ is endowed with the inner product

$$< v, w > = \int_a^b < v(t), w(t) > dt = \int_a^b \left(\sum_{i=1}^n v_i(t) w_i(t) \right) dt.$$

If γ is a critical point of $F : \Omega(p, q) \to \mathbb{R}$, then $dF_\gamma.v = 0$ for all $v \in T_\gamma \Omega(p, q)$. To describe the differential $dF_\gamma.v$, we proceed the same way as we did to obtain the directional derivative in Chap. 1. For all $v \in T_\gamma(p, q)$, the curve $\gamma_s : (-\epsilon, \epsilon) \to \mathbb{R}^n$, given by $s \to \gamma_s(t) = \gamma(t) + s.v(t)$ is contained in $\Omega(p, q)$ ($\gamma_0 = \gamma$ and $\frac{d\gamma_s}{ds} = v$) and defines a variation of γ. The Lagrangian

$$L = L(t, x_1, \ldots, x_n, y_1, \ldots, y_n)$$

is a function on $(2n + 1)$-parameters and $y_i = \dot{x}_i$, $1 \le i \le n$. Considering $L \circ \Gamma : (-\epsilon, \epsilon) \times [a, b] \to \mathbb{R}$, $L \circ \Gamma(s, t) = L(t, \gamma_s(t), \dot{\gamma}_s(t))$, we have

$$dF_\gamma.v = \frac{dF(\gamma_s)}{ds} \Big|_{s=0} = \int_a^b \frac{d(L \circ \Gamma)}{ds} \Big|_{s=0} dt = \int_a^b dL_{\Gamma(t)}.\frac{d\Gamma(\gamma_s)}{ds} \Big|_{s=0} dt =$$

$$= \int_a^b < \nabla L(\Gamma(t)), \frac{d\Gamma(\gamma_s)}{ds} \Big|_{s=0} > dt = \int_a^b < \nabla L(\Gamma(t)), (0, v, \dot{v}) > dt =$$

$$= \int_a^b \sum_{i=1}^n \left(\frac{\partial L}{\partial x_i}.v_i + \frac{\partial L}{\partial y_i}.\dot{v}_i \right) dt = \int_a^b \sum_{i=1}^n \left[\frac{\partial L}{\partial x_i} - \frac{d}{dt}\left(\frac{\partial L}{\partial y_i} \right) \right].v_i dt +$$

$$+ \underbrace{\frac{\partial L}{\partial y_i}(b)v_i(b) - \frac{\partial L}{\partial y_i}(a)v_i(a)}_{=0}.$$

Let the gradient of F be the operator

$$\nabla F = \sum_{i=1}^{n} \left[\frac{\partial L}{\partial x_i} - \frac{d}{dt} \left(\frac{\partial L}{\partial y_i} \right) \right] e_i.$$

Therefore the expression above can be written as

$$dF_\gamma.v =< \nabla F(\gamma), v >= \int_a^b < \nabla F(\gamma(t)), v(t) > dt.$$

Consequently, γ is a critical point of F if

$$\int_a^b < \nabla F(\gamma), v > dt = \int_a^b \sum_{i=1}^{n} \left[\frac{\partial L}{\partial x_i} - \frac{d}{dt} \left(\frac{\partial L}{\partial y_i} \right) \right].v_i dt = 0, \qquad (14)$$

for all $v \in T_\gamma \Omega$.

Lemma 2 (du Bois-Reymond) *Let $f \in C^0([a, b])$ be an admissible function. If $\int_a^b f(t)v(t)dt = 0$ for all $v \in C^0([a, b])$ satisfying $v(a) = v(b) = 0$, then $f = 0$ in $[a, b]$.*

Proof Let $g(t) = \int_a^t f(\xi)d\xi$. It follows from the hypothesis that

$$0 = \int_a^b \frac{d(g(t).v(t))}{dt} dt = \underbrace{\int_a^b g'(t).v(t)dt}_{=0} + \int_a^b g(t).v'(t)dt.$$

So $\int_a^b g(t).v'(t)dt = 0$ for all v. Consider $v(t) = \int_a^t (f(\xi) - c)d\xi, t \in [a, b]$ and $c = \frac{1}{b-a} \int_a^b f(\xi)d\xi$. By construction, we have $v(a) = 0$ and $v(b) = 0$. Therefore

$$0 \le \int_a^b (f(t) - c)^2 dt = \int_a^b (f(t) - c).v'(t)dt = \int_a^b f(t)v'(t)dt - c(v(b) - v(a)) = 0.$$

Now we have $f(t) = c$, but this yields $f(t) = 0$.   □

Taking $v(t) = v_k(t)e_k, k \in \{1, \ldots, n\}$ in Eq. (14), we get

$$\int_a^b < \nabla F(\gamma), v > dt = \int_a^b \left[\frac{\partial L}{\partial x_k} - \frac{d}{dt} \left(\frac{\partial L}{\partial y_k} \right) \right] (t).v_k(t)dt = 0.$$

Applying Lemma 2, the Euler-Lagrange equations of the functional F are

$$\frac{\partial L}{\partial x_k} - \frac{d}{dt} \left(\frac{\partial L}{\partial y_k} \right) = 0, \ k = 1, \ldots, n. \qquad (15)$$

Since $\dot{\gamma}(t) = (\dot{x}_1(t), \ldots, \dot{x}_n(t))$, the ordinary differential equations (15) above are most often written as

$$\frac{\partial L}{\partial x_k} - \frac{d}{dt}\left(\frac{\partial L}{\partial \dot{x}_k}\right) = 0, \ k = 1, \ldots, n. \tag{16}$$

Therefore the E-L equations are 2nd-order ODEs. For an admissible curve $\gamma(t) = \big(x_1(t), \ldots, x_n(t)\big)$, we have

$$L(t, \gamma(t), \dot{\gamma}(t)) = L(t, x_1, \ldots, x_n, \dot{x}_1, \ldots, \dot{x}_n).$$

There are particular cases where the E-L equations can be reduced to 1st-order ODEs; this is done analyzing the L dependence on the $(2n + 1)$-variables. We denote $L_y = \frac{\partial L}{\partial y}$.

(i) Assume $L = L(t, \dot{x}_1, \ldots, \dot{x}_n)$;
   It follows from the E-L equations that

$$\frac{d}{dt}\left(\frac{\partial L}{\partial \dot{x}_k}\right) = 0.$$

(ii) $L = L(x_1, \ldots, x_n, \dot{x}_1, \ldots, \dot{x}_n)$;
   In this case, L does not depend on the parameter t, so

$$\frac{dL_{x_i}}{dx} = (L_{\dot{x}_i})_{y_i}\dot{x}_i + (L_{y_i})_{\dot{x}_i}\dot{x}_i = L_{\dot{x}_i y_i}\dot{x}_i + L_{\dot{x}_i \dot{x}_i}\ddot{x}_i.$$

Dealing with the E-L , we get

$$L_{x_i} - L_{\dot{x}_i x_i}\dot{x}_i - L_{\dot{x}_i \dot{x}_i}y\ddot{x}_i = 0.$$

After multiplying by $\dot{x}_i$, we have

$$L_{x_i}\dot{x}_i - L_{\dot{x}_i x_i}(\dot{x}_i)^2 - L_{\dot{x}_i \dot{x}_i}\ddot{x}_i\dot{x}_i = L_{x_i}\dot{x}_i + L_{\dot{x}_i}\ddot{x}_i - [L_{\dot{x}_i}\ddot{x}_i + \dot{x}_i L_{\dot{x}_i x_i}\dot{x}_i + \dot{x}_i L_{\dot{x}_i \dot{x}_i}\ddot{x}_i] =$$
$$= \frac{dL}{dt} - \frac{d}{dt}(\dot{x}_i L_{\dot{x}_i}) = \frac{d}{dt}[L - \dot{x}_i L_{\dot{x}_i}] = 0.$$

Therefore we have a constant $c \in \mathbb{R}$ such that

$$L - \dot{x}_i L_{\dot{x}_i} = c_i, \ i = 1, \ldots, n; \tag{17}$$

(iii) $L = L(t, x_1 \ldots, x_n)$;
   Since $L_{\dot{x}_i} = 0, i = 1, \ldots, n$, the E-L equations become

$$\frac{\partial L}{\partial x_i} = 0, \ i = 1, \ldots, n.$$

5.2 *Examples*

We will introduce some classical examples. Let $\Omega(p, q)$ be the space of continuous curves in $\mathbb{R}^2$ connecting p to q, i.e.,

$$\Omega(p, q) = \{\gamma : [0, 1] \to \mathbb{R}^2 \mid \gamma \in C^0([0, 1]), \ \gamma(0) = p, \ \gamma(1) = q\}.$$

The space of admissible functions to be considered is a subset of the space

$$\Omega^{gr}(p, q) = \{\gamma \in \Omega(p, q) \mid \gamma(x) = (x, y(x))\} \subset \Omega(p, q),$$

with $y : [0, 1] \to \mathbb{R}$ a C^1-function.

(1) Brachistochrone curve (*brachystos*=minimum, *chronos*=time).
Starting from the point $p = (a, A)$, a heavy body slides to the point $q = (b, B)$. Find the curve $\gamma \in \Omega^g(p, q)$ so that the body reaches the point q in a minimum amount of time.
variational formulation:
Let us consider that the trajectory of the particle is given as $\gamma(t) = (x, y(x))$, $y \in C^1([a, b])$. We need to find the time spent by the body to go through the trajectory described by γ. The arc-length element of γ is $ds = \sqrt{1 + [y'(t)]^2}dx$. The velocity of the body at time t is $\vec{v} = \dot{\gamma}(t)$ and the module of the velocity is $v = |\vec{v}| = |\dot{\gamma}| = \sqrt{1 + [y'(x)]^2}\frac{dx}{dt}$. Assuming gravity's acceleration is constant and equal to g, we have $v = \sqrt{2gy}$, and so $dt = \frac{\sqrt{1+[y'(x)]^2}}{\sqrt{2gy(x)}}dx$. Therefore

$$T = \frac{1}{\sqrt{2g}} \int_a^b \sqrt{\frac{1 + [y'(x)]^2}{y(x)}} dx. \tag{18}$$

The variational problem to be addressed is

$$\begin{cases} \text{minimize } T(\gamma) = \frac{1}{\sqrt{2g}} \int_a^b \sqrt{\frac{1+[y'(x)]^2}{y(x)}}dx, \\ \text{constrain to } \gamma \in \Omega^g(p, q) = \{y : [a, b] \to \mathbb{R} \mid f \in C^1([a, b])\} \subset (C^1([a, b]), \|\cdot\|_1). \end{cases} \tag{19}$$

solution:
The Lagrangian does not depend on x, so it satisfies Eq. 17;

$$L - y'L_{y'} = \sqrt{\frac{1 + (y')^2}{y}} - \frac{1}{\sqrt{2g}}\frac{y'}{\sqrt{1 + (y')^2}} = \frac{1}{\sqrt{2g}}\frac{1}{\sqrt{y(1 + (y')^2)}} = c.$$

Taking $A = \frac{1}{c\sqrt{2g}}$, we have $y' = \sqrt{\frac{A^2 - y}{y}}$, i.e.,

$$dx = \sqrt{\frac{y}{A^2 - y}}dy \Rightarrow x = \int \sqrt{\frac{y}{A^2 - y}}dy.$$

Changing the variable of integration to $y = A^2 \sin^2(\theta/2)$, we get

$$x(\theta) = \frac{A^2}{4}(\theta - \sin(\theta)) + c, \quad y(\theta) = \frac{A^2}{4}(\theta - \cos(\theta)).$$

The parametrized curve $\gamma(\theta) = \left(\frac{A^2}{4}(\theta - \sin(\theta)) + c, \frac{A^2}{4}(\theta - \cos(\theta))\right)$ is a cycloid.

(2) Geodesics.

Let $S \subset \mathbb{R}^3$ be a surface. A Riemannian metric defined in S is a C^∞-map that associates to every $x \in S$ an inner product $g_x : T_x S \times T_x S \to \mathbb{R}$. Taking a frame $\alpha_x = \{e_1(x), e_2(x)\}$ at each point $x \in S$, the metric g_x evaluated on vectors $u, v \in T_x S$ can be written as the product of matrices

$$g_x(u, v) = u^t g(x)v.$$

The entries of matrix $g(x)$ are $[g(x)]_{ij} = g_x(e_i, e_j)$. The length of a curve $\gamma : [a, b] \to S$ is

$$L(\gamma) = \int_a^b \sqrt{g(\gamma'(t), \gamma'(t))}dt. \tag{20}$$

Since the interval is $[a, b]$, consider

$$\Omega(p, q) = \{\gamma \in C^0([a, b]; S) \mid \gamma(a) = p, \ \gamma(b) = q\}.$$

Definition 7 Let $p, q \in S$. A geodesic connecting the point p to the point q is a curve $\gamma \in \Omega(p, q)$ such that

$$L(\gamma) = \min_{\delta \in \Omega(p,q)} L(\delta). \tag{21}$$

The distance $d(p, q)$ between p and q is

$$d(p, q) = \min_{\delta \in \Omega(p,q)} L(\delta). \tag{22}$$

Indeed, the distance between p and q is the length of the geodesic connecting p to q. Then to find the distance between two points $p, q \in S$, we need to solve the variational problem

$$\begin{cases} \text{minimize: } L(\gamma) = \int_a^b \sqrt{\gamma'(t)^t . g(\gamma(t)) . \gamma'(t)}dt, \\ \text{constrain to: } \gamma \in \Omega(p, q). \end{cases} \tag{23}$$

Next, we will work out the geodesic variational problem in Euclidean, spherical and hyperbolic geometries.

(a) **Euclidean Geometry**
In this case, the surface is $S = \mathbb{R}^2$ and

$$g_p(u, v) = u^t \begin{pmatrix} 1 & 0 \\ 0 & 1 \end{pmatrix} v.$$

Let's consider the space of admissible curves as the curves defined by the graph of a function $\Omega^g(p, q) = \{y : [a, b] \to \mathbb{R} \mid y \in C^1([a, b])\}$. The length of a curve $\gamma(t) = (t, y(t))$ is

$$L(\gamma) = \int_a^b \sqrt{1 + (\dot{y}(t))^2} dt.$$

variational formulation:
To find the distance between two points $p, q \in \mathbb{R}^2$, we need to solve the problem

$$\begin{cases} \text{minimize: } \int_a^b \sqrt{1 + (\dot{y}(t))^2} dt., \\ \text{constrain to: } \gamma \in \Omega^g(p, q). \end{cases} \tag{24}$$

Solution:

$$L(\gamma) = \int_a^b \sqrt{1 + (y')^2} dt.$$

The Lagrangian $L(t, y, y') = \sqrt{1 + (y')^2}$ does not depend on t and y, so it satisfies the equation

$$L - y' L_{y'} = \sqrt{1 + (y')^2} - \frac{(y')^2}{\sqrt{1 + (y')^2}} = \frac{1}{\sqrt{1 + (y')^2}} = c.$$

In this way, we have a constant $k \in \mathbb{R}$ such that $y' = k$, and consequently, we get $y(t) = kt + k'$. In Euclidean Geometry, tracing a geodesic involves straight lines.

(b) **Spherical Geometry**
Let the surface be the sphere $S = S^2 = \{(\cos(\theta) \sin(\psi), \cos(\theta) \sin(\psi), \cos(\psi)) \mid (\theta, \psi) \in (0, 2\pi) \times (0, \pi)\}$. A curve $\gamma : [a, b] \to S^2$ is given by

$$\gamma(t) = \Big(\cos\big(\theta(t)\big) \sin\big(\psi(t)\big), \cos\big(\theta(t)\big) \sin\big(\psi(t)\big), \cos\big(\psi(t)\big) \Big).$$

The Riemannian metric at $p = (\cos(\theta) \sin(\psi), \cos(\theta) \sin(\psi), \cos(\psi))$ is

$$g_p(u, v) = u^t \cdot \begin{pmatrix} \sin^2(\psi) & 0 \\ 0 & 1 \end{pmatrix} . v, \quad \text{for all } u, v \in T_p S^2.$$

The length of γ is

$$L(\gamma) = \int_a^b \sqrt{\sin^2(\psi(t))\dot{\theta}^2 + \dot{\psi}^2} dt.$$

Let's consider $\theta = \theta(\psi)$ as a function of ψ and denote $\theta' = \frac{d\theta}{d\psi}$. Now the length of γ is

$$L(\gamma) = \int_a^b \sqrt{\sin^2(\psi(t))\theta'^2 + 1} d\psi.$$

variational formulation:

$$\begin{cases} \text{minimize: } \int_a^b \sqrt{\sin^2(\psi(t))\theta'^2 + 1} d\psi, \\ \text{constrain to: } \gamma \in \Omega^g(p, q) = \{\theta : [a, b] \to [0, 2\pi] \mid \theta \in C^1([a, b]; U)\}. \end{cases}$$
(25)

Solution:
The Euler-Lagrange equation is

$$\frac{d}{d\theta}(L_{\theta'}) - \frac{\partial L}{\partial \theta} = 0.$$

Since the Lagrangian L does not depend on θ, we have $\frac{\partial L}{\partial \theta} = 0$, and so there is a constant $c \in \mathbb{R}$ such that

$$\frac{d}{d\theta}(L_{\theta'}) = \frac{d}{d\theta}\left(\frac{\sin^2(\psi)\theta'}{\sqrt{\sin^2(\psi)\theta'^2 + 1}}\right) = 0 \quad \Rightarrow \quad \frac{\sin^2(\psi)\theta'}{\sqrt{\sin^2(\psi)\theta'^2 + 1}} = c.$$

Therefore we have

$$\theta' = \frac{k}{\sin(\psi)\sqrt{\sin(\psi) - k^2}} \quad \Rightarrow \quad \theta = \int \frac{k}{\sin(\psi)\sqrt{\sin(\psi) - k^2}} d\psi.$$

Changing the variable of integration to $u = \text{cotg}(\ )$, we get $\sin(\psi) = \frac{1}{\sqrt{1+u^2}}$ and

$$d\theta = -\frac{du}{1 + u^2}, \quad \sqrt{\sin^2(\psi) - c^2} = c.\sqrt{\frac{1 - c^2}{c^2} - u^2}.$$

Taking $k = \frac{1-c^2}{c^2}$, the integral is written as

$$\theta = -\int \frac{du}{\sqrt{k^2 - u^2}} = \arccos\left(\frac{u}{k}\right) + \psi_0 \quad \Rightarrow \quad \cos(\theta - \psi_0) = \frac{\text{cotg}(\psi)}{k},$$

so

$$\cos(\psi_0).\cos(\theta) - \sin(\psi_0).\sin(\theta) - \frac{\cos(\phi)}{k.\sin(\psi)} = 0$$

$$\frac{\cos(\psi_0).\cos(\theta).\sin(\psi) - \sin(\psi_0).\sin(\theta)\sin(\psi) - \cos(\phi)}{k.\sin(\psi)} = 0.$$

Since coordinates of the curve are $x = \cos(\theta).\sin(\psi)$, $y = \sin(\theta).\sin(\psi)$ and $z = \cos(\phi)$, then the curve is contained in the plane

$$\cos(\psi_0).x - \sin(\psi_0).y - z = 0.$$

Therefore the trace of a spherical geodesic is a segment of a great circle of the sphere.

(c) Hyperbolic Geometry.

Let the surface be $S = \mathbb{H}^2 = \{(x, y \in \mathbb{R}^2 \mid y > 0\}$. The Riemannian metric at $p = (x, y)$ is

$$g_p(u, v) = u^t \begin{pmatrix} \frac{1}{y^2} & 0 \\ 0 & \frac{1}{y^2} \end{pmatrix} v.$$

The length of $\gamma(t) = (x(t), y(t)) \in \Omega(p, q)$ is

$$L(\gamma) = \int_a^b \frac{\sqrt{((x'(t))^2 + (y'(t))^2}}{y(t)} dt.$$

Considering that $\gamma(t) = (t, y(t)) \in \Omega^g(p, q)$, the length is

$$L(\gamma) = \int_a^b \frac{\sqrt{(1 + (y'(t))^2}}{y(t)} dt.$$

variational formulation:

$$\begin{cases} \text{minimize: } \int_a^b \frac{\sqrt{1+(y'(t))^2}}{y(t)} dt, \\ \text{constrain to: } \gamma \in \Omega^g(p, q). \end{cases} \tag{26}$$

Solution:

As in the last examples, we have

$$L - y'L_{y'} = \frac{\sqrt{1 + (y')^2}}{y} - \frac{(y')^2}{y\sqrt{1 + (y')^2}} = \frac{1}{y\sqrt{1 + (y')^2}} = c.$$

Taking $k = \frac{1}{\sqrt{c}}$, we get $y\sqrt{1 + (y')^2} = k^2$, and so

$$\left(\frac{dy}{dx}\right)^2 = \frac{k^2 - y^2}{y^2}, \quad \Rightarrow \quad \frac{dy}{dx} = \sqrt{\frac{k^2 - y^2}{y^2}}.$$

By integrating the equation $dx = \frac{y}{\sqrt{k^2 - y^2}} dy$, we get a constant $c \in \mathbb{R}$ such that

$$(x - c)^2 + y^2 = k^2.$$

Then the trace of a hyperbolic geodesic, defined by the graph of a function, is an arc contained in the upper part ($y > 0$) of a circumference centered on the x-axis. When the points p, q are on a vertical line, say $p = (x_0, a)$ and $q = (x_0, b)$, the geodesic is a curve $\gamma(t) = (0, y(t))$ with a trace that is a vertical segment of the line with length

$$L(\gamma) = \int_a^b \frac{y'}{y} dt = \ln\left(\frac{y(b)}{y(a)}\right).$$

(3) **Minimum area of Surface of Revolution**

Given a curve $\gamma \subset \mathbb{R}^2$ and $\gamma \in \Omega(p, q)$, we would like to find a surface $S \subset \mathbb{R}^3$ with the minimum area among all the surfaces obtained by rotating a curve $\gamma \in \Omega(p, q)$ around some fixed axis. Given the curve $\gamma \in \Omega^g(p, q)$, $\gamma(t) = (t, f(t))$, consider the surface S_γ obtained by rotating γ around the axis-x. A parametrization of S_γ is given by

$$\begin{pmatrix} x(t) \\ y(t) \\ z(t) \end{pmatrix} = \begin{pmatrix} 1 & 0 & 0 \\ 0 & \cos(\theta) & -\sin(\theta) \\ 0 & \sin(\theta) & \cos(\theta) \end{pmatrix} \cdot \begin{pmatrix} t \\ f(t) \\ 0 \end{pmatrix} = \begin{pmatrix} t \\ f(t) . \cos(\theta) \\ f(t) . \sin(\theta) \end{pmatrix}.$$

The area of S_γ is

$$A(f) = 2\pi \int_a^b f(t)\sqrt{1 + [f'(t)]^2} dt.$$

variational formulation:

$$\begin{cases} \text{minimize: } A(f) = 2\pi \int_a^b f(t)\sqrt{1 + [f'(t)]^2} dt, \\ \text{constrain to: } f \in \Omega^g(p, q). \end{cases} \quad (27)$$

(4) **Lagrangian Formulation of Classical Mechanics**

Joseph Louis Lagrange observed that the trajectory of a particle, subject to a force field $\vec{F}$, is the one minimizing the action of the motion. Let the cinematic energy of the particle, traversing a trajectory $\gamma(t)$, be $K(\gamma) = \frac{1}{2}m |\dot{\gamma}|^2$, and let the potential energy be $U \in C^1(\mathbb{R}^3)$. The Lagrangian function is defined by $L(t, \gamma, \dot{\gamma}(t)) = K(\gamma) - U(\gamma)$ and the Action of the movement is defined by the integral

$$S(\gamma) = \int_a^b \big(K(\gamma) - U(\gamma)\big)dt.$$

Then to determine the trajectory of the part subject to the field $\vec{F} = -\nabla U$, we have to solve the following variational problem:

variational formulation:

$$\begin{cases} \text{minimize: } S(\gamma) = \int_a^b (K(\gamma) - U(\gamma))\, dt., \\ \text{constrain to: } \gamma \in \Omega(p, q) = \{\gamma \in C^0([a, b]; \mathbb{R}^n) \mid \gamma(0) = p, \gamma(b) = q\}. \end{cases}$$
$$(28)$$

In Classical Mechanics, the dynamics of a body of mass m, subject to a force field $\vec{F}$, is obtained by studying the properties of the solutions of Newton's $\vec{F} = m\vec{a}$ as time evolves. Lagrange noted that Newton's equation is equivalent to minimizing a functional. Let's assume the force field $\vec{F} = -\nabla U$ is defined by the potential $U : \mathbb{R}^3 \to \mathbb{R}$, $U = U(\gamma)$. Given a curve $\gamma(t) = \big(\gamma_1(t), \ldots, \gamma_n(t)\big)$, the Lagrangian $L : \mathbb{R}^3 \times \mathbb{R}^3 \times \mathbb{R} \to \mathbb{R}$, $L\big(t, \gamma(t), \dot\gamma(t)\big) = L\big(t, x_1(t), \ldots, x_n(t), \dot{x}_1(t), \ldots, \dot{x}_n(t)\big)$, is

$$L(t, \gamma(t), \dot\gamma(t)) = \frac{m}{2} \sum_{i=1}^{n} |\, \dot\gamma_i(t)\,|^2 - U(\gamma(t)).$$

The Euler-Lagrange equations define the ODE system

$$m\ddot\gamma_i = -\frac{\partial U}{\partial x_i}, \quad i = 1, \ldots, n.$$

The Euler-Lagrange equations turn out to be Newton's equation $\vec{F} = m\ddot\gamma(t)$.

(5) **Euclidean Geometry II**

Now we consider the space of admissible functions as the space of parametrized curves in $\mathbb{R}^2$:

$$\Omega(p, q) = \{\gamma : [a, b] \to \mathbb{R}^2 \mid \gamma(t) = (x(t), y(t)) \in C^1([a, b], \mathbb{R}^2), \; \gamma(a) = p, \; \gamma(b) = q\}.$$

The variational problem is

$$\begin{cases} \text{minimize: } L(\gamma) = \int_a^b \sqrt{\dot{x}^2 + \dot{y}^2} \\ \text{constrain to: } \gamma \in \Omega(p, q). \end{cases}$$

Consider the space $\mathcal{V} = \{\Lambda : [a, b] \to \mathbb{R}^2 \mid \Lambda \in C^1, \Lambda(a) = \Lambda(b) = 0\}$ and let $\gamma_t(s) : (-\epsilon, \epsilon) \to \Omega(p, q)$ be the curve in the space of admissible functions given by

$$\gamma_s(t) = \gamma(t) + s\Lambda(t), \quad \Lambda \in V,$$

and such that $\gamma_0(t) = \gamma(t)$. If $\gamma \in \Omega(p, q)$ and $\Lambda \in V$, then

$$\gamma_0(t) = \gamma(t), \quad \frac{\partial \gamma_s(t)}{\partial s}\Big|_{s=0} = \Lambda(t).$$

Let $\gamma \in \Omega(p, q)$, $\dot{\gamma} = \frac{d\gamma}{dt}$ and $\gamma' = \frac{d\gamma}{ds}$.

(a) γ is regular at $t \in [0, 1]$ when it is differentiable at t and $\gamma'(t) \neq 0$. We say that γ is regular if it is regular for all $t \in [0, 1]$.

(b) The normal vector to γ is

$$N = \frac{\ddot{\gamma} - \frac{<\dot{\gamma},\ddot{\gamma}>}{|\dot{\gamma}|^2}\dot{\gamma}}{\left|\ddot{\gamma} - \frac{<\dot{\gamma},\ddot{\gamma}>}{|\dot{\gamma}|^2}\dot{\gamma}\right|}.$$

(c) The curvature of γ is the function $k_\gamma : [0, 1] \to \mathbb{R}$ given by

$$k_\gamma = \frac{|\dot{\gamma} \wedge \ddot{\gamma}|}{|\dot{\gamma}|^3}.$$

Let $\gamma \in \Omega(p, q)$ be a regular curve with a curvature k_γ and normal vector N. The length of $\gamma_s(t)$ defines a function $L : (-\epsilon, \epsilon) \to \mathbb{R}$,

$$L(s) = L(\gamma_s) = \int_0^1 |\dot{\gamma}_s|\, dt.$$

Therefore we have

$$L(s) = \int_0^1 |\dot{\gamma} + s\dot{\Lambda}|\, ds = \int_0^1 \sqrt{|\dot{\gamma}|^2 + 2t <\dot{\gamma}, \dot{\Lambda}> + s^2 |\dot{\Lambda}|^2}\, dt.$$

The variational derivative of $L(s)$ with respect to the parameter s is

$$\frac{dL}{ds}\Big|_{s=0} = \lim_{\Delta s \to 0} \frac{L(\Delta s) - L(0)}{\Delta t} = dL_\gamma.\Lambda.$$

Since

$$\frac{dL}{ds} = \int_0^1 \frac{<\dot{\gamma}, \dot{\Lambda}> + s|\dot{\Lambda}|^2}{\sqrt{|\dot{\gamma}|^2 + 2s <\dot{\gamma}, \dot{\Lambda}> + s^2 |\dot{\Lambda}|^2}}\, dt,$$

we have

$$\frac{dL}{ds}\Big|_{s=0} = \int_0^1 <\frac{\dot{\gamma}}{|\dot{\gamma}|}, \dot{\Lambda}>\, dt. \tag{29}$$

We integrate the expression by parts aware of the following identities:

$$(i) \quad \frac{d < \gamma, \Lambda >}{dt} = < \frac{d}{dt} \left(\frac{\dot{\gamma}}{|\dot{\gamma}|} \right), \Lambda > + < \frac{\dot{\gamma}}{|\dot{\gamma}|}, \dot{\Lambda} >,$$

$$(ii) \quad \frac{d}{dt} (\frac{\dot{\gamma}}{|\dot{\gamma}|}) = \frac{1}{|\dot{\gamma}|} \left[\ddot{\gamma} - \frac{< \dot{\gamma}, \ddot{\gamma} >}{|\dot{\gamma}|^2} \dot{\gamma} \right],$$

$$(iii) \quad \left| \ddot{\gamma} - \frac{< \dot{\gamma}, \ddot{\gamma} >}{|\dot{\gamma}|^2} \dot{\gamma} \right|^2 = |\ddot{\gamma}|^2 - \frac{< \dot{\gamma}, \ddot{\gamma} >^2}{|\dot{\gamma}|^2} = \frac{|\dot{\gamma} \wedge \ddot{\gamma}|^2}{|\dot{\gamma}|^2}.$$

So we have

$$\frac{dL}{ds} = < \dot{\gamma}(1), \Lambda(1) > - < \dot{\gamma}(0), \Lambda(0) > + \int_0^1 < -\frac{|\dot{\gamma} \wedge \ddot{\gamma}|}{|\dot{\gamma}|^3} N, \Lambda > dt.$$

The space $\mathcal{V}$ is endowed with the inner product

$$< \Lambda_1, \Lambda_2 >= \int_a^b < \Lambda_1(t), \Lambda_2(t) > dt.$$

Since $\Lambda(1) = \Lambda(0) = 0$ and $k_\gamma = \frac{|\dot{\gamma} \wedge \ddot{\gamma}|}{|\dot{\gamma}|^3}$, the equation becomes

$$\frac{dL}{ds} |_{s=0} = \int_0^1 < -k_\gamma(t) N(t), \Lambda(t) > dt = < -k_\gamma N, \Lambda > .$$

Hence $dL_\gamma.\Lambda = < -k_\gamma N, \Lambda >$.

Theorem 9 *Let $p, q \in \mathbb{R}^2$. A curve $\gamma : [a, b] \to \mathbb{R}^2$ is a geodesic from Euclidean Geometry if and only if the trace of γ is a straight line.*

Proof Suppose $\gamma \in \Omega(p, q)$ is a critical point. Therefore $dL_\gamma.\Lambda = 0$ for all $\Lambda \in V$. This is equivalent to

$$< \nabla L(\gamma), \Lambda >= 0 \quad \text{for all } \Lambda \in V.$$

The du Bois-Reymond Lemma implies that $\nabla L(\gamma) = -k_\gamma N = 0$. Since $N \neq 0$, then $k_\gamma = 0$. Consequently, $\dot{\gamma} \wedge \ddot{\gamma} = 0$, and so one of the following cases may occur:

(a) $\ddot{\gamma} = 0$.
     In this case, we have $\gamma(t) = t(q - p) + p$. Hence the trace is a straight segment.
(b) We have a function $f : [0, 1] \to \mathbb{R}$ such that for all $t \in [0, 1]$, $\ddot{\gamma}(t) = f(t).\dot{\gamma}(t)$. The general solution of the ODE

$$\begin{cases} \ddot{\gamma} - f\dot{\gamma} = 0, \\ \gamma(0) = p, \\ \gamma(1) = q, \end{cases}$$

is given as

$$\gamma(t) = \frac{\int_0^t e^{\int_0^\sigma f(\tau)d\tau} d\sigma}{\int_0^1 e^{\int_0^\sigma f(\tau)d\tau} d\sigma}(q - p) + p.$$

Hence the trace of γ is a straight line in $\mathbb{R}^2$.

$\square$

6   Fredholm Maps

A differentiable map $f : E \to F$ between Banach spaces is a Fredholm map if $df_x : E \to F$ is a linear Fredholm operator for all $x \in E$. Since the Fredholm index is invariant by a continuous perturbation, the df_x index does not depend on x. Let $\mathrm{ind}(f)$ be the index of f. Some theorems for differentiable maps between finite dimensional spaces extend to the case when the map is Fredholm. A point $y \in F$ is a regular value of f if df_x is surjective and has a right-inverse for all $x \in f^{-1}(y)$; otherwise y is a critical value. Under these conditions, we will prove that $f^{-1}(y)$ is a differentiable $n = \mathrm{ind}(f)$ dimensional manifold.

Theorem 10 *Let E, F be Banach spaces and let $U \subset E$ be an open subset. If $f : U \to F$ is a Fredholm C^k-map and $y \in F$ is a regular value of f, then*

$$\mathcal{M} = f^{-1}(y) \subset X$$

is a C^k-differentiable Banach manifold and $T_x\mathcal{M} = \mathrm{Ker}(df_x)$ for all $x \in \mathcal{M}$. Moreover, $\dim(\mathcal{M}) = \mathrm{ind}(f)$ is finite.

The proof of the above theorem follows the arguments outlined in the Preamble of the Inverse Function Theorem. Again, the Contraction Lemma is the backbone of the proof, which is a task requiring a clear goal and also a lot of perseverance.

Before proceeding with the proof, we will discuss the strategy we will use. We consider $y = 0$ a regular value of f and $\mathcal{M} = f^{-1}(0)$. Let $x_0 \in U$ be a point, not necessarily in $\mathcal{M}$, such that df_{x_0} is a surjective Fredholm operator for which we define $E_0 = \mathrm{Ker}(df_{x_0})$. Then we have a closed subspace $E_1 \subset E$, such that $E = E_0 \oplus E_1$ and $df_{x_0} : E_1 \to F$ is an isomorphism. Let $Q_{x_0} = df_{x_0}^{-1} : F \to E_1$ be the linear operator given by $Q_{x_0}(y) = df_{x_0}^{-1}(y)$. Q_{x_0} is injective and is the right inverse of df_{x_0}. For this reason, for every $x \in E$, we have $\xi \in \mathrm{Ker}(df_0)$ and $\eta \in F$ such that

$$x = (x_0 + \xi) + Q_{x_0}(\eta).$$

For the following, we fix the point x_0, so that $f(x_0)$ is very small and define $df_0 = df_{x_0}$ and $Q_{x_0} = Q$. The goal is to show that we have open sets $V \subset \mathrm{Ker}(df_0)$, $W \subset \mathcal{M}$ and a diffeomorphism $\Phi : V \to W$. Consequently $\mathcal{M}$ is locally diffeomorphic to $\mathrm{Ker}(df_0)$. Given $x_1 = x_0 + \xi$, such that $f(x_1)$ is sufficiently small and $\xi \in \mathrm{Ker}(df_0)$, therefore the strategy is to prove the solution of $f(x) = 0$ in the equivalence class $x_1 + \mathrm{Im}(Q)$. There is no need for x_0 and x_1 to be solutions of $f(x) = 0$; we just need them to be near to a solution. For this purpose, let $U \subset E$ be an open subset containing x_0, such that $U \cap \mathcal{M} \neq \varnothing$, and consider the differentiable maps

$$h : U \to F \qquad\qquad\qquad \psi : U \to E$$
$$h(x) = f(x) - df_0(x - x_1) \,, \quad \psi(x) = x + Qh(x).$$

Proposition 10 *Let $f : U \to F$ be a C^k-map and let x_0 be such that $df_0 : E \to F$ is surjective. So $\psi(x) = x_1$ if and only if $f(x) = 0$ and $x - x_1 \in \mathrm{Im}(Q)$.*

Proof Let's assume $\psi(x) = x_1$, so

$$Q(f(x) - df_0(x - x_1)) = x_1 - x \;\Rightarrow\; x - x_1 \in \mathrm{Im}(Q).$$

Let $u \in F$ be such that $Q(u) = x - x_1$;

$$x_1 = \psi(x) = x + Q(f(x) - df_0 Q(u)) = x + Q(f(x) - u)$$
$$= x + Q(f(x)) - x + x_1 \;\Rightarrow\; Q(f(x)) = 0.$$

Therefore we have $f(x) = 0$ since Q is injective. The reverse assertion is trivial (Fig. 7).   □

We need to show that the equation $\psi(x) = x_1$ has a unique solution in the neighborhood of x_0. We take $\delta > 0$ and $c > 0$ such that $B_\delta(x_0) \subset U$, $\mid Q \mid \leq c$ and

$$x \in B_\delta(x_0) \;\Rightarrow\; \| \, df_x - df_0 \, \| \leq \frac{1}{2c}.$$

For every $y \in U$, we consider the map $\phi_y : B_\delta(x_0) \to E$, $\phi_y(x) = x + (y - \psi(x))$. Then

$$\mid d(\phi_y)_x \mid \, = \mid I - d\psi_x \mid \, = \mid Q.(df_x - df_{x_0} \mid \, \leq \frac{1}{2},$$

for all $x \in B_\delta(x_0)$. Applying the Mean Value inequality, it follows that ϕ_y is a contraction since

$$\| \, \phi_y(x_2) - \phi_y(x_1) \, \| \leq \frac{1}{2} \, \| \, x_2 - x_1 \, \| . \tag{30}$$

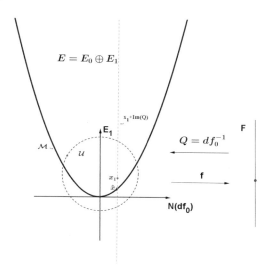

Fig. 7 $f : E \to F$ is Fredholm

The map $\psi : B_\delta(x_0) \to E$ is injective since

$$\| \psi(x_2) - \psi(x_1) \| = \| (x_2 - x_1) - (\phi_y(x_2) - \phi_y(x_1)) \| \geq \| \| x_2 - x_1 \| - \| \phi(x_2) - \phi(x_1) \| \| \geq$$

$$\geq \| \| x_2 - x_1 \| - \frac{1}{2} \| x_2 - x_1 \| \| \geq \frac{1}{2} \| x_2 - x_1 \|,$$

$$(31)$$

for any $x_1, x_2 \in B_\delta(x_0)$. By restricting the domain of ϕ_y to $B_\delta = B_\delta(x_0)$, we have the map $\psi : B_\delta \to U_\delta = \psi(B_\delta)$. Therefore we assert $\psi(x) = y$ has a solution in B_δ for $y \in U_\delta$.

Lemma 3 *Let* $y_0 \in U_\delta$ *and* $y_0 = \psi(x_0)$. *Consider* $0 < \epsilon \leq \delta$ *such that* $\overline{B_\epsilon(x_0)} \subset B_\delta(x_0)$ *and* $B_{\epsilon/2}(y_0) = \{y \in E; \| y - y_0 \| < \epsilon/2\}$. *So*
(i) $U_\delta \subset E$ *is open. Indeed, if* $y \in B_{\epsilon/2}(y_0)$, *then* $y \in U_\delta$.
(i) $\psi : B_\delta(x_0) \to U_\delta$ *is a diffeomorphism.*

Proof (i) For any $x \in B_\epsilon(x_0)$ and $y \in B_{\epsilon/2}(y_0)$, we have

$$\| \phi_y(x) - x_0 \| \leq \| \phi_y(x) - \phi_y(x_0) \| + \| \phi_y(x_0) - x_0 \| \leq$$

$$\leq \frac{1}{2} \| x - x_0 \| + \| y - y_0 \| \leq \epsilon.$$

So $\phi_y(x) \in B_\epsilon(x_0)$ and $\phi_y(B_\epsilon(x_0)) \subset \overline{B_\epsilon(x_0)}$. Since $\overline{B_\epsilon(x_0)} \subset E$ is a complete metric space, it follows from the Contraction Lemma that $\phi_y : \overline{B_\epsilon(x_0)} \to \overline{B_\epsilon(x_0)}$ has a single fixed point. Let $x_y \in \overline{B_\epsilon(x_0)}$ be the fixed point of ϕ_y. Therefore $y_0 \in U_\delta$ since $\psi(x_y) = y$. Moreover, since $B_{\epsilon/2}(y_0) \subset U_\delta$, the subset U_δ is open.

(ii) In particular, $\phi_y : \overline{B_\delta(x_0)} \to \overline{B_\delta(x_0)}$ is a contraction. Due to the uniqueness of the fixed point, the map $\psi : B_\delta \to U_\delta$ is injective. Also, we have $d\psi_{x_0} = I$. Hence $\psi : B_\delta \to U_\delta$ is a diffeomorphism. $\qquad\qquad\square$

Theorem 11 *Let $x_0 \in \mathcal{M} = f^{-1}(0)$ and suppose that df_{x_0} is surjective. Let δ and c be such that $B_\delta(x_0) \subset U$, $\mid Q \mid \le c$ and*

$$x \in B_\delta(x_0) \Rightarrow \| df_x - df_{x_0} \| \le \frac{1}{2c}.$$

If $x_1 \in E$ satisfies $\| x_1 - x_0 \| < \frac{\delta}{8}$ and $\| f(x_1) \| \le \frac{\delta}{4c}$, then we have a unique $\hat{x}_1 \in U_\delta$ such that
(i) $f(\hat{x}_1) = 0$, and $\hat{x}_1 - x_1 \in \mathrm{Im}(Q)$.
(ii) $\| \hat{x}_1 - x_1 \| \le 2c \| f(x_1) \|$.

Proof (i) We need to check that $x_1 \in U_\delta$. Since $\psi(x_1) = x_1 + Qf(x_1)$, we have

$$\| x_1 - \psi(x_0) \| = \| \psi(x_1) - \psi(x_0) - Qf(x_1) \| \le \| \psi(x_1) - \psi(x_0) \| + \mid Q \mid . \| f(x_1) \|$$
$$\le 2 \| x_1 - x_0 \| + c. \| f(x_1) \| < \frac{\delta}{2}.$$

So $x_1 \in B_{\delta/2}(\psi(x_0)) \subset U_\delta$. From Lemma 3 we have $\hat{x}_1 \in B_\delta(x_0)$ such that $f(\hat{x}_1) = 0$ and $\hat{x}_1 - x_1 \in \mathrm{Im}(Q)$.

(ii) It follows from the Inequality (31) that

$$\frac{1}{2} \| \hat{x}_1 - x_1 \| \le \| (\hat{x}_1 - x_1) - (\phi(\hat{x}_1) - \phi(x_1)) \| = \| \psi(\hat{x}_1) - \psi(x_1) \| .$$

Since $\psi(\hat{x}_1) = x_1$, $h(x_1) = f(x_1)$, we have

$$\| \hat{x}_1 - x_1 \| \le 2 \| \psi(\hat{x}_1) - \psi(x_1) \| \le 2c \| f(x_1) \| .$$

$\qquad\qquad\square$

Therefore for every $x \in B_{\delta/2}(x_0)$, we have a unique element $\hat{x} \in B_\delta(x_0)$ such that $\psi(\hat{x}) = x$, or equivalently, such that $f(\hat{x}) = 0$ and $\hat{x} - x \in \mathrm{Im}(Q)$. Let $\xi \in \mathrm{Ker}(df_0)$ be such that $\| \xi \| < \delta/2$ and $x_\xi = x_0 + \xi$. We have a unique $\hat{x}_\xi \in B_\delta$ such that $\psi(\hat{x}_\xi) = x_\xi$, and so $f(\hat{x}_\xi) = 0$ and $\hat{x}_\xi - x_\xi \in \mathrm{Im}(Q)$. Let $\eta : \mathrm{Ker}(df_0) \to E_1$ be the map

$$\eta(\xi) = \hat{x}_\xi - x_\xi.$$

The map $\sigma = Q^{-1}\eta : \mathrm{Ker}(df_0) \to F$ is well-defined since $Q : E_1 \to F$ is an isomorphism. Let $\mathcal{F} : \mathrm{Ker}(df_0) \to \mathcal{M}$ be the local C^k-map

$$\xi \overset{\mathcal{F}}{\to} \hat{x}_\xi = x_0 + \xi + Q\sigma(\xi) = x_\xi + Q\sigma(\xi).$$

Proposition 11 $d\mathcal{F}_0 = I$.

Proof The differential of $\mathcal{F}$ at $\xi = 0$ is

$$d\mathcal{F}_0 = I + Qd\sigma_0.$$

Since $x_\xi = \psi(\hat{x}_\xi) = \hat{x}_\xi + Qh(\hat{x}_\xi) = \hat{x}_\xi - df_0(\hat{x}_\xi - x_\xi)$ and $df_0(x_\xi - x_0) = 0$, we have

$$x_\xi = \hat{x}_\xi - Qdf_0(\hat{x}_\xi - x_\xi) + Qdf_0(x_\xi - x_0) = \hat{x}_\xi + Qdf_0(\hat{x}_\xi - x_0) \Rightarrow$$
$$\hat{x}_\xi = x_\xi - Qdf_0(\hat{x}_\xi - x_\xi) = x_\xi - Qh(\hat{x}_\xi),$$

and so $\sigma(\xi) = -h(\hat{x}_\xi)$. Therefore

$$\sigma(\xi) = -h(\hat{x}_\xi) = -f(\hat{x}_\xi) + df_0(\hat{x}_\xi - x_\xi) = df_0(\hat{x}_\xi) - df_0(x_\xi)$$
$$= df_0\psi^{-1}(x_0 + \xi) - df_0(x_\xi).$$

Consequently the differential of σ at $\xi = 0$ is

$$d\phi_0 = df_0.d\psi_0^{-1} - df_0 = 0 \; (d\psi_0 = I).$$

Hence $d\mathcal{F}_0 = I$. $\qquad\square$

Given the InFT, there are opens sets $V \subset \mathrm{Ker}(df_0)$, $W \subset \mathcal{M} \cap B_{\delta/2}$ such that $\mathcal{F} : V \to W$ is a diffeomorphism. Identifying $\mathcal{M} \cap B_{\delta/2}$ with the graph $\xi \overset{\Sigma}{\mapsto} (\xi, Q\sigma(\xi))$, we get that $\mathcal{M}$ is a differentiable manifold and $\dim(\mathcal{M}) = \dim(\mathrm{Ker}(df_0))$.

The manifold $\mathcal{M}$ can be locally described as $\hat{f}^{-1}(0)$, with $\hat{f} : \mathrm{Ker}(df_0) \to \mathrm{CoKer}(df_0)$ a differentiable map between finite dimensional spaces.

Proposition 12 (Kuranishi model) *Let $f : E \to F$ be a Fredholm map, $0 \in F$ a regular value of f and $f(0) = 0$. There are open sets $\mathcal{U}_0 \subset \mathrm{Ker}(df_0)$ and $\mathcal{V}_0 \subset \mathrm{CoKer}(df_0)$, $0 \in \mathcal{V}_0$, and a differentiable map $\hat{f} : \mathcal{U}_0 \to \mathcal{V}_0 \subset \mathrm{CoKer}(df_0)$ such that $\mathcal{M} \cap \mathcal{U}_0 = \hat{f}^{-1}(0)$.*

Proof Let $df_0 : E \to F$ be the differential of f at $x = 0$ and let $P : F \to E$ be the pseudo-inverse of df_0. Let $U \subset E$ and let $\vartheta : U \to E$ be the differentiable map $\vartheta(x) = x + P(f(x) - df_0(x))$ such that $\vartheta(0) = 0$ and $d\vartheta_0 = I$. By the Inverse Function Theorem, there are open sets $V_0, U_0 \subset E$ such that $\vartheta : V_0 \to U_0 = \vartheta(V_0)$ is a diffeomorphism. Now we define the differentiable maps $\rho = \vartheta^{-1} : U_0 \to V_0$ and

$$\hat{f} : U_0 \rightarrow \mathrm{CoKer}(df_0),$$
$$\hat{f}(u) = (I - df_0.P) \circ f \circ \rho(u). \tag{32}$$

The operator $df_0.P : F \rightarrow \mathrm{Im}(df_0)$ is a projection according to Proposition 11, Chap. 2. Now we have $f \circ \rho : U_0 \rightarrow F$ as follows:

$$f \circ \rho(u) = f \circ \rho(u) - df_0 \circ P \circ f \circ \rho(u) + df_0 \circ P \circ f \circ \rho(u)$$
$$= \hat{f}(u) + df_0 \circ P \circ f \circ \rho(u).$$

Since $P.df_0.P = P$, we have $P(\hat{f}(u)) = 0$, and so $\mathrm{Im}(\hat{f}) \subset \mathrm{CoKer}(df_0)$. The expression

$$df_0(\vartheta(x)) = df_0(x) + df_0.P[f(x) - df_0)(x)] = df_0.P(f(x))$$

yields $df_0 = df_0.P \circ f \circ \vartheta^{-1}$, and so

$$f \circ \rho = f \circ \vartheta^{-1} = (I - df_0.P) \circ f \circ \vartheta^{-1} + df_0 = \hat{f} + df_0.$$

Considering the open sets $\mathcal{U}_0 = U_0 \cap \mathrm{Ker}(df_0)$ and $\mathcal{V}_0 = V_0 \cap \mathrm{CoKer}(df_0)$, the map $\hat{f}$ is equal to the restriction $\hat{f} = f : \mathcal{U}_0 \rightarrow \mathcal{V}_0$. Taking $x = \rho(u)$, we have

$$f(x) = \hat{f}(u) + df_0(u) = 0 \;\Leftrightarrow\; \hat{f}(u) = 0 \text{ and } df_0(u) = 0.$$

So $x \in \mathcal{M}$ if and only if $\hat{f}(u) = 0$ and $u \in \mathcal{U}_0$. Hence $\hat{f} : \mathcal{U}_0 \rightarrow \mathcal{V}_0$ is a map between finite dimensional spaces and

$$\mathcal{M} = \hat{f}^{-1}(0).$$

$\square$

6.1 Final Comments and Examples

In Differential Topology, there are many results with a proof that relies on the properties of differentiable maps on differentiable manifolds. In the infinite dimensional case, Fredholm maps form the category for which some of the techniques are extended. Some properties of the Fredholm maps are listed below; the proofs can be found in Ref. [43].

The index of a Fredholm operator is computed in many examples applying the Atiyah-Singer Theorem, one of the most profound and broad theorems in mathematics.

In the following theorems, let M and N be a second countable Banach manifold, and assume M is connected. Therefore the index of a Fredholm map $f : M \to N$ does not depend on the choice of $p \in M$. A point $y \in N$ is said to be a regular value of f if the linear map $df_x : T_x M \to T_y N$ is surjective for all $x \in f^{-1}(y)$; otherwise, y is a critical value. The first interesting result that extended to Banach manifolds is the Sard-Smale theorem which extends the Sard Theorem in the finite dimensional context according to the following statement.

Theorem 12 (Sard-Smale) *Let E and F be separable Banach spaces and let $U \subset E$ an open subset. Assume $f : U \to F$ is a Fredholm C^k-map, and $k \geq \max\{1, \text{index}(f) + 1\}$. So the set of regular values*

$$F_{\text{reg}}(f) = \{y \in F \mid x \in U, \ y = f(x), \ df_x \ surjective\}$$

is a Baire 2nd category set in F.

The theorem extends to Banach manifolds.

Theorem 13 *Suppose that $f : M \to N$ is a C^k-map between Fredholm Banach manifolds, $k \geq \max\{1, \text{index}(f) + 1\}$. Then the regular values are a Baire 2nd category set in N or the critical values are 1st category. If f is proper (i.e., K compact $\Rightarrow f^{-1}(K)$ compact), then the regular values are open and dense.*

Theorem 14 (Domain Invariance) *Let $f : M \to N$ be a locally injective Fredholm C^k-map such that $\text{ind}(f) = 0$. So f is an open map.*

The following theorem can be interpreted as the nonlinear version of the Fredholm alternative;

Theorem 15 *Let $f : M \to N$ be a proper, locally injective Fredholm C^k-map such that $\text{ind}(f) = 0$. If N is connected, then f is a covering map.*

Example 4 The following examples are seminal examples of applications to Gauge Theory and to the study of invariants of 4-manifolds. Although they require information beyond the scope of this text, we will make a brief presentation for illustrative purposes. These examples will become more understandable after the study of differential forms and Maxwell's equations in Chap. 7.

(1) **Moduli Spaces of Instantons** [9].

Let (M, g) be a closed Riemannian 4-manifold, $\pi : P_k \to M$ a principal bundle on M with structural group $SU(2)$ and $c_2(\xi) = k$. Let $\Omega^p(\mathfrak{su}_2)$ be the space of differential forms taking values in the Lie algebra $\mathfrak{su}_2$. Consider $\mathcal{A}$ the affine space of the connections defined over P. Fixing a connection $A_0 \in \mathcal{A}$, we have $\mathcal{A} - A_0 = \Omega^1(\mathfrak{su}_2)$. For every connection $A \in \mathcal{A}$, we associate the exterior derivative operator $d_A : \Omega^p(\mathfrak{su}_2) \to \Omega^{p+1}(\mathfrak{su}_2)$ and its curvature to be

$$F_A = d_A \circ d_A : \Omega^0(\mathfrak{su}_2) \to \Omega^2(\mathfrak{su}_2).$$

Every connection A can be identified with a 1-form in $\Omega^1(\mathfrak{su}_2)$; its curvature is a 2-form $F_A \in \Omega^2(\mathfrak{su}_2)$. In dimension 4, Hodge's star-operator $*$: $\Omega^2(\mathfrak{su}_2) \to \Omega^2(\mathfrak{su})$ satisfies $*^2 = \mathrm{id}$. Therefore the $*$-eigenvalues are ± 1 and the corresponding eigenspaces $\Omega^2_\pm(\mathfrak{su}_2)$ induce the decomposition $\Omega^2(\mathfrak{su}_2) = \Omega^2_+(\mathfrak{su}_2) \oplus \Omega^2(\mathfrak{su}_2)_-$. The curvature 2-form decomposes as $F_A = F_A^+ + F_A^-$, and

$$*F_A^+ = F_A^+, \quad *F_A^- = -F_A^-.$$

F_A^+ is the self-dual component and F_A^- is the anti-self-dual component. The L^2-energy of the curvature, known in Physics as the Yang-Mills functional, is

$$YM(A) = \int_M \| F_A \|^2 .$$

The Yang-Mills equation, which is the Euler-Lagrange equation of the functional is

$$*d_A * F_A = 0.$$

The gauge group $\mathcal{G} = \{g : P \to P \mid \pi \circ g = g\}$ is the group of the fiber-preserving automorphism of the bundle P. The gauge group acts on the spaces of connections and 2-forms as follows;

$$\mathcal{G} \times \mathcal{A} \to \mathcal{A} \qquad\qquad \mathcal{A} \times \Omega^2(\mathfrak{su}_2) \to \Omega^2(\mathfrak{su}_2)$$
$$g.A = g^{-1}dg + g^{-1}Ag \, , \quad g.\omega = g^{-1}.\omega.g.$$

The connections with the trivial isotropy group are irreducible and the reducible are those for which we have a $g \in \mathcal{G}$, such that $g \neq I$ and $g.A = A$. The isotropic group of a reducible connection is isomorphic to U_1. On the spaces $\mathcal{A}$, $\mathcal{G}$ and $\Omega^p(\mathfrak{su}_2)$, we consider the topology induced by the $L^{k,2}$-norms

$$\mathcal{A} = L^{1,2}(\mathcal{A}), \ \mathcal{G} = L^{2,2}(\mathcal{G}), \ \Omega^p(\mathfrak{su}_2) = L^{1,2}(\Omega^p(\mathfrak{su}_2)).$$

We see that these are Hilbert manifolds and the actions are differentiable. The Yang-Mills functional is invariant by the $\mathcal{G}$ action. Since $YM(A) < \infty$ for all $A \in \mathcal{A}$, the functional is well-defined. The moduli space $\mathcal{B}_k = \mathcal{A}_k/\mathcal{G}$ is a differentiable Banach manifold in a neighborhood of an irreducible point $[A]$. Due to the Bianchi identity $d_A F_A = 0$, if the curvature F_A of a connection A satisfies either $F_A = F_A^+$ or $F_A = F_A^-$, then A is a solution to the Yang-Mills equation. The space of the instantons $\mathcal{M}_k \subset \mathcal{B}_k$ is the space of the anti-self-dual connections (solutions), that is,

$$\mathcal{M}_k = \{A \in \mathcal{A} \mid F_A = F_A^-\}/\mathcal{G}.$$

Defining the map

$$F^+ : \Omega^1(\xi) \to \Omega^2_+(\xi)$$
$$A \to F^+_A,$$

we have $\mathcal{M}_k = (F^+)^{-1}(0)/\mathcal{G}$. Up to the gauge action, F^+ defines a differentiable Fredholm map. To avoid technical issues and to keep the exposition accessible, we assume A is an irreducible connection. To define the differential $d(F^+)_A$ of F^+ at A, we consider the projection $\pi_+ : \Omega^2(\mathfrak{su}_2) \to \Omega^2_+(\mathfrak{su}_2)$,

$$d(F^+)_A = d^+_A = \pi^+ \circ d_A : \Omega^1(\mathfrak{su}_2) \to \Omega^2_+(\mathfrak{su}_2).$$

So the following items are valid; (see [9]).
(i) Let $d_A : \Omega^0(\mathfrak{su}_2) \to \Omega^1(\mathfrak{su}_2)$ be the exterior derivative; A is a reducible point, i.e., fixed by the $\mathcal{G}$-action if and only if $\mathrm{Ker}(d_A) \neq 0$. The tangent space to the orbit $\mathcal{G}.A$ at A is $T_A\mathcal{G}.A = \mathrm{Im}(d_A)$.
(ii) The tangent plane $T_{[A]}\mathcal{M}_k$ is $\mathrm{Ker}(d^+_A)/\mathrm{Im}(d_A)$.
(iii) If $\Omega^2_+(\mathfrak{su}_2)/\mathrm{Im}(d_A) = 0$, then the differential $d(F^+)_A$ is surjective.
Considering the diagram

$$0 \longrightarrow \Omega^0(\mathfrak{su}_2) \xrightarrow{\ d_A\ } \Omega^1(\mathfrak{su}_2) \xrightarrow{\ d^+_A\ } \Omega^2_+(\mathfrak{su}_2),$$

we define the vector spaces

$$H^0_A = \mathrm{Ker}(d_A),\ H^1_A = \mathrm{Ker}(d_A)/\mathrm{Im}(d_A),\ H^+_A = \Omega^2(\mathfrak{su}_2)/\mathrm{Im}(d_A).$$

We now have that $\mathcal{M}_k$ is a differentiable manifold in the neighborhood of A if $H^0_A = \mathrm{Ker}(d_A) = \{0\}$, $T_{[A]}\mathcal{M}_k = H^1_A$ and $H^+_A = \{0\}$. Using the Hilbert space structure of $\Omega^1(\mathfrak{su}_2)$, we have that the complement of $\mathrm{Im}(d_A)$ in $\Omega^1(\mathfrak{su}_2)$ is $\mathrm{Ker}(d^*_A)$, and $d^*_A : \Omega^1(\mathfrak{su}_2) \to \Omega^0(\mathfrak{su}_2)$. That is, the tangent space to $\mathcal{M}_k$ at A is

$$H^1_A = \{\omega \in \Omega^1(\mathfrak{su}_2) \mid d^*_A\omega = 0,\ d^+_A\omega = 0\}.$$

However, the operator $d(F^+)_A$ restricted to $T_{[A]}\mathcal{M}_k$ is a Fredholm operator. Further, it follows from the Sard-Smale Theorem that under the condition $b^+_2 = \dim(H^2_+(M)) \geq 1$, we can perturb the Riemannian metric g defined in M to obtain $H^+_A = \{0\}$. In this way, the dimension of $\mathcal{M}_k$ is equal to the index of $d(F^+)_A$, which, by the Atiyah-Singer theorem, is

$$\dim(\mathcal{M}_k) = 8k - 3(1 - b_1 + b^+_2).$$

The integers $b_1 = \dim(H^1(M; \mathbb{R}))$, $b^+_2 = \dim(H^2_+(M; \mathbb{R}))$ are topological invariants of M. Using Kuranishi's model, we have $\mathcal{M}_k = \hat{f}^{-1}(0)$, and

$$\hat{f} : \mathrm{Ker}(d_A) \to \mathrm{CoKer}(d^+_A).$$

(2) **Seiberg-Witten Monopoles Moduli Space** [40].

Let $(M, g, \mathfrak{c})$ be a closed Riemannian 4-manifold with metric g and endowed with a Spin^c structure $\mathfrak{c}$. Associated to $\mathfrak{c}$ we have a pair of vector bundles $(\mathcal{L}(E)_{\mathfrak{c}}, S_{\mathfrak{c}})$;

(i) $\mathcal{L}(E)_{\mathfrak{c}}$ is a complex line bundle with the 1st Chern class $c_1(\mathcal{L}(E)_{\mathfrak{c}}) = \alpha$,

(ii) $S_{\mathfrak{c}}$ is a complex spinorial bundle with fibers isomorphic to $\mathbb{C}^4$, they are $\mathbb{C}l_4$-modules.[3] The spinor bundle $S_{\mathfrak{c}}$ admits a decomposition in vector sub-bundles

$$S_{\mathfrak{c}} = S_{\mathfrak{c}}^+ \oplus S_{\mathfrak{c}}^- .$$

$S_{\mathfrak{c}}^{\pm}$ are the positive and negative spinor bundles, respectively, with fibers that are isomorphic to $\mathbb{C}^2$. Let $\mathcal{A}_{\mathfrak{c}}$ be the space of connections defined on $\mathcal{L}(E)_{\mathfrak{c}}$ and let $\Gamma_{\mathfrak{c}}^{\pm} =$ be the space of sections of each bundle $S_{\mathfrak{c}}^{\pm}$, respectively. The configuration space is $\mathcal{C}_{\mathfrak{c}} = \mathcal{A}_{\mathfrak{c}} \times \Gamma_{\mathfrak{c}}^+$. In Seiberg-Witten theory, the gauge group $\mathcal{G} = \mathrm{Map}(M; U_1)$ is an abelian group and acts as

$$\mathcal{G} \times \mathcal{C}_{\mathfrak{c}} \to \mathcal{C}_{\mathfrak{c}}$$
$$g.(A, \phi) = (A + 2g^{-1}dg, g^{-1}.\phi).$$

For every $g \in \mathcal{G}$, we have $g.(A, \phi) = (A, \phi)$ if and only if $\phi = 0$ and $g : M \to U_1$ is constant. Therefore every point (A, ϕ), $\phi \neq 0$ is irreducible, while every point $(A, 0)$ is reducible. The isotropic group of a reducible point is isomorphic to the group U_1. The moduli space is $\mathcal{B}_{\mathfrak{c}} = \mathcal{C}_{\mathfrak{c}}/\mathcal{G}$. With a connection $A \in \mathcal{A}_{\mathfrak{c}}$ and a local frame $\{e_1, \ldots, e_4\}$ on M, we have the positive linear Dirac operator $D_A^+ :$ $\Gamma_{\mathfrak{c}}^+ \to \Gamma_{\mathfrak{c}}^-$ acting in the space of the sections. In 1994, Ed. Witten introduced the system of 1st-order coupled equations,

$$\begin{aligned} D_A^+\phi &= 0, &\quad (2.1) \\ F_A^+ &= \sigma(\phi), &\quad (2.2) \end{aligned} \tag{33}$$

in which F_A^+ is the self-dual component of the curvature 2-form F_A and $\sigma :$ $\Gamma_{\mathfrak{c}}^+ \to \Omega_+^2(i\mathbb{R})$ is the self-dual 2-form

$$\sigma(v)(X, Y) = <X.Y.v, v> + \frac{1}{2} <X, Y>|v|^2$$

coupling the connection A to the spinor ϕ. The Seiberg-Witten map is

$$\mathcal{F} : \mathcal{A} \times \Gamma_{\mathfrak{c}}^+ \to \Omega_+^2(\mathfrak{u}_1) \times \Gamma_{\mathfrak{c}}^-$$
$$\mathcal{F}(A, \phi) = \left(F_A^+ - \sigma(\phi), D_A^+\phi\right). \tag{34}$$

The Seiberg-Witten Monopole moduli space is $\mathcal{M}_{\mathfrak{c}} = \mathcal{F}^{-1}(0)/\mathcal{G}$. Let $T_{(A,\phi)}\mathcal{G}$ be the tangent plane to the orbit $\mathcal{G}.(A, \phi)$ and $T_{(A,\phi)}\mathcal{C}_{\mathfrak{c}}$ the tangent plane to $\mathcal{C}_{\mathfrak{c}}$

---

[3] $\mathbb{C}l_4$ is a Clifford Algebra.

at the point (A, ϕ). The differential of $\mathcal{F}$ at (A, ϕ) defines the linear operator $d\mathcal{F}_{(A,\phi)} : \Omega^1(\mathfrak{u}_1) \times \Gamma_c^+ \to \Omega^1(\mathfrak{u}_1) \times \Gamma_c^-$. Using a diagram, we get

$$0 \longrightarrow \Omega^0(\mathfrak{u}_1) \xrightarrow{d_1} \Omega^1(\mathfrak{u}_1) \times \Gamma_c^+ \xrightarrow{d\mathcal{F}_{(A,\phi)}} \Omega_+^2(\mathfrak{u}_1) \times \Gamma_c^-.$$

Likewise in the Instantons example, we introduce the vector spaces $H^0_{(A,\phi)} = \text{Ker}(d_1)$, $H^1_{(A,\phi)} = \text{Ker}(d\mathcal{F}_{(A,\mathbb{E})})/\text{Im}(d_1)$ and $H^+_{(A,\phi)} = \frac{\Omega_+^2(\mathfrak{u}_1) \times \Gamma_c^+}{\text{Im}(d\mathcal{F}_{(A,\mathbb{E})})}$. Analogously,
(i) (A, ϕ) is irreducible if and only if $H^0_{(A,\phi)} = 0$.
(ii) The tangent plane to $\mathcal{M}_c$ at (A, ϕ) is $H^1_{(A,\phi)}$.
(iii) The differential $d\mathcal{F}_{(A,\phi)}$ is surjective if and only if $H^+_{(A,\phi)} = 0$.

Theorem 16 *If $b_2^+(M) \geq 1$, then the space $\mathcal{M}_c$ is either empty or it is a differentiable manifold for a generic metric g. Moreover, $\mathcal{M}_c$ is compact and orientable with dimension*

$$\dim(\mathcal{M}_c) = \frac{1}{2}\left[c_1^2(\mathcal{L}(E)_c) - 2\chi(M) - 3\sigma(M)\right], \tag{35}$$

$\chi(M)$ is the Euler characteristic of M and $\sigma(M)$ is the signature of the intersection form of M.

7 An Application of the Inverse Function Theorem to Geometry

In the previous section, we introduced the concept of geodesic, and showed how to find a geodesic connecting two points p, q by solving a variational problem. When we consider an open subset $U \subset \mathbb{R}^n$ endowed with a Riemannian metric g, it is very difficult to find a geodesic connecting two points $p, q \in U$. In this section, we will prove the existence of a geodesic if p, q are non-conjugate points, which means they are close enough. The proof is an application of InFT.

We remind the reader that the Riemannian metric g induces an internal product $g : T_xU \times T_xU \to \mathbb{R}$ on the tangent plane $T_xU \simeq \mathbb{R}^n$. Once a frame $\beta = \{e_i \mid 1 \leq i \leq n\}$ is fixed on U, the entries of the matrix associated to the metric g, with respect to the frame β, are functions $g_{ij} : U \to \mathbb{R}$ given by $g_{ij}(x) = g_x(e_i, e_j)$ ($g_{ij} = g_{ji}$). The tangent vector to a curve $\gamma : [a, b] \to U$, $\gamma(t) = (\gamma_1(t), \ldots, \gamma_n(t))$, at the instant t, is $\dot{\gamma}(t) = (\dot{\gamma}_1(t), \ldots, \dot{\gamma}_n(t)) = \sum_{ij} \dot{\gamma}_i e_i$. The length of γ is

$$L(\gamma) = \int_a^b \sqrt{g(\dot{\gamma}(t)), g(\dot{\gamma}(t))} = \int_a^b \sqrt{\sum_{i,j} g_{ij}(t)\dot{\gamma}_i(t)\dot{\gamma}_j(t)},$$

$g_{ij}(t) = g_{\gamma(t)}(e_i, e_j)$. The Euler-Lagrange equations are

$$\frac{d\gamma_k^2}{dt^2} + \sum_{i,j} \Gamma_{ij}^k(\gamma(t))\frac{d\gamma_i}{dt}\frac{d\gamma_j}{dt} = 0, \; k = 1, \ldots, n.$$

The functions $\Gamma_{ij}^k : U \to \mathbb{R}$ are Christoffel's symbol of the metric g. They are given by

$$\Gamma_{ij}^m = \frac{1}{2}\sum_k \left(\frac{\partial g_{jk}}{\partial x_i} + \frac{\partial g_{ki}}{\partial x_j} - \frac{\partial g_{ij}}{\partial x_k}\right).g^{km}; \; i, j, k \in \{1, \ldots, n\}.$$

The Banach space $C^2([a,b]^n) = \{f : [a,b]^n \to \mathbb{R}^n \mid f \in C^2\}$ is endowed with the norm,

$$\| f \|_{C^2} = \sup_{0 \le i \le 2} \max_{t \in [a,b]} \left| \frac{d^i f}{dt^i}(t) \right|.$$

On the open subset $\tilde{U} = \{f \in C^2([a,b]^n) \mid f([a,b]) \subset U\}$ of $C^2([a,b]^n)$, consider the map

$$\mathcal{G} : \tilde{U} \to C^0([a,b]^n)$$

$$\gamma \to \left(\frac{d\gamma_1^2}{dt^2} + \sum_{i,j}\Gamma_{ij}^1(\gamma(t))\frac{d\gamma_i}{dt}\frac{d\gamma_j}{dt}, \; \ldots, \; \frac{d\gamma_n^2}{dt^2} + \sum_{i,j}\Gamma_{ij}^n(\gamma(t))\frac{d\gamma_i}{dt}\frac{d\gamma_j}{dt}\right).$$

The differentiable map that will matter to solve our question is the following;

$$\mathcal{P} : \tilde{U} \to C^0([a,b]^n) \times \mathbb{R}^n \times \mathbb{R}^n$$

$$\gamma \to (\mathcal{G}(\gamma), \gamma(a), \gamma(b)).$$

With respect to the frame β, we have

$$\mathcal{G}(\gamma) = \sum_\lambda \left(\frac{d^2\gamma_\lambda}{dt^2} + \sum_{i,j}\Gamma_{ij}^\lambda(\gamma(t))\frac{d\gamma_i}{dt}\frac{d\gamma_j}{dt}\right)e_\lambda = \sum_\lambda \mathcal{G}_\lambda e_\lambda.$$

In local coordinates, the differential $d\mathcal{G}_\gamma : C^2([a,b]^n) \to C^0([a,b]^n)$ is given by

$$(d\mathcal{G}_\gamma.h)^\lambda = \frac{d^2 h^\lambda}{dt^2} + 2\sum_{i,j}\Gamma_{ij}^\lambda(\gamma)\frac{d\gamma_i}{dt}\frac{dh_j}{dt} + \sum_l\sum_{i,j}\frac{\partial\Gamma_{ij}^\lambda}{\partial x_l}\frac{d\gamma_i}{dt}\frac{d\gamma_j}{dt}h^l.$$

Let's consider the decomposition

$$(d\mathcal{G}_\gamma.h)^\lambda = D_2 h^\lambda + K_1^\lambda(h) + K_2^\lambda(h) \tag{36}$$

and

$$D_2 h^\lambda = \frac{d^2 h^\lambda}{dt^2}, \quad K_1^\lambda(h) = \sum_{i,j} \Gamma_{ij}^\lambda(\gamma) \frac{d\gamma_i}{dt} \frac{dh_j}{dt}, \quad K_2^\lambda(h) = \sum_l \sum_{i,j} \frac{\partial \Gamma_{ij}^\lambda}{\partial x_l} \frac{d\gamma_i}{dt} \frac{d\gamma_j}{dt} h^l.$$

The differential of $\mathcal{P} : \tilde{U} \to C^0([a,b]^n) \times \mathbb{R}^n \times \mathbb{R}^n$ at γ is

$$d P_\gamma : C^2([a,b]^n) \to C^0([a,b]^n) \times \mathbb{R}^n \times \mathbb{R}^n,$$
$$d P_\gamma . h = (d\mathcal{G}_\gamma . h, h(a), h(b))$$
$$d P_\gamma . h = D_2 h^\lambda + K_1^\lambda(h) + K_2^\lambda(h)$$

and

$$\mathcal{D}_2 h^\lambda = (D_2 h^\lambda, h(a), h(b))$$

$$K_1^\lambda(h) = \left(\sum_{i,j} \Gamma_{ij}^\lambda(\gamma) \frac{d\gamma_i}{dt} \frac{dh_j}{dt}, 0, 0 \right), \quad K_2^\lambda(h) = \left(\sum_l \sum_{i,j} \frac{\partial \Gamma_{ij}^\lambda}{\partial x_l} \frac{d\gamma_i}{dt} \frac{d\gamma_j}{dt} h^l, 0, 0 \right).$$

We will show that the operator $\mathcal{D}_2$ is a Fredholm operator with index 0 and the operators K_1^λ and K_2^λ are compact. Once this is done, we will conclude that the operator $d P_\gamma$ is a Fredholm operator with index 0.

(i) $\mathcal{D}_2$ is a Fredholm operator with index 0:
   The assertion follows from the following items;

   (a) $\text{Ker}(\mathcal{D}_2) = \{0\}$. Let $h \neq 0 \in \text{Ker}(\mathcal{D}_2)$, so $h''(t) = 0$ and $h(a) = h(b) = 0$. Therefore $h'(t) = u_0 \in \mathbb{R}^n$ and $h(t) = u_0 t + u_1, u_0, u_1 \in \mathbb{R}^n$. Since $h(a) = h(b) = 0$, then $u_0 = u_1 = 0$.

   (b) $\mathcal{D}_2 : C^2([a,b]^n) \to C^0([a,b]^n) \times \mathbb{R}^n \times \mathbb{R}^n$ is surjective, i.e., $\text{CoKer}(\mathcal{D}_2) = \{0\}$.
   Let $(g, p, q) \in \tilde{U} \times \mathbb{R}^n \times \mathbb{R}^n$ and define

$$h(t) = p + \left(q - p - \int_a^b \int_a^s g(\theta) d\theta ds \right) \cdot \frac{t-a}{b-a} + \int_a^t \int_a^s g(\theta) d\theta ds.$$

   Therefore $h(a) = p$, $h(b) = q$ and $h''(t) = g(t)$. Also the linear operator $\mathcal{D}_2$ is surjective and $\text{ind}(\mathcal{D}_2) = 0$.

   Remark: The bounded functions

$$a_{ij}^\lambda = \sum_i \Gamma_{ij}^\lambda(\gamma) \frac{d\gamma_i}{dt}, \quad \vartheta_l^\lambda = \sum_{ij} \frac{\partial \Gamma_{ij}^\lambda}{\partial x_l} \frac{d\gamma_i}{dt} \frac{d\gamma_j}{dt}$$

allow us to consider the constants $C_a^\lambda = \sup_j | a_j^\lambda |$ and $C_\vartheta^\lambda = \sup_l | \vartheta_l^\lambda |$. So the operators

$$\hat{K}_1 : V \to C^0\big((a,b)^n; U\big) \quad \hat{K}_2 : V \to C^0\big((a,b)^n; U\big)$$
$$\hat{K}_1(v) = \sum_j a_j^\lambda v_j \qquad\qquad \hat{K}_2(v) = \sum_l \vartheta^\lambda(\gamma) v_l$$

are bounded since $\| \hat{K}_1(v) \| \le C_a^\lambda \| v \|$ and $\| \hat{K}_2(v) \| \le C_\vartheta^\lambda \| v \|$.

(ii) K_1^λ, for all λ, is a linear compact operator.
The operator $D_1 : C^2\big((a,b); \mathbb{R}\big) \to C^0\big((a,b); \mathbb{R}\big)$, $D_1(h) = h'$, is compact. Fix the Banach spaces E, F, G, the composition $T \circ K$, $T \in L(F,G)$ and $K \in \mathcal{K}(E,F)$ is always a compact operator. So the compactness of K_1^λ follows from the composition

$$K_1^\lambda(h) = \hat{K}_1(D_1(h)) = \sum_j a_j^\lambda D_1(h_j).$$

(iii) The compactness of K_2^λ for all λ is a trivial consequence of

$$K_2^\lambda(h) = \sum_l \vartheta_l^\lambda h_l.$$

The linear operator $dP_\gamma : C^2\big([a,b]^n\big) \to C^0\big([a,b]^n\big) \times \mathbb{R}^n \times \mathbb{R}^n$ is a Fredholm operator and $\mathrm{ind}(dP) = 0$. It follows from a Riemannian Geometry Theorem that if p and q are not conjugated points, then the kernel of the differential dP_γ is trivial. The kernel of dP_γ contains the Jacobi Fields (see [7]). Therefore dP_γ is injective. Since $\mathrm{ind}(dP) = 0$, then it is also surjective. Hence $dP_\gamma : C^2\big([a,b]^n\big) \to C^0\big([a,b]^n\big) \times \mathbb{R}^n \times \mathbb{R}^n$ is an isomorphism at γ.

By the InFT, there are open neighborhoods $V_\gamma \subset \widetilde{U}$ and $W \subset C^0\big([a,b]^n\big) \times \mathbb{R}^n \times \mathbb{R}^n$ such that $P : V_\gamma \to W$ is a diffeomorphism. Set γ as a constant geodesic, let's say $\gamma(t) = x_0 \in U$, so $P(\gamma) = (0, x_0, x_0)$, and take $W = V_\gamma \times V_0 \times V_0$, with $V_\gamma \subset \widetilde{U}$ a neighborhood of γ and V_0 is a neighborhood of x_0 in U. Therefore every pair of points $(p,q) \in V_0 \times V_0$ is connected by a geodesic.

Theorem 17 *Let $U \subset \mathbb{R}^n$ be an open subset endowed with a Riemannian metric g. If the points $p, q \in U$ are non-conjugated, then we have a unique geodesic in U joining them.*

Chapter 4
Vector Fields

Vector fields arise naturally in physics where several variables are of a vectorial nature. We will look at examples in which a physical system is modeled by an ordinary differential equation (ODE). In Classical Mechanics, Newton's 2nd law imposes the differential equation $\vec{F} = m\frac{d\vec{v}}{dt}$. An understanding of the analytical, algebraic and geometric properties of vector fields is the core of the study to understand the evolution of a system governed by an ODE. We will take an easy approach to cover what is needed in the text. Whenever appropriate, we will also consider vector fields $X : E \to E$ defined in the space of Banach E. We will denote the Vector Fields by the capital letters F, X, Y, V.

1 Vector Fields in $\mathbb{R}^n$

A vector field in $U \subset \mathbb{R}^n$ is a map $F : U \to T\mathbb{R}^n$, such that at each point $p \in \mathbb{R}^n$ associates a vector $F(p) \in T_pU = \mathbb{R}^n$. Let $\Gamma(U)$ be the space of the vector fields defined in U. In the following examples, it is understood that an orthonormal frame $\beta_c = \{e_1, \ldots, e_n\}$ is fixed on U. A C^k-vector field F defined on an open subset $U \subset \mathbb{R}^n$ is a map $F \in C^k(U; \mathbb{R}^n)$ which we write in local coordinates $x = (x_1, \ldots, x_n)$ as

$$F(x) = \sum_{i=1}^{n} f_i(x)e_i, \quad f_i \in C^k(U).$$

On many occasions, we may denote just $F(x) = (f_1(x), \ldots, f_n(x))$. Unless otherwise stated, we assume that the fields are C^∞.

© Springer Nature Switzerland AG 2021
C. M. Doria, *Differentiability in Banach Spaces, Differential Forms and Applications*,
https://doi.org/10.1007/978-3-030-77834-7_4

Example 1 Examples of vector fields.

1. In $\mathbb{R}^2$;

    (a) $F(x, y) = (1, 0)$ (constant vector field).
    (b) $F(x, y) = (x, y)$.
        Using polar coordinates, we have $F(r, \theta) = r\hat{r}$ with $\hat{r} = \frac{x}{\sqrt{x^2+y^2}}e_1 + \frac{y}{\sqrt{x^2+y^2}}e_2$.
    (c) $F(x, y) = (-y, x)$.
        Using polar coordinates, we have $F(r, \theta) = r\hat{\theta}$, with $\hat{\theta} = \frac{-y}{\sqrt{x^2+y^2}}e_1 + \frac{x}{\sqrt{x^2+y^2}}e_2$.
    (d) $F(x, y) = \left(x^2 + \cos(xy), \frac{1}{y^2} + e^{\sec(x^2 \cdot y)}\right)$.

2. In $\mathbb{R}^3$,

$$(a)\ F(x, y, z) = \frac{(x, y, z)}{\sqrt{x^2 + y^2 + z^2}}, \quad (b)\ F(x, y, z) = (xy, yz, zx).$$

3. In $\mathbb{R}^n$. Let $A \in M_n(\mathbb{R})$ be a matrix and define the linear vector field $F : \mathbb{R}^n \to \mathbb{R}^n$, $F(x) = A.x$;

$$F(x, y, z) = \begin{pmatrix} a_{11} & a_{12} & \cdots & a_{1n} \\ a_{21} & a_{22} & \cdots & a_{2n} \\ \vdots & \vdots & \vdots & \vdots \\ a_{n1} & a_{n2} & \cdots & a_{nn} \end{pmatrix} \cdot \begin{pmatrix} x_1 \\ x_2 \\ \vdots \\ x_n \end{pmatrix}.$$

4. The Euler-Lagrange equations (15) in Chap. 3 associated with the problem

$$\begin{cases} \text{Minimize: } F(\gamma) = \int_a^b L(t, \gamma(t), \dot{\gamma}(t))dt, \\ \text{constraint to: } \gamma \in E \end{cases} \tag{1}$$

defines a vector field in E.

Integral curves of a vector field $F : U \to \mathbb{R}^n$ are those curves $\gamma : (a, b) \to U$ satisfying the equation

$$\gamma'(t) = F(\gamma(t)), \tag{2}$$

i.e., they are curves with a tangent vector at the point $\gamma(t)$ that is $F(\gamma(t))$. Considering $\gamma = (\gamma_1, \ldots, \gamma_n)$ and $F(x) = \sum_{i=1}^n f_i(x)e_i$, Eq. (2) is written as a system of ODE

$$\begin{cases} \gamma_1'(t) = f_1(\gamma(t)), \\ \vdots \\ \gamma_n'(t) = f_n(\gamma(t)). \end{cases} \tag{3}$$

Once a point $x_0 \in U$ is fixed, the IVP (initial value problem) associated with Eq. (2) is

$$\begin{cases} \gamma'(t) = F(\gamma(t)), \\ \gamma(0) = x_0 . \end{cases} \quad (4)$$

Example 2 Examples.

1. The ODE $x'' + x = 0$ governs the simple harmonic oscillator without friction and without external force. If the initial conditions are $x(0) = a$ and $x'(0) = b$, the IVP solution is

$$x(t) = a \cos(t) + b \sin(t) = \sqrt{a^2 + b^2} \cos(t - \theta_0),$$

with $\cos(\theta_0) = \frac{a}{\sqrt{a^2+b^2}}$ and $\sin(\theta_0) = \frac{b}{\sqrt{a^2+b^2}}$. Let's introduce a vector field associated with the ODE by considering the variable $y = x'$. The ODE becomes the linear system

$$\begin{pmatrix} x' \\ y' \end{pmatrix} = \begin{pmatrix} 0 & 1 \\ -1 & 0 \end{pmatrix} \cdot \begin{pmatrix} x \\ y \end{pmatrix} = \begin{pmatrix} -y \\ x \end{pmatrix},$$

associated with the linear vector field $F(x, y) = (-y, x)$.
2. The nonlinear pendulum is governed by the ODE,

$$\theta'' + \frac{k}{m}\theta' + \frac{1}{l} \sin(\theta) = 0. \quad (5)$$

In introducing the variable $\omega = \theta'$, the ODE becomes the linear system

$$\begin{cases} \theta' = \omega, \\ \omega' = -\frac{k}{m}\theta' - \frac{1}{l} \sin(\theta). \end{cases} \quad (6)$$

In this way the parameter is governed by the ODE associated with the planar vector field $F(\theta, \omega) = (\omega, -\frac{k}{m}\theta' - \frac{1}{l} \sin(\theta))$.

Definition 1 Let $U \subset \mathbb{R}^n$ be an open subset and let $F : U \to \mathbb{R}^n$ be a vector field. A point $p \in U$ is a singularity or a critical point of F if $F(p) = 0$.

A singularity is a stationary point in the vector field F since the only solution to the IVP $\gamma'(t) = F(\gamma(t)) \gamma(0) = p$ is $\gamma(t) = p$.

Example 3 Examples of singularities of linear vector fields on $\mathbb{R}^2$;

1. $F(x, y) = (x, y)$. The only singularity is at $p = (0, 0)$. For the point (x_0, y_0), the solution to the IVP:

$$\begin{pmatrix} x' \\ y' \end{pmatrix} = \begin{pmatrix} 1 & 0 \\ 0 & 1 \end{pmatrix} \cdot \begin{pmatrix} x \\ y \end{pmatrix} = \begin{pmatrix} x \\ y \end{pmatrix}, \quad \begin{pmatrix} x(0) \\ y(0) \end{pmatrix} = \begin{pmatrix} x_0 \\ y_0 \end{pmatrix},$$

is the integral curve $\gamma(t) = e^t(x_0, y_0)$. Taking the limit $t \to \infty$, we have $\gamma(t) \to \infty$. This is an example of an unstable singularity, also called a source because any point $x \neq p$ in a neighborhood of p moves away from p, while the parameter t evolves to ∞, as shown in Fig. 1.

2. $F(x, y) = (-x, -y)$. In this case, $p = (0, 0)$ is the only singularity. For each point (x_0, y_0), the solution to the IVP

$$\begin{pmatrix} x' \\ y' \end{pmatrix} = \begin{pmatrix} -1 & 0 \\ 0 & -1 \end{pmatrix} \cdot \begin{pmatrix} x \\ y \end{pmatrix} = \begin{pmatrix} -x \\ -y \end{pmatrix}, \quad \begin{pmatrix} x(0) \\ y(0) \end{pmatrix} = \begin{pmatrix} x_0 \\ y_0 \end{pmatrix}$$

is the integral curve $\gamma(t) = e^{-t}(x_0, y_0)$. Taking the limit $t \to \infty$, we have $\gamma(t) \to p$. This is an example of a stable singularity called a sink because any point $x \neq p$ in a neighborhood of p moves towards p, while the parameter t evolves to ∞, as shown in Fig. 2.

3. $F(x, y) = (x, -y)$. The only singularity is at $p = (0, 0)$. For each point (x_0, y_0), the solution to the IVP

$$\begin{pmatrix} x' \\ y' \end{pmatrix} = \begin{pmatrix} 1 & 0 \\ 0 & -1 \end{pmatrix} \cdot \begin{pmatrix} x \\ y \end{pmatrix} = \begin{pmatrix} x \\ -y \end{pmatrix}, \quad \begin{pmatrix} x(0) \\ y(0) \end{pmatrix} = \begin{pmatrix} x_0 \\ y_0 \end{pmatrix},$$

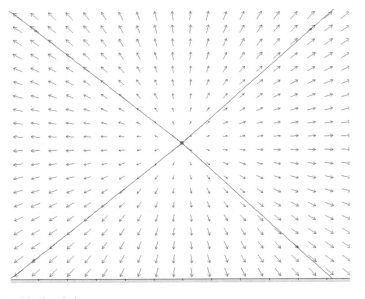

Fig. 1 Unstable singularity

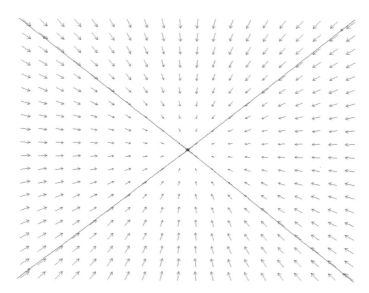

Fig. 2 Stable singularity

is the integral curve $\gamma(t) = (e^t x_0, e^{-t} y_0)$. As we can see in Fig. 3, there are directions in which the trajectory approaches p and also directions moving away from p. In this case, p is a saddle point singularity.

4. $F(x, y) = (-y, x)$. The point $p = (0, 0)$ is the only singularity. For each point (x_0, y_0), it follows from item (1) in Example 2 that the solution to the IVP

$$\begin{pmatrix} x' \\ y' \end{pmatrix} = \begin{pmatrix} 0 & 1 \\ -1 & 0 \end{pmatrix} \cdot \begin{pmatrix} x \\ y \end{pmatrix} = \begin{pmatrix} -y \\ x \end{pmatrix}, \quad \begin{pmatrix} x(0) \\ y(0) \end{pmatrix} = \begin{pmatrix} x_0 \\ y_0 \end{pmatrix},$$

is the integral curve

$$\gamma(t) = (A \cos(t + \theta_0), A \sin(t + \theta_0)), \quad A = \sqrt{x_0^2 + y_0^2}.$$

Therefore $\gamma(t)$ is a closed orbit for which the trace is the circumference $x^2 + y^2 = A^2$. Note in this example that $p = (0, 0)$ is neither a sink nor a source.

5. $F(x, y) = (y - \frac{x}{2}, -x - \frac{y}{2})$. The point $p = (0, 0)$ is the only singularity at $p = (0, 0)$ and it is stable, as shown in Fig. 5.

We highlight below some elementary questions in studying the vector field $F : U \to \mathbb{R}^n, U \subset \mathbb{R}^n$.

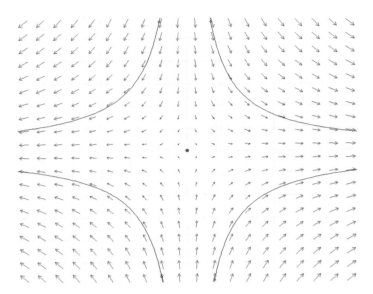

Fig. 3 Saddle singularity

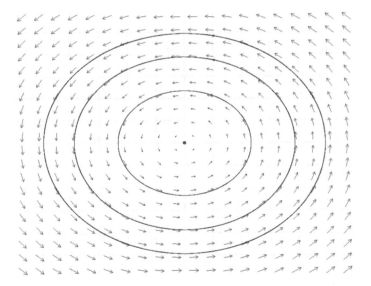

Fig. 4 Periodic orbit

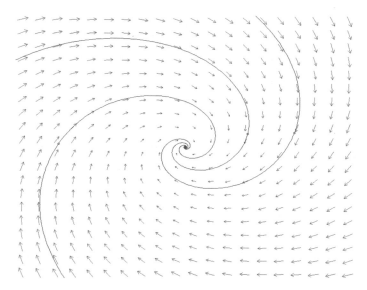

Fig. 5 $F(x, y) = (y - \frac{x}{2}, -x - \frac{y}{2})$

Question 1:   Let $p \in U \subset E$ and let $F : U \to E$ be a vector field. The IVP

$$\begin{cases} \gamma'(t) = F(\gamma(t)), \\ \gamma(0) = p \end{cases} \tag{7}$$

admits a solution? How many?

Question 2:   Find all the singularities of a vector field F and classify them in terms of their stability (easy for linear fields, hard for nonlinear).

Question 3:   Assuming that the IVP (7) has a unique solution $\gamma_p(t)$ (deterministic problem), is it possible to describe the evolution and the limit of p along its trajectory (integral curve)? In $\mathbb{R}^2$, the Bendixon-Poincaré theorem asserts that if $\overline{U}$ is compact, then either $\lim_{t \to \infty} \gamma_p(t)$ is a singularity of F or it is asymptotic to a closed orbit (see [31]).

Question 4:   Are there periodic orbits?

Example 4 Nonlinear Pendulum.

Let's look at the linear system of the Linear Pendulum with an edge measuring l, as shown in Fig. 6. We will assume that the gravitational constant is $g = 1$, the mass of the edge is negligible, and there is a frictional force F_f proportional to the speed at the end of the pendulum. Let θ be the angle formed by the edge of the pendulum and the vertical line, so the angular velocity is $\theta' = \frac{d\theta}{dt}$ and, consequently, the velocity

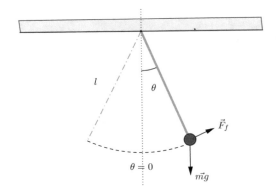

Fig. 6 Pendulum

is $v = l.\theta'$ and $F_f = kl.\theta'$ $(k > 0)$. It follows from Newton's Law that the resulting force is $F = -F_f - m.\sin(\theta)$, that is, the ODE governing the system is

$$\theta'' + \frac{k}{m}\theta' + \frac{1}{l}\sin(\theta) = 0. \tag{8}$$

As seen before, the pendulum is governed by the ODE associated with the planar vector field $F(\theta, \omega) = (\omega, -\frac{k}{m}\theta' - \frac{1}{l}\sin(\theta))$. Let's study the possibilities;

(i) the free pendulum with small oscillations ($F_f = 0$ and $\sin(\theta) \sim \theta$).
The vector field $F(\theta, \omega) = (\omega, -\frac{1}{l}\theta)$ has a unique singularity at $p = (0, 0)$. So it is equal to the harmonic oscillator, and so the integral curves are centered at the origin as shown in Fig. 4.
(ii) pendulum without friction ($F_f = 0$).
In this case, the field $F(\theta, \omega) = (\omega, -\frac{1}{l}\sin(\theta))$ has singularities at $p_k = (k\pi, 0)$, $k \in \mathbb{N}$. The phase space is shown in Fig. 7. We stress the existence of closed orbits and unstable singularities.
(iii) pendulum with friction.
The force field in this case is $F(\theta, \omega) = (\omega, -\frac{k}{m}\omega - \frac{1}{l}\sin(\theta))$. Again, the only singularities are at $p_k = (k\pi, 0)$, $k \in \mathbb{N}$. There is no closed orbit, as shown in Fig. 8.

The phase portrait of a vector field $F : U \to \mathbb{R}^n$ ($U \subset \mathbb{R}^n$) is a copy of U in which a point represents a possible physical state of a system governed by ODE $\gamma'(t) = F(\gamma(t))$. The integral curves passing through the points represent the trajectory of each point while the parameter evolves, therefore showing the evolution of the states of the system. Figures 4, 7 and 8 show the phase portrait of each pendulum system under consideration. In describing the phase portrait of a vector field, we can make a qualitative analysis of the evolution of each state, i.e., the evolution of each point along the integral curve $\gamma_p(t)$ when $t \to \infty$. It is also important, though a difficult

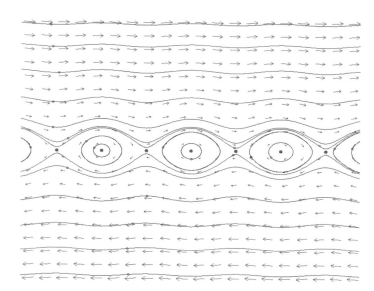

Fig. 7 Frictionless pendulum

Fig. 8 Pendulum with friction

task, to find the invariant sets by the integral curves, i.e., for any integral curve γ, find those sets $V \subset U$ such that if $\gamma \cap V \neq \varnothing$, then $\gamma(t) \subset V$ for all $t \in \mathbb{R}$.

Exercises

1. Show that the system

$$\begin{cases} x' = -y + x(x^2 + y^2 - 4), \\ y' = x + y(x^2 + y^2 - 4) \end{cases} \tag{9}$$

has a periodic orbit in which all other integral curves accumulate. Conclude that the phase space of the system is the one in Fig. 9 (hint: use polar coordinates):

2. Show the phase portrait of the system below is the one in Fig. 10:

$$\begin{cases} x' = 2x - y + 3(x^2 - y^2) + 2xy, \\ y' = x - 3y - 3(x^2 - y^2) + 3xy \ . \end{cases} \tag{10}$$

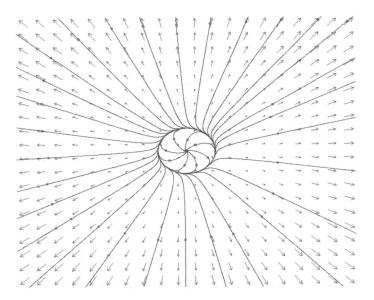

Fig. 9 Attracting orbit

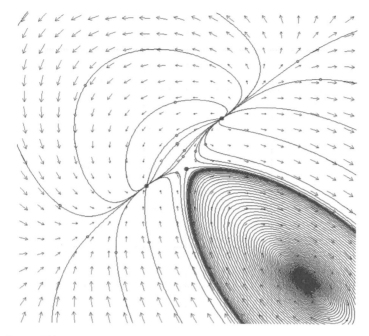

Fig. 10 System (10)

2 Conservative Vector Fields

Let $U \subset \mathbb{R}^n$ be a star-shaped open subset. A vector field $F : U \to \mathbb{R}^n$ is conservative, if for any closed continuous curve $\gamma : [0, 1] \to \mathbb{R}^n$, $\gamma(0) = \gamma(1)$, the path integral over γ is null, i.e.,

$$\int_\gamma F = \int_0^1 < F(\gamma(t)), \gamma'(t) > dt = 0.$$

As a result, for any pair of points $p, q \in U$ and a C^0 curve $\alpha : [0, 1] \to \mathbb{R}^n$ such that $\alpha(0) = p$ and $\alpha(1) = q$, if F is conservative, then the integral $\int_\alpha F$ does not depend on α. In the next section, assuming U is star-shaped and F is conservative, we will prove that we have a function $V : U \to \mathbb{R}$ such that $F = -\nabla V$. The function V is the potential function of F. Conservative vector fields are also known as gradient vector fields; they are of great importance since they emerge in many applications. In Classical Mechanics, a body of mass m displaced under the effects of a conservative vector field $F = -\nabla V$ has its total energy conserved. Given Newton's 2nd Law, a force field acting on a body of mass m and velocity v satisfies the identity $F = m\frac{dv}{dt}$. In this case, let α be the curve defined by the trajectory along the displacement from p to q $(v = \alpha')$. The work done by F along the displacement is

$$\int_\alpha F = \int_0^1 < m\frac{d\alpha'}{dt}, \alpha'(t) > dt = m \int_0^1 \frac{d}{dt}\left(|\alpha'(t)|^2\right) dt = \frac{1}{2}m \mid v(q) \mid^2 -\frac{1}{2}m \mid v(p) \mid^2.$$

For a body of mass m, the kinetic energy is $K(v) = \frac{1}{2}m \mid v \mid^2$. Since F is conservative, we have $F = -\nabla V$, and so

$$\int_\alpha F = -\int_0^1 < \nabla V(\alpha(t)), \alpha'(t) > dt = \int_0^1 \frac{dV}{dt} dt = -V(q) + V(p).$$

Consequently, $\frac{m}{2} \mid v(q) \mid^2 -\frac{m}{2} \mid v(p) \mid^2 = -V(q) + V(p)$. Therefore

$$\frac{m}{2} \mid v(q) \mid^2 +V(q) = \frac{m}{2} \mid v(p) \mid^2 +V(p).$$

The total energy of the system at the point $x \in U$ is given by the function

$$E(x) = \frac{m}{2} \mid v(x) \mid^2 +V(x).$$

So we have proved the conservation of the total energy.

Theorem 1 *Let $F = -\nabla V$ be a vector field and suppose a body of mass m is under the action of F. If $\gamma : (-a, a) \to U$ is an integral curve of the equation $\gamma'(t) = F(\gamma(t))$, then the energy function $E_\gamma = E(\gamma(t)) : (-a, a) \to \mathbb{R}$ is constant along γ.*

A vector field $F : U \to \mathbb{R}^n$ is a central vector field if we have a function $\lambda : U \to \mathbb{R}$ such that $F(x) = \lambda(x).x$.

Exercises

1. Show that a central vector field is conservative.
2. Let $k > 0$ be a constant. Consider $F(x) = -kx$ and show it is conservative. Apply Newton's Second Law to find the integral curves of F.
3. Find necessary conditions for a vector field F to be conservative.
4. Find sufficient conditions for a vector field F to be conservative. (hint: read Chap. 5).

3 Existence and Uniqueness Theorem for ODE

The cornerstone of this chapter is the Existence and Uniqueness Theorem, which answers Question (1) above. Let E be a Banach space and let $U \subset E$ be an open subset. We will assume that $X : U \to E$ is a Lipschitz function with constant $K > 0$, that is, for every pair $x, y \in U$, we have

$$| X(y) - X(x) | \le K. | y - x | .$$

Let $\mathcal{I} : C^0([-a, a]); E) \to C^1([-a, a]); E)$ be the operator given by

$$\mathcal{I}(\alpha)(t) = x_0 + \int_0^t X(\alpha(s))ds. \tag{11}$$

If γ is an integral curve of X and $\gamma(0) = x_0$ (IVP (2)), then

$$\gamma(t) = x_0 + \int_0^t X(\gamma(s))ds. \tag{12}$$

Then the identity $\mathcal{I}(\gamma) = \gamma$ implies that γ is a fixed point of $\mathcal{I}$. To prove that the IVP

$$\gamma'(t) = X(\gamma(t)),$$
$$\gamma(0) = x_0,$$

has a solution, we need the hypothesis of X to be Lipschitz. The strategy will be to show the existence of a fixed point of $\mathcal{I}$. Consider $I_a = (-a, a)$ to be the open interval and $\mathcal{I}^n(\alpha) = \mathcal{I}(\mathcal{I}^{n-1}(\alpha))$.

Proposition 1 *Let $K > 0$ be the Lipschitz constant of $X : I_a \times U \to E$. So*

$$| \mathcal{I}^n(\alpha)(t) - \mathcal{I}^n(\beta)(t) | \le \frac{K^n}{n!} | t |^n . \| \alpha - \beta \|_0, \tag{13}$$

for all $\alpha, \beta \in C^0(I_a, E)$, $n > 0$ and $t \in I_a$.

Proof The statement is immediate for $n = 0$. Let $\alpha, \beta \in C^0(I_a, E)$. From the definition of $\mathcal{I}$, we have

$$\| \mathcal{I}(\alpha) - \mathcal{I}(\beta) \|_0 \le \int_0^t | X(\alpha(s)) - X(\beta(s)) | \, ds \le K. \int_0^t | \alpha(s) - \beta(s) | \, ds \le$$
$$\le K | t | . \| \alpha - \beta \|_0 .$$

The identity is true for $n = 1$. Let's check the case $n = 2$;

$$\| \mathcal{I}^2(\alpha) - \mathcal{I}^2(\beta) \|_0 \leq K \int_0^t | \mathcal{I}(\alpha)(s) - \mathcal{I}(\beta)(s) | \, ds \leq K^2 \, \| \alpha - \beta \|_0 \int_0^t | s | \, ds \leq$$

$$\leq K^2 \frac{| t |^2}{2} \cdot \| \alpha - \beta \|_0 \,.$$

By induction, we assume the inequality (13) is true for n. By an analogous procedure, we obtain the inequality (13) for $n + 1$. $\qquad\square$

Lemma 1 *The operator* $\mathcal{I} : C^0(I_a; E) \to C^1(I_a; E)$ *has a unique fixed point.*

Proof Since the series $e^{K|t|}$ is convergent for any $t \in I_a$, we have $\lim \frac{K^n}{n!} | t |^n = 0$. Setting $c \in (0, 1)$ and taking n_0 so that for every $n > n_0$, we have $\frac{K^n}{n!} | t |^n < c < 1$, then we have

$$\| \mathcal{I}^n(\alpha) - \mathcal{I}^n(\beta) \|_0 < c \cdot \| \alpha - \beta \|_0 \,.$$

Therefore for all $n > n_0$, $\mathcal{I}^n$ is a contraction. $\qquad\square$

Let $\overline{B}_r(x_0) = \{x \in U; \, | x - x_0 | \leq r\}$ be the closed ball.

Theorem 2 (Existence and Uniqueness) *Let E be a Banach space and assume* $X : U \subset E \to E$ *is a Lipschitz map with the constant* $K > 0$ *such that* $\sup_{x \in B_r(x_0)} | X(x) | \leq M$. *So there is* $\epsilon > 0$ *such that the IVP (4) admits a unique solution* $\gamma : (-\epsilon, \epsilon) \to E$ *of class* C^1.

Proof Consider the Banach space $\mathcal{P}_r = C^0 \left(I_a; \overline{B}_r(x_0) \right)$. Let's check that the hypothesis implies the existence of $\epsilon > 0$, such that $\mathcal{I} : \mathcal{P}_r \to \mathcal{P}_r$ has a unique fixed point. Since

$$| \mathcal{I}(\alpha)(t) - x_0 | \leq \int_0^t | X(\alpha(s)) | \, ds \leq M | t |,$$

we have that if $| t | \leq \frac{r}{M}$, then $\mathcal{I}(\alpha)(t) \in \overline{B}_r(x_0)$. Taking n_0 as prescribed in Lemma 1 and $\epsilon = \min\{\frac{c.n!}{K^n}, \frac{r}{M}\}$, the operator $\mathcal{I}^n : \mathcal{P}_r \to \mathcal{P}_r$ is well-defined and is a contraction for $n > n_0$. Therefore from Lemma 1, we now have a unique $\gamma \in \mathcal{P}_r$ such that $\mathcal{I}(\gamma) = \gamma$. As previously noted, γ is the only solution to the IVP. $\qquad\square$

Remark 1 The vector field $F : \mathbb{R} \to \mathbb{R}$, given by $F(x) = 3x^{2/3}$, is continuous but it is not Lipschitz. The IVP with $x(0) = 0$ admits two solutions; they are $x(t) = t^3$ and $x(t) = 0$.

Theorem 2 suggests an interaction process to extend the solution $\gamma : (-\epsilon, \epsilon) \to E$ for an interval greater than $(-\epsilon, \epsilon)$, as we show next: let $\gamma_0 : (-\epsilon_0, \epsilon_0) \to E$ be a solution to the IVP (4) and $x_1 = \gamma(t_1)$, $t_1 \in (-\epsilon_0, \epsilon_0)$. When we change the initial condition of the IVP to $\gamma(0) = x_1$, we obtain a new IVP whose solution we call $\gamma_1 : (-\epsilon_1, \epsilon_1) \to E$. Due to the uniqueness, γ_0 and γ_1 coincide in a neighborhood of x_1. By combining the solutions, we obtained an extension of γ_0. This process can be applied so that either γ is defined for every $t \in \mathbb{R}$; in this case we say that the integral curve $\gamma : \mathbb{R} \to E$ is complete, or that there is a maximal interval $J(x_0) \subset \mathbb{R}$ in which γ is well-defined.

Example 5 Incomplete vector fields.
(1) Let $X(x) = x^2$ be a vector field defined on $\mathbb{R}$. The IVP

$$\gamma'(t) = X(\gamma(t)) = \gamma^2(t),$$
$$\gamma(0) = x_0,$$

has a solution given that $\gamma(t) = \frac{x_0}{1-tx_0}$. Therefore the maximal interval is $J(x_0) = \mathbb{R}\setminus\{x_0\}$.

(2) Let $X(x, y) = (0, 1)$. The integral curve passing by $p = (x_0, y_0)$ is $\gamma(t) = (x_0, y_0 + t)$. If we consider $X : \mathbb{R}^2 \to \mathbb{R}^2$ and $p = (0, -1)$, then $J(p) = \mathbb{R}$. If we consider $X : \mathbb{R}^2\setminus\{0\} \to \mathbb{R}^2$ and $p = (0, -1)$, then $J(p) = (-\infty, 1)$.

An important fact arising from this uniqueness is that if two integral curves $\gamma_1 : I \to U$ and $\gamma_2 : I_2 \to U$ pass through $p \in U$, then they coincide in the intersection of their domains. To prove this, we consider the intersection $I = I_1 \cap I_2$ and the set $C = \{t \in I \mid \gamma_1(t) = \gamma_2(t)\}$. C is connected; let's check whether it is open and closed. It is a closed set because the integral curves are continuous. It is open because the interactive process mentioned above shows we can always extend a solution near $t \in C$.

A vector field $X : U \to E$ is complete, if for all $x \in U$, we have $J(x) = \mathbb{R}$. If X has compact support, then it is complete. The following theorem tells us about the continuous dependence of the solution with respect to the initial condition.

Theorem 3 *Let $X : U \to E$ be a vector field satisfying the hypothesis of Theorem 2. Let $\alpha(t)$, $\beta(t)$ be solutions of the ODE $\gamma'(t) = X(\gamma(t))$ on the closed interval $[t_0, t_1]$. So we have*

$$\mid \alpha(t) - \beta(t) \mid \leq \mid \alpha(t_0) - \beta(t_0) \mid .\exp(K(t - t_0)),$$

for all $t \in [t_0, t_1]$.

Proof See in [31].   □

4 Flow of a Vector Field

In this section, we consider only complete C^∞-vector fields. Let $U \subset \mathbb{R}^n$ be an open subset and let $X : U \to \mathbb{R}^n$ be a vector field. In the previous section, we showed the existence and uniqueness of an integral curve of X passing by each point $x \in U$. In this way, we associate the integral curve $\gamma(t, x) = \gamma_x(t)$ with every $x \in U$. The flow of X is defined by the map $\Phi^X : \mathbb{R} \times U \to E$ given by $\Phi^X(t, x) = \gamma_x(t)$. Φ^X is a C^∞ map (see [31]).

Example 6 Linear flows

For a fixed matrix $A \in M_n(\mathbb{R})$, consider the linear vector field $F : \mathbb{R}^n \to \mathbb{R}^n$ given by $F(x) = A.x$. These are the simplest non-constant fields to be studied. Simplicity is revealed in the fact that it is possible to describe analytically all of its integral curves. We set the point $x_0 \in \mathbb{R}^n$ and consider the trajectory $\gamma(t) = \exp(At).x_0$, with $\exp(A)$ the exponential operator of A. We will show that $\gamma(t) = \exp(At).x_0$ is the solution of the IVP

$$\begin{cases} \gamma'(t) = A.\gamma(t)), \\ \gamma(0) = x_0. \end{cases} \tag{14}$$

It is straightforward to verify that $\gamma(0) = x_0$ and $\gamma \in C^\infty$. The trajectory $\gamma(t)$ satisfies the ODE since

$$\gamma'(t) = \frac{d}{dt}\left(\sum_{i=0}^{\infty}\frac{A^n}{n!}t^n\right) = \sum_{i=1}^{\infty}\frac{A^n}{(n-1)!}t^{n-1} = A.\sum_{i=1}^{\infty}\frac{A^{n-1}}{(n-1)!}t^{n-1} = A.\exp(tA) = A.\gamma(t)$$

and $\gamma(0) = x_0$. F is complete, since $\gamma(t)$ is defined for all $t \in \mathbb{R}$. The flow of $F(x) = A.x$ is the map

$$\Phi^F : \mathbb{R} \times \mathbb{R}^n \to \mathbb{R}^n,$$
$$\Phi^F(t, x) = \gamma_x(t) = \exp(tA).x .$$

The flow $\gamma_x(t) = x.\exp(tA)$ satisfies the following conditions;
(i) $\gamma_x(0) = x$ and (ii) $\gamma_{\gamma_x(t)}(s) = \gamma_x(t + s)$.
In terms of the flow, these identities mean that $\Phi^F(s, \Phi^F(t, x)) = \Phi^F(t + s, x)$.

The flow properties obtained in the example above generalizes to linear vector fields $F : E \to E$, $F(x) = A.x$, defined in a Banach space E by an operator $A \in \mathcal{L}(E)$. Indeed, the properties are true in general. Any complete C^∞ vector field $X : U \to E$, satisfying the hypothesis of Theorem 2, defines a C^∞ flow $\Phi^X : \mathbb{R} \times U \to E$.

$$\Phi^X : \mathbb{R} \times U \to E$$
$$(t, x) \to \gamma_t(x). \tag{15}$$

Of course, $\Phi^X(0, t) = x$. To prove the identity $\Phi^X(s, \Phi^X(t, x)) = \Phi^X(t + s, x)$, we consider the curves $\alpha(s) = \gamma_u(t + s)$ and $\beta(s) = \gamma_{\gamma_u(t)}(s)$, and note that

$$\begin{cases} \alpha(0) = \gamma_u(t), \\ \alpha'(s) = \gamma_u'(t + s) = X(\gamma_u(t + s)) \end{cases} , \quad \begin{cases} \beta(0) = \gamma_u(t), \\ \beta'(s) = \gamma_{\gamma_u(t)}'(s) = X(\gamma_{\gamma_u(t)}(s)). \end{cases}$$

Since α and β satisfy the same IVP, they must be equal. Therefore the flux satisfies the conditions

$$\Phi^X(0, x) = x, \quad \text{and} \quad \Phi^X(t + s, u) = \Phi^X(s, \Phi^X(t, u)). \tag{16}$$

For fixed $t \in \mathbb{R}$, the flow Φ^X induces the diffeomorphism $\phi_t : U \to U$, $\phi_t(x) = \Phi^X(t, x)$. By varying the parameter t, we get the 1-parameter subgroup $\{\phi_t : U \to U \mid t \in \mathbb{R}\}$ for the diffeomorphism group $\mathrm{Diff}(X)$.

Definition 2 Let $X \in \Gamma(U)$, $V \in \Gamma(V)$ and let $\phi : U \to V$ be differentaible maps.

(i) The vector fields X and Y are ϕ-related ($X \overset{\phi}{\sim} Y$) if $d\phi.X = Y \circ \phi$. Equivalently, the relation $X \overset{\phi}{\sim} Y$ means the diagram below commutes;

$$
\begin{array}{ccc}
TU & \xrightarrow{\ d\phi\ } & TV \\[4pt]
\big\uparrow{\scriptstyle X} & & \big\uparrow{\scriptstyle Y} \\[4pt]
U & \xrightarrow{\ \phi\ } & V\,.
\end{array}
\tag{17}
$$

(ii) The flows Φ^X, Φ^Y are conjugated if $\phi \circ \Phi^X = \Phi^Y \circ \phi$. If ϕ is a diffeomorphism, then $\Phi^Y = \phi \circ \Phi^X \circ \phi^{-1}$. In particular, Φ^X is conjugated to itself.

The ϕ-related vector fields share the following algebraic properties;

(i) If $X_i \overset{\phi}{\sim} Y_i$, $i = 1, 2$, and $a, b \in \mathbb{R}$, then $aX_1 + bX_2 \overset{\phi}{\sim} aY_1 + bY_2$.

(ii) If $X \in \Gamma(U)$, $Y \in \Gamma(V)$ and $Z \in \Gamma(W)$, then

$$
\begin{cases}
X \overset{\phi}{\sim} Y \\[4pt]
Y \overset{\phi}{\sim} Z
\end{cases}
\Rightarrow X \overset{\psi \circ \phi}{\sim} Z.
\tag{18}
$$

Proposition 2 *Given the vector fields $X \in \Gamma(U)$ and $V \in \Gamma(V)$, let $\psi : U \to V$ be a differentiable map and let Φ^X, Φ^Y be their flows, respectively. So $X \overset{\psi}{\sim} Y$ if and only if their flows are conjugated. In particular, if ψ is a diffeomorphism, then $Y = \psi_*(X)$ if and only if $\Phi^Y = \psi \circ \Phi^X \circ \psi^{-1}$ ($(\Phi^X)_*(X) = X$).*

Proof Let's prove each direction of the statements separately. For $t \in \mathbb{R}$, consider the diffeomorphism $\phi_t = \Phi^X(t, .) : U \to U$ and fix $p \in U$.

(i) ($\Rightarrow$) Let $c(t) = \phi_t(p)$, so $c'(t) = X(p)$. It follows that

$$
\frac{d}{dt}\left(\psi(c(t))\right) = d\psi_{c(t)}.c'(t) = d\psi_{\Phi_t^X(p)}.X(p) = \left(Y \circ \psi\right)(\Phi_t^X(p)) =
$$
$$
= \left(Y \circ \psi\right)(c(t)) = Y\left((\psi \circ c)(t)\right).
$$

Therefore $\left(\psi \circ c\right)(t)$ is an integral curve of Y with the initial condition $\psi\left(c(0)\right) = \psi(p)$. The uniqueness implies that $\left(\psi \circ \Phi_t^X\right)(p) = \left(\psi \circ c\right)(t) = \Phi_t^Y\left(\psi(p)\right)$.

(ii) ($\Leftarrow$) Differentiating the expression $\left(\psi \circ \Phi_t^X\right)(p) = \left(\Phi_t^Y \circ \psi\right)(p)$ with respect to t, we get

$$
d\psi_{\Phi_t^X(p)}.\frac{d\Phi_t^X}{dt} = \frac{d\Phi_t^Y}{dt}\left(\psi(p)\right) \Rightarrow d\psi_{\Phi_t^X}.X = Y\left(\psi(p)\right).
$$

$\square$

Definition 3 Let G be a group with product operation $\cdot : G \times G \to G$ and let e be the identity element. A G-action on a topological space X is a map $G \bullet : G \times X \to X$ satisfying the following items; for all $x \in X$,

(i) $e \bullet x = x$,

(ii) $g' \bullet (g \bullet x) = (g' \cdot g) \bullet x$.

By denoting $\gamma(t, x) = t \bullet x$, it follows from the identities in Eq. (16) that a flow defines an action of the additive group $G = (\mathbb{R}, +)$ on E since it satisfies the following items;

(i) $0 \bullet x = x$;

(ii) $s \bullet (t \bullet x) = (s + t) \bullet x$, for any $t, s \in \mathbb{R}$ and $x \in E$.

The orbit of an element $x \in E$ generated by the $\mathbb{R}$ action is the set $\mathcal{O}_x = \{t \bullet x \mid t \in \mathbb{R}\}$, which is the trajectory of the point x with the flow Φ^X. If x_0 is a singularity of X, $X(x_0) = 0$, then x_0 is a fixed point for the action, that is, its orbit is the constant trajectory $t \bullet x_0 = x_0$. In general, the $\mathbb{R}$-action induced by a vector field X induces the 1-parameter subgroup of the diffeomorphism $\phi_t : U \to U$, $\phi_t(x) = t \bullet x$.

Let G be a group. A dynamical system defined on an open subset $U \subset E$ is the flow spanned by the action $G \times U \to E$, $(g, x) \to g \bullet x$. When $G = \mathbb{R}$ and the action is induced by the flow of a vector field $X \in \Gamma(U)$ (EDO), we have $t \bullet x = \gamma_x(t)$. The dynamical system can be continuous or discrete, depending on whether G is continuous or discrete as a topological space. When $X(x) = A \cdot x$, the 1-parameter subgroup of the diffeomorphism is $\{\exp(tA) \mid t \in \mathbb{R}\}$.

Exercises

1. Show that the system below is ϕ-related to the system defined in Eq. (9). $\Big($hint: ϕ is the diffeomorphism $\phi : \mathbb{R}^2 \setminus \{0\} \to \mathbb{R}^2 \setminus \{0\}$, $\phi(\theta, r) = (r \cdot \cos(\theta), r \cdot \sin(\theta))\Big)$.

$$\begin{cases} r' = r(r^2 - 4), \\ \theta' = 1. \end{cases} \qquad (19)$$

2. Show that the solution to the IVP given by the ODE in the item above with the initial condition $(x(0), y(0)) = (a, b)$ is

$$\begin{cases} (x(t), y(t)) = 2\sqrt{\frac{k}{k - e^{8\theta}}}\big(\cos(\theta), \sin(\theta)\big), & \text{if } r > 2 \\ (x(t), y(t)) = 2\big(\cos(\theta), \sin(\theta)\big), & \text{se } r = 2, \\ (x(t), y(t)) = 2\sqrt{\frac{k}{e^{8\theta} - k}}\big(\cos(\theta), \sin(\theta)\big), & \text{if } r < 2. \end{cases}$$

The constants are $k = \frac{a^2 + b^2}{a^2 + b^2 - 4} \cdot e^{8\theta_0}$ and $\theta_0 = \mathrm{tg}^{-1}\big(\frac{b}{a}\big)$.

5 Vector Fields as Differential Operators

Let $U \subset \mathbb{R}^n$ be an open subset and let $TU = U \times \mathbb{R}^n$ be its tangent bundle. A vector field is an element of the space $\Gamma(U)$, that is, a section of the tangent bundle of U as described in Chap. 2. Consider a fixed orthonormal frame $\beta_c = \{e_i \mid 1 \leq i \leq n\}$ of $\mathbb{R}^n$, then a C^∞-vector field $X : U \to TU$ can be written as

$$X(x) = \sum_{i=1}^{n} a_i(x)e_i, \quad a_i \in C^\infty(U).$$

The field X induces the $C^\infty(U)$ map $X : C^\infty(U) \to C^\infty(U)$,

$$X(f) = (df.X)(x) = \sum_{i=1}^{n} a_i(x)\frac{\partial f}{\partial x_i}(x) = \left(\sum_{i=1}^{n} a_i(x)\frac{\partial}{\partial x_i}\right)(f(x)), \qquad (20)$$

and so the bijection

$$X = \sum_{i=1}^{n} a_i(x)e_i \longleftrightarrow \sum_{i=1}^{n} a_i\frac{\partial}{\partial x_i}.$$

We have identified the vector field with a 1st-order linear differential operator. As differential operators, vector fields have the following properties: let $f, g \in C^\infty(U)$, $a, b \in \mathbb{R}$ and $X \in \Gamma(U)$;
(i) $X(af + bg) = aX(f) + bX(g)$ (linear).
(ii) $X(f.g) = X(f).g + f.X(g)$ (Leibniz's rule).

Definition 4 The Lie bracket $[X, Y]$, or commutator, of two smooth vector fields $X, Y \in \Gamma(U)$ is the differentiable vector field $[X, Y] : C^\infty(U) \to C^\infty(U)$,

$$[X, Y](f) = X\big(Y(f)\big) - Y\big(X(f)\big). \qquad (21)$$

Proposition 3 *Let* $X, Y \in \Gamma(U)$ *and* $f, g \in C^\infty(U)$; *so*

1. $[fX, gY] = fg[X, Y] + fX(g)Y - gY(f)X$.
2. $[Y, X] = -[X, Y]$.
3. $[[X, Y], Z] + [[Z, X], Y] + [[Y, Z], X] = 0$ *(Jacobi's identity)*.

Proof It is left as an exercise since it is straightforward from the definition. □

We note that for any $X \in \Gamma(U)$ and $f \in C^\infty(V)$, we have

$$[d\phi(X)](f)(x) = df_{\phi(x)}.d\phi_x.X(x) = d_x(f \circ \phi).X(x) = X\big(f \circ \phi\big)(x).$$

If $X \overset{\phi}{\sim} Y$, then

$$\big(d\phi(X)\big)(f)(x) = df_{\phi(x)}.d\phi_x.X(x) = d_{\phi(x)}f.Y\big(\phi(x)\big) = Y(f) \circ \phi.$$

Proposition 4 *Let $U \subset \mathbb{R}^n$ and $V \subset \mathbb{R}^m$ be open subsets and let $\phi : U \to V$ be a C^∞ map. If $X, X' \in \Gamma(U)$ and $Y, Y' \in \Gamma(V)$ are vector fields such that $X \overset{\phi}{\sim} Y$ and $X' \overset{\phi}{\sim} Y'$, then $[X, Y] \overset{\phi}{\sim} [X', Y']$.*
Proof

$$d\phi([X, X']_x)(x) = [X, X'](f \circ \phi)(x) = X(X'(f \circ \phi)) - X'(X(f \circ \phi)) =$$
$$= X(d\phi.X'(f)) - X'(d\phi.X(f)) = X(Y'(f) \circ \phi) - X'(Y(f) \circ \phi) =$$
$$= d\phi(X)(Y'(f)) - d\phi(X')(Y(f)) = Y(Y'(f)) - Y'(Y(f)) = [Y, Y'](f).$$

$\blacksquare$

$\square$

Definition 5 A Lie algebra is a vector space $\mathfrak{g}$ endowed with a bilinear map $[., .] : \mathfrak{g} \to \mathfrak{g}$ satisfying the following identities: for any $X, Y, Z \in \mathfrak{g}$;
(i) $[Y, X] = -[X, Y]$.
(ii) $[[X, Y], Z] + [[Z, X], Y] + [[Y, Z], X] = 0$ (Jacobi's identity).

Therefore the space $\Gamma(U)$ of vector fields defined on an open subset $U \subset \mathbb{R}^n$ is a Lie algebra.

Remark 2 Let's show a useful interpretation for the bracket $[X, Y]$ of $X, Y \in \Gamma(U)$. Consider the points $p_0, p_1, p_2, p_3, p_4 \in U$, as illustrated in Fig. 11 and let $\alpha_0, \alpha_1, \beta_0, \beta_1$ be the integral curves such that:

- α_0, α_1 are integral curves of X, i.e., $\alpha_i'(t) = X(\alpha(t))$. Besides, α_0 satisfies $\alpha_0(0) = p_0$ and $\alpha_0(h) = p_2$, and α_1 satisfies $\alpha_1(0) = p_1$ and $\alpha_1(h) = p_4$.
- β_0, β_1 are integral curves of Y, i.e., $\beta_i'(t) = X(\beta(t))$. β_0 satisfies $\beta_0(0) = p_0$ and $\beta_0(0) = p_1$, and β_1 satisfies $\beta_1(0) = p_2$ and $\beta_1(h) = p_3$.

For any $f \in C^\infty(U)$, let's find the value of $f(p_4) - f(p_3)$. We make use of the approximations in the Taylor series, as follows:

(a)   Let p_i, p_j be α_i-adjacent vertices in the polygon illustrated in Fig. 12. Then we have

$$\alpha_i(h) = \alpha_i(0) + h\alpha_i'(0) + o(h) = \alpha_i(0) + hX(\alpha_i(0)) + o(h),$$

and $\lim \frac{o(h)}{h} = 0$. Also we have $p_2 = p_0 + hX(p_0)$ and $p_4 = p_1 + hX(p_1)$.
(b)   Let p_i, p_j be β_i-adjacent vertices in the polygon illustrated in Fig. 12. Then we have

$$\beta_i(h) = \beta_i(0) + h\beta_i'(0) + o(h) = \beta_i(0) + hX\big(\beta_i(0)\big) + o(h),$$

and $\lim \frac{o(h)}{h} = 0$. So we have $p_1 = p_0 + hY(p_0)$ and $p_3 = p_2 + hX(p_2)$.

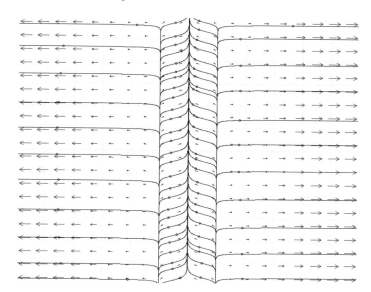

Fig. 11   Phase portrait of system (19)

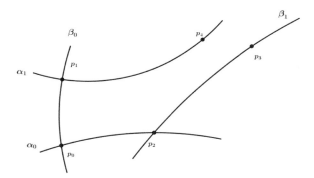

Fig. 12   Coordinates curves

(c)   For any $f \in C^\infty(U)$, let H_f be the Hessian operator of f. So we have the identity.

$$d(df).X.Y = H_f(X, Y).$$

(d)   For any $X, Y \in \Gamma(U)$, we have

$$[X, Y](f) = X\big(Y(f)\big) - Y\big(X(f)\big) = d(df.Y).X - d(df.X).Y =$$
$$= \big(d(df).Y + df.dY\big).X - \big(d(df).X + df.dX\big).Y =$$
$$= (dY.X - dX.Y)(f).$$

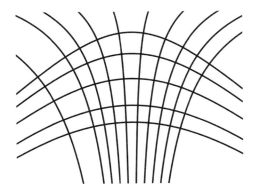

Fig. 13 Coordinate system

Now we can proceed to compute $f(p_4) - f(p_3)$;

$$f(p_4) - f(p_3) = [f(p_4) - f(p_1)] + [f(p_1) - f(p_0)] + [f(p_0) - f(p_2)] + [f(p_2) - f(p_3)] =$$
$$= (df.X)_{p_1} + o(h) + (df.Y)_{p_0} + o(h) - (df.X)_{p_0} + o(h) - (df.Y)_{p_2} + o(h) =$$
$$= [(df.X)_{p_1} - (df.X)_{p_0}] - [((df.Y)_{p_2} - ((df.Y)_{p_0}] + o(h) =$$
$$= [d(df.X).Y]_{p_0} - [d(df.Y).X]_{p_0} + o(h) = df.(dX).Y)_{p_0} + H_f(X, Y) - df.(dY.X)]_{p_0} - H_f(X, Y)$$
$$= df.[dX.Y - dY.X] + o(h) = df.[X, Y] + o(h) = [X, Y](f) + o(h).$$

Since the identity obtained holds for any f, we conclude that the paths close if and only if $[X, Y] = 0$. If $\{(x_1, \ldots, x_n) \in U\}$ defines a coordinate system in U, then $[\frac{\partial}{\partial x_i}, \frac{\partial}{\partial x_j}] = 0$, for all i, j.

Definition 6 Let $U \subset \mathbb{R}^n$. A set of vector fields $\{X_1, \ldots, X_n\} \subset \Gamma(U)$ defines a coordinate system in U if the following conditions are satisfied:
(i) for all $x \in U$, the set $\{X_1(x), \ldots, X_n(x)\}$ is a basis of $T_x U$,
(ii) $[X_i, X_j] = 0$, for all pairs (i, j).
In this case, the integral curves are called coordinate curves , as illustrated in Fig. 13.

Exercise

1. Prove the identities in Proposition 3.

6   Integrability, Frobenius Theorem

Let $U \subset \mathbb{R}^n$ be an open subset and let $\mathcal{D}_x \subseteq T_x U$ be a k-dimensional vector subspace for all $x \in U$. We say that $\mathcal{D} = \cup_{x \in U} \mathcal{D}_x$ is a k-distribution. $\mathcal{D}$ is differentiable if we have a set of k C^∞-vector fields $\beta = \{X_1, \ldots, X_k\} \subset \Gamma(U)$, such that at $x \in U$, the set $\beta(x) = \{X_1(x), \ldots, X_k(x)\}$ is a base of $\mathcal{D}_x$. A field X belongs to the distribution $\mathcal{D}$ if $X(x) \in \mathcal{D}_x$ for all $x \in U$. The set β is a frame on U when $\beta(x)$ is a basis of $T_x U$ for all $x \in U$.

Definition 7 A distribution $\mathcal{D}$ is involutive if $[X, Y] \in \mathcal{D}$ for all pairs of vector fields $X, Y \in \mathcal{D}$.

Now we will consider the integrability of a distribution $\mathcal{D}$. Indeed, this is a generalization of the case $k = 1$, for which Theorem 2 ensures the response is positive. In the previous section, we considered the case $k = 2$ and showed that if the distribution satisfies $[X_1, X_2] = 0$, then locally these fields define a surface. The Lie bracket plays a central role in answering the question.

Definition 8 A k-distribution $\mathcal{D}$ is integrable if there is a submanifold $M^k(x) \subset U$ for all $x \in U$, such that
(i) $x \in M^k(x)$,
(ii) $T_x M^k(x) = \mathcal{D}_x$.

$M^k(x)$ is a leaf of the foliation defined by the distribution $\mathcal{D}$. The distribution $\mathcal{D}$ is maximal if $M^k(x)$ is connected and it is not contained in a larger leaf of the foliation for all $k \in \mathbb{N}$; i.e., if $N^k(x)$ is another leaf containing x, then either $N^k(x) \subset M^k(x)$ or $N^k(x) = M^k(x)$ for all $x \in U$.

For the k-dimensional case, the integrability of a distribution defines a local foliation as illustrated in Figs. 14 and 15. The necessary and sufficient condition for the integrability of a distribution is provided by the Frobenius theorem. Before proving the Frobenius theorem, we will prove two lemmas.

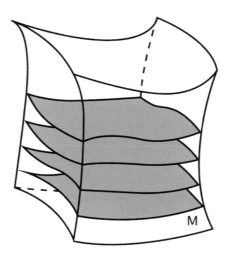

Fig. 14 Foliation A

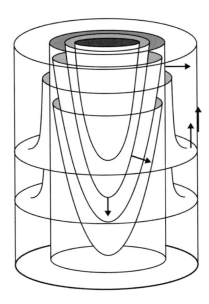

Fig. 15 Foliation B

Lemma 2 *Let $\mathcal{D}$ be an involutive k-distribution in U of class C^∞. So we have a local frame $\beta = \{Y_i \mid 1 \leq i \leq k\}$ such that $[Y_i, Y_j] = 0$ for every pair i, j and for all $x \in U$.*

Proof Let $\beta = \{X_1, \ldots, X_k\} \subset \Gamma(U)$ be a k-frame defining the distribution $\mathcal{D}$ and let $\beta_c = \{e_i \mid 1 \leq i \leq n\}$ be the canonical frame composed by the vector fields $e_i = \frac{\partial}{\partial x_i}$. Let $X_i = \sum_{i=1}^{n} t_{ji} e_j$ be the representation given by the coordinates in basis β_c. The matrix $T = (t_{ij})$ is equivalent by elementary row operations to

$$\begin{pmatrix} I_{k \times k} & A_{(n-k) \times k} \\ 0_{(n-k) \times k} & B_{(n-k) \times (n-k)} \end{pmatrix}.$$

Therefore we have a unique k-frame $\beta' = \{Y_i \mid 1 \leq i \leq k\}$ given by

$$Y_i = e_i + \sum_{i=k+1}^{n} \alpha_{ji} e_j$$

also spanning $\mathcal{D}$. Let $\hat{\mathcal{D}}$ be the distribution generated by the vector fields $\{e_{k+1}, \ldots, e_n\}$. Since $[e_i, e_j] = 0$, this yields that $[Y_i, Y_j] \in \hat{\mathcal{D}}$. Since $\mathcal{D} \cap \hat{\mathcal{D}} = \{0\}$, we have $[Y_i, Y_j] \in \mathcal{D}$, and so $[Y_i, Y_j] = 0$, $\forall i, j$.   □

Lemma 3 *The vector fields* $X, Y \in \Gamma(U)$ *commute* ($[X, Y] = 0$) *if and only if the flows* Φ^X *and* Φ^Y *are also commutative, i.e.,*

$$\Phi^X_t \circ \Phi^Y_s = \Phi^Y_s \circ \Phi^X_t, \; \forall s, t \in \mathbb{R}.$$

Proof (i) ($\Rightarrow$) To prove this direction, the procedure is the same as in Remark 2; we have to show that $[X, Y]$ measures the difference $f(p_4) - f(p_3)$. Assuming t and s are small enough such that

$$\Phi^X_t(x) = x + X(x)h + r_1(h), \; \Phi^Y_s(x) = x + Y(x)h + r_2(h),$$

and $\lim_{h \to 0} \frac{r_i(h)}{h} = 0, i = 1, 2$, we have the identity

$$\Phi^X_t(\Phi^Y_s(x)) - \Phi^Y_s(\Phi^X_t(x)) = [X, Y].h + r(h), \quad \lim_{h \to 0} \frac{r(h)}{h} = 0.$$

(ii) ($\Leftarrow$) Consider the surface $\xi(t, s) = \Phi^X_t(\Phi^Y_s) = \Phi^X(t, \Phi^Y_s)$. Then

$$\frac{\partial \xi}{\partial t} = \frac{\partial \Phi^X(t, \Phi^Y_s)}{\partial t} = X(\Phi^Y_s)$$

and

$$\frac{\partial^2 \xi}{\partial s \partial t} = \frac{\partial X(\Phi^Y_s)}{\partial s} = dX_{\Phi^Y_s}.\frac{\partial \Phi^Y_s}{\partial s} = dX_{\Phi^Y_s}.Y(\Phi^Y_s).$$

Analogously, using the fact that $\xi(t, s) = \Phi^Y_s \circ \Phi^X_t$, we get $\frac{\partial^2 \xi}{\partial t \partial s} = dY_{\Phi^X_t}.X(\Phi^X_t)$. From the hypothesis, we have $\frac{\partial^2 \xi}{\partial s \partial t} = \frac{\partial^2 \xi}{\partial t \partial s}$, and therefore $[X, Y] = dX.Y - dY.X = 0$. $\qquad\square$

Theorem 4 (Frobenius) *A k-dimensional C^∞-distribution $\mathcal{D}$ in U is integrable if and only if it is involutive.*

Proof Let $\beta = \{X_1, \ldots, X_k\} \subset \Gamma(U)$ be a k-frame defining $\mathcal{D}$;
(i) integrable $\Rightarrow$ involutive.
Consider $\iota : M \to U$ the embedding of M such that $\iota(m_0) = x_0$. For $m \in M$, $d\iota : T_m M \to T_{\iota(m)} U$ is an isomorphism, so we have vector fields $\widetilde{X}_1, \ldots, \widetilde{X}_k \in T_m M$, such that $d\iota(\widetilde{X}_i) = X_i \circ \iota$ for all i. $\widetilde{X}_i, \widetilde{X}_j$ are ι-related to X_i, X_j, respectively; therefore $[X_i, X_j] = d\iota([\widetilde{X}_i, \widetilde{X}_j])$ for all pairs i, j. Hence $\mathcal{D}$ is involutive.

(ii) involutive $\Rightarrow$ integrable.
Fix a point $x_0 \in U$. Consider $\beta = \{X_i \mid 1 \le i \le k\}$ a k-frame generating $\mathcal{D}$ in a neighborhood $U_0 \subset U$ of x_0 and satisfying the condition $[X_i, X_j] = 0, \forall i, j$ (Lemma 2). Let $\epsilon > 0$ be such that the flow Φ^{X_i} is defined for $(-\epsilon, \epsilon) \times U_0 \subset \mathbb{R} \times U$ for all i. Considering the k-cube $(-\epsilon, \epsilon)^k$, we define the map

$$\Phi : (-\epsilon, \epsilon)^k \to U,$$

$$(t_1, \ldots, t_k) \to \Phi(t_1, \ldots, t_k) = (\Phi_{t_1}^{X_1} \circ \cdots \circ \Phi_{t_k}^{X_k})(x_0). \tag{22}$$

Let $\hat{t} = (t_1, \ldots, t_k)$. Applying Lemma 3, we get

$$\frac{\partial \Phi}{\partial t_i} \Big|_{t=\hat{t}} = \frac{d[(\Phi_{t_1}^{X_1} \circ \cdots \circ \Phi_{t_k}^{X_k})(x_0)]}{dt} \Big|_{t=\hat{t}} = \frac{d[\Phi_{t_i}^{X_i} \circ (\Phi_{t_1}^{X_1} \circ \cdots \circ \Phi_{t_k}^{X_k})(x_0)]}{dt} =$$

$$= X_i(\Phi_{t_i}^{X_i} \circ [(\Phi_{t_1}^{X_1} \circ \cdots \circ \Phi_{t_k}^{X_k})(x_0)]) = X_i(\Phi(\hat{t})). \tag{23}$$

So Φ is an immersion. Given the Local Immersion Theorem, we can reduce ϵ to $\epsilon_1, 0 < \epsilon_1 < \epsilon$, so that the image of the map $\Phi : (-\epsilon_1, \epsilon_1)^k \to U$ is a k-submanifold $M^k(x_0) \subset U$. Therefore $\mathcal{D}$ is integrable. $\qquad\Box$

7   Lie Groups and Lie Algebras

In Appendix B, we introduce the concept of a Lie Group G; now we will associate G to a Lie Algebra $\mathfrak{g}$. As an application of the Frobenius theorem, we will use the SO(n) example to show that a subgroup H of a group G can be reconstructed when we know its subalgebra $\mathfrak{h} \subset \mathfrak{g}$. Let us first recapitulate some assertions.

Definition 9 Let $g \in G$. A left translation by an element $g \in G$ is the diffeomorphism $L_g : G \to G$, $L_g(h) = g.h$. The right translation by g is the diffeomorphism $R_g : G \to G$, $R_g(h) = h.g$.

That we have left and right translations make the Lie groups a very special class of spaces to be studied.

Definition 10 A vector field X is left-invariant if $dL_\sigma . X = X \circ L_\sigma$ ($X \overset{L_\sigma}{\sim} X$).

Let $e \in G$ be the identity. For $g \in G$, the diffeomorphism $L_g : G \to G$, $L_g(h) = g.h$, induces the isomorphism $dL_g : T_e G \to T_g G$, $dL_g(X) = g.X$. For every $X_e \in T_e G$, the vector field $X(g) = dL_g . X_e$ is left-invariant.

$$(dL_\sigma X)(g) = dL_\sigma . X(g) = dL_\sigma . dL_g . X_e = dL_{\sigma.g} . X_e = X(\sigma.g) = (X \circ L_\sigma)(g).$$

Proposition 5 *Let $\mathfrak{g}$ be the vector space of left-invariant vector fields on a Lie group G. So*
(i) $\mathfrak{g}$ is isomorphic to $T_e G$.
(ii) $\mathfrak{g}$ is a Lie algebra endowed with a Lie bracket.

Proof (i) Since every field $X \in \mathfrak{g}$ is defined by the relation $X(g) = dL_g.X_e$, we consider the map $\theta : \mathfrak{g} \to T_eG$ given by $\theta(X) = X_e$. So θ is a vector space isomorphism.

(ii) The Lie bracket $[X, Y]$ of two fields $X, Y \in \mathfrak{g}$ also belongs to $\mathfrak{g}$ since

$$dL_g.[X, Y] = [dL_gX, dL_gY] = [X \circ L_g, Y \circ L_g] = [X, Y] \circ L_g.$$

$\square$

Definition 11 The Lie algebra of a Lie group G is the vector space $\mathfrak{g}$ of left-invariant vector fields defined on G. A subspace $\mathfrak{h} \subset \mathfrak{g}$ is a Lie subalgebra of $\mathfrak{g}$ if $[X, Y] \in \mathfrak{h}$ whenever $X, Y \in \mathfrak{h}$.

The group $GL_n(\mathbb{R})$ of invertible matrices is a differentiable manifold locally modelled on the Euclidean space $\mathbb{R}^{n^2}$. Indeed, $GL_n(\mathbb{R})$ is an open subset of $M_n(\mathbb{R})$. The tangent plane at identity $I \in GL_n(\mathbb{R})$ is $T_IG_0 = M_n(\mathbb{R})$. Let $\mathfrak{gl}_n(\mathbb{R})$ be the Lie algebra of $GL_n(\mathbb{R})$. The vector space $M_n(\mathbb{R})$ is a Lie algebra endowed with the Lie bracket $[A, B] = A.B - B.A$, so $\mathfrak{gl}_n(\mathbb{R})$ and $M_n(\mathbb{R})$ are isomorphic Lie algebras, since

$$[X, Y](g) = dL_g.[X(I), Y(I)].$$

In Example 7, Chap. 1, we show that the group of orthogonal matrices

$$SO(n) = \{X \in GL_n(\mathbb{R}) \mid X.X^t = X^t.X = I, \det(X) = 1\}$$

is a Lie group of dimension $d(n) = \frac{n(n-1)}{2}$. The tangent plane $T_ISO(n) = \{A \in M_n(\mathbb{R}) \mid A^t = -A\}$ at the identity is exactly the subspace $A_n(\mathbb{R})$ of the real skew-symmetric matrices. We note that $A_n(\mathbb{R})$ is a Lie subalgebra of $\mathfrak{gl}_n(\mathbb{R})$ with the Lie bracket $[A_1, A_2] = A_1A_2 - A_2A_1$, since

$$[A_1, A_2]^t = (A_1A_2 - A_2A_1)^t = A_2^t A_1^t - A_1^t A_2^t = A_2A_1 - A_1A_2 = [A_2, A_1] = -[A_1, A_2].$$

Therefore $[A_1, A_2] \in \mathfrak{so}_n$ satisfies the following items:
(i) $[A_2, A_1] = -[A_1, A_2]$,
(ii) $[[A_1, A_2], A_3] + [[A_3, A_1], A_2] + [[A_2, A_3], A_1] = 0$.

Moreover the Lie algebra $\mathfrak{so}_n$ of $SO(n)$ is isomorphic to the Lie algebra $A_n(\mathbb{R})$. A basis of $\mathfrak{so}_n$ is given by $\beta = \{E_{ij} = (e_{ij}) \mid 1 \le i, j \le n\}$, and the entries of E_{ij} are

$$e_{\alpha\beta} = \begin{cases} 1, & \text{if } \alpha < \beta \text{ and } \alpha = i, \beta = j, \\ -1, & \text{if } \alpha > \beta \text{ and } \alpha = j, \beta = i, \\ 0, & \text{if } \{\alpha, \beta\} \ne \{i, j\}. \end{cases}$$

The dimension of $\mathfrak{so}_n$ is $d(n) = \frac{n(n-1)}{2}$. Let's construct an involutive $d(n)$-distribution $\mathcal{D}$ in $GL_n(\mathbb{R})$. The fields $X_{ij}(g) = dL_g.E_{ij}$ form a basis for $\mathfrak{so}_n$. Set $\mathcal{D}(g)$ to be the subspace $T_g GL_n(\mathbb{R})$ generated by the base $\beta(g) = \{X_{ij}(g) \mid 1 \leq i < j \leq n\}$ and $\mathcal{D} = \{\mathcal{D}(g) \mid g \in GL_n(\mathbb{R})\}$. According to the Frobenius theorem, I is in a maximal leaf M(integral subvariety) of $\mathcal{D}$, which is invariant by left translation $(L_{g^{-1}}(M) = M)$ since the distribution $\mathcal{D}$ is invariant by left translation. We have to show that $M = SO(n)$ is a group, but this follows directly from the fact that $L_{g^{-1}}(M) = M$ since $g^{-1}h \in M$ for all $g, h \in M$. Consequently, M is a subgroup of $GL_n(\mathbb{R})$ with Lie algebra $\mathfrak{so}_n$. And $SO(n)$ as well is an integral submanifold (leaf) of $\mathcal{D}$, and a Lie subgroup of $GL_n(\mathbb{R})$. To verify that $SO(n)$ is maximal, we note that the dimension of $SO(n)$ is $d(n)$ and is a connected, compact topological space.

Exercises

1. Show that $\mathbb{R}^3$ provided with the Lie bracket $[u, v] = u \times v$ is a Lie algebra, isomorphic to $A_3(\mathbb{R})$ given that $\times$ is the vector cross product.
2. Show that there is a bijection between the sets

$$\{\text{Lie subalgebras of } \mathfrak{g}\} \longleftrightarrow \{\text{subgroups of } G\}.$$

3. Let $A \in A$. Consider the ODE $\gamma'(t) = A\gamma(t)$ defined in $\mathbb{R}^n$. Show that the 1-parameter subgroup of diffeomorphisms generated by the flow is $\{\exp(tA) \mid t \in \mathbb{R}\}$.
4. Show that $SO(n)$ is compact and connected.
5. Show that the exponential map $\exp : A_n(\mathbb{R}) \to SO(n)$, $\exp(A) = \sum_{i=0}^{\infty} \frac{A^n}{n!}$, is well-defined and surjective.
6. Show that $\exp : A \to O(n)$ is not surjective.
7. For all $A \in SO(n)$, show that the 1-parameter subgroup of diffeomorphisms of the linear field $F(x) = A.X$ is a subgroup of $SO(n)$.
8. Baker-Campbell-Hausdorff formula: let A, B and C be matrices in $M_n(\mathbb{C})$, such that $\exp(C) = \exp(A).\exp(B)$. Show that there are homogeneous polynomials F_n of degree n such that

$$C = \sum_{i=1}^{\infty} F_n(A, B).$$

Moreover, if $[A, B] = 0$, then $F_n(A, B) = 0$ for all $n \geq 2$. Show that

$$C = A + B + \frac{1}{2}[A, B] + \frac{1}{12}([A, [A, B]] + [[A, B], B]) + \ldots.$$

8 Variations over a Flow, Lie Derivative

Let Φ^X be the flux of the field X and let $\{\phi_t \mid t \in \mathbb{R}\}$ be the 1-parameter subgroup of diffeomorphisms generated by the flow. We want to study the variations of functions and vector fields along the flow of X. We will consider all functions, maps, and vector fields to be C^∞.

Definition 12 Let $U \subset \mathbb{R}^n$ and $V \subset \mathbb{R}^m$ be open sets and let $\phi : U \to V$ be a map in $C^\infty(U, V)$;
(i) Let $f \in C^\infty$, $X \in \Gamma(U)$;
(i.1) the pullback of f by $\phi \in C^\infty(V)$ is $\phi^* f(x) = f(\phi(x)) \in C^\infty(U)$. The pullback induces a map $\phi^* : C^\infty(V) \to C^\infty(U)$.
(i.2) Assuming ϕ is a diffeomorphism, the pushforward of $f \in C^\infty(U)$ induced by ϕ is

$$\phi_* f = f \circ \phi^{-1} \in C^\infty(V). \text{ We have the map } \phi_* : C^\infty(U) \to C^\infty(V).$$

(ii) Let $\phi : U \to V$ be a diffeomorphism.
(ii.1) The pullback of $Y \in \Gamma(V)$ induced by ϕ is $\phi_*(X) = d\phi^{-1} \circ X \circ \phi \in \Gamma(U)$. The pullback induces the map $\phi^* : \Gamma(V) \to \Gamma(U)$.
(ii.2) The pushforward of X by ϕ is $\phi_*(X) = d\phi \circ X \circ \phi^{-1}$. The pushforward induces the map $\phi^* : \Gamma(U) \to \Gamma(V)$.

A diffeomorphism $\psi : U \to V$ also induces the pushforward of flows. The pushforward $\psi^* \Phi^X = \psi \circ \Phi^X \circ \psi^{-1}$ of Φ^X by ψ induces the group homomorphism $\mathrm{Diff}(U) \to \mathrm{Diff}(V)$, $\Phi_t^X \to \psi \circ \Phi_t^X \circ \psi^{-1}$. The pushforward of Φ^X with the diffeomorphisms $\phi_t = \Phi_t^X$ is Φ^X itself.
We fix an orthonormal basis $\beta = \{e_1, \ldots, e_n\}$ of R^n and let $X = \sum_i x_i e_i$. Consider $\phi : U \to V$ a diffeomorphism and $q = \phi(p)$. Then

$$d\phi_p(X(p)) = \sum_i d\phi_p(e_i) = \sum_i x_i(p) \frac{\partial \phi}{\partial x_i}(p) = \sum_i \left[\sum_j x_j(p) \frac{\partial \phi^i}{\partial x_j}(p) \right] e_i.$$

Two fields can be related by a diffeomorphism, for example by a coordinate change; therefore it is necessary to describe the action of the diffeomorphisms on functions and maps.

Proposition 6 *The pullback and the pushforward of functions enjoy the following algebraic properties;*
(i) the pullback and the pushforward of functions are linear operators and

$$\phi^*(f.g) = \phi^*(f).\phi^*(g), \quad \phi_*(f.g) = \phi_*(f).\phi_*(g). \tag{24}$$

(ii) Let $\phi : U \to V$ and $\psi : V \to W$, so $(\psi \circ \phi)^* = \phi^* \circ \psi^*$ and $(\psi \circ \phi)_* = \phi_* \circ \psi_*$.

Proof The proofs are only immediate from the definitions, so they will be left as an exercise.

$\square$

Definition 13 Let $f \in C^\infty(U)$ and $X \in \Gamma(U)$. The Lie derivative of f in the direction of the vector field X is the operator

$$\mathcal{L}_X : C^\infty(U) \to C^\infty(U),$$
$$(\mathcal{L}_X f)(x) = X(f)(x) = df_x.X(x). \tag{25}$$

Therefore the Lie derivative is a derivative operator satisfying the following identities; for all $a, b \in \mathbb{R}$ and $f, g \in C^\infty(U)$,
(i) $\mathcal{L}_X(af + bg) = a\mathcal{L}_X(f) + b\mathcal{L}_X(g)$ ($\mathbb{R}$-linear),
(ii) $\mathcal{L}_X(f.g) = \mathcal{L}_X(f).g + f.\mathcal{L}_X(g)$ (Leibniz's rule).

Proposition 7 *Let $\psi : U \to V$ be a diffeomorphism. So $\mathcal{L}_X$ is natural with respect to the pushforward by ψ, i.e., for all $f \in C^\infty(U)$, we have $\mathcal{L}_{\psi_* X}(\psi_* f) = \psi_*(\mathcal{L}_X(f))$. Equivalently, the diagram below commutes;*

$$
\begin{array}{ccc}
C^\infty(U) & \xrightarrow{\psi_*} & C^\infty(V) \\
\downarrow{\scriptstyle \mathcal{L}_X} & & \downarrow{\scriptstyle \mathcal{L}_{\psi_*}} \\
C^\infty(U) & \xrightarrow{\psi_*} & C^\infty(V).
\end{array}
\tag{26}
$$

Proof Let $x \in U$,

$$\mathcal{L}_{\psi_* X}(\psi_* f)(x) = d(f \circ \psi^{-1}).(\psi_* X)(x) = d(f \circ \psi^{-1}).(d\psi.X.\psi^{-1})(x) =$$
$$= df_{\psi^{-1}(x)}.d\psi_x^{-1}.d\psi.X(\psi^{-1}(x)) = df_{\psi^{-1}(x)}.X(\psi^{-1}(x)) = \psi_*(\mathcal{L}_X(f)).$$

$\square$

For maps, we have the following assertion;

Proposition 8 *Let $\psi : U \to V$ be a map, $X \in \Gamma(U)$ and $Y \in \Gamma(V)$. If $X \overset{\psi}{\sim} Y$, then $\mathcal{L}_X(\psi^* f) = \psi^*(\mathcal{L}_Y(f))$ for all maps $f : V \to F$ given that F is a Banach space.*

$$
\begin{array}{ccc}
C^\infty(V, F) & \xrightarrow{\psi^*} & C^\infty(U, F) \\
\downarrow{\scriptstyle \mathcal{L}_Y} & & \downarrow{\scriptstyle \mathcal{L}_X} \\
C^\infty(V, F) & \xrightarrow{\psi^*} & C^\infty(U, F).
\end{array}
\tag{27}
$$

Proof For all $x \in U$,

$$\mathcal{L}_X(\psi^* f)(x) = d(f \circ \psi)_x . X(x) = df_{\psi(x)}.d\psi_x . X(x) = df_{\psi(x)}.Y(\psi(x)) = \psi^*(\mathcal{L}_Y f)(x).$$

□

Definition 14 Let $X, Y \in \Gamma(U)$. The Lie derivative of Y with respect to a vector field X at $x \in U$ is

$$\mathcal{L}_X Y = \lim_{t \to 0} \frac{d\Phi^X_{-t}\left(Y(\Phi^X_t)\right) - Y(x)}{t}. \tag{28}$$

In the definition above, we need to apply the isomorphism $d\Phi^X_{-t} = (d\Phi^X_t)^{-1}$: $T_{\Phi^X_t(x)}U \to T_x U$ (Fig. 16) to compute $\mathcal{L}_X Y$ since the vectors $Y(\Phi^X_t(x))$ and $Y(x)$ do not belong to the same tangent plane (Fig. 17).

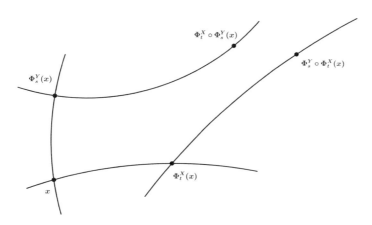

Fig. 16 Comparing $\Phi^X_t \circ \Phi^Y_s(x)$ with $\Phi^Y_s \circ \Phi^X_t(x)$

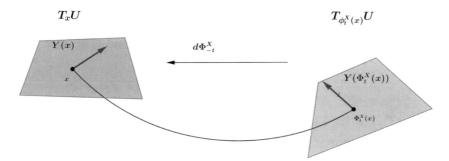

Fig. 17 Lie derivative

Proposition 9 *Let* $X, Y \in \Gamma(U)$, *so*

$$\mathcal{L}_X Y = [X, Y].\tag{29}$$

Proof According to the definition,

$$(\mathcal{L}_X Y)(f) = \left(\lim_{t \to 0} \frac{d\Phi_{-t}^X(Y(\Phi_t^X)) - Y(x)}{t}\right)(f) = \frac{d}{dt}\left[d\Phi_{-t}^X(Y(\Phi_t^X))(f)\right]|_{t=0} =$$

$$= \frac{d}{dt}\left[Y(\Phi_t^X)(f \circ \Phi_{-t}^X)\right]|_{t=0} .\tag{30}$$

For any $x \in U$, when we consider the function $H(t, u) = f\big((\phi_{-t}^X \circ \Phi_u^Y \circ \Phi_t^X)(x)\big)$, while setting $q = (\phi_{-t}^X \circ \Phi_u^Y \circ \Phi_t^X)(x)$, we have

$$\frac{\partial H}{\partial u}\,|_{(t,0)} = df_q . \frac{\partial}{\partial u}\left(\Phi_{-t}^X(\Phi_u^Y(\Phi_t^X(x)))\right)|_{(t,0)} = df_q . d\Phi_{-t}^X . Y(\Phi_t^X(x)) = Y(\Phi_t^X(x))(f \circ \Phi_{-t}^X).$$

Therefore we get $\frac{\partial^2 H}{\partial t \partial u}\,|_{(0,0)} = (\mathcal{L}_X Y)(f)$. We have to find $\frac{\partial^2 H}{\partial t \partial u}\,|_{(0,0)}$. Consider the function $K(t, u, s) = f(\Phi_s^X \circ \Phi_u^Y \circ \Phi_t^X(x))$, so $H(t, u) = K(t, u, -t)$. In this way,

$$\frac{\partial H}{\partial u} = \frac{\partial K}{\partial u}, \quad \frac{\partial^2 H}{\partial t \partial u}\Big|_{(0,0)} = \frac{\partial^2 K}{\partial t \partial u}\Big|_{(0,0,0)} - \frac{\partial^2 K}{\partial s \partial u}\Big|_{(0,0,0)}.$$

Since $K(t, u, 0) = f(\Phi_u^Y \circ \Phi_t^X(x))$, and considering $\hat{q} = (\Phi_u^Y \circ \Phi_t^X)(x)$, it follows that the identity $\frac{\partial K}{\partial u}\,|_{(0,0,0)} = df_{\hat{q}} . Y(\Phi_t^X(x))(f)$ and, consequently, we have

$$\frac{\partial^2 K}{\partial t \partial u}\,|_{(0,0,0)}(f) = \{d(df)_{\hat{q}} . (d\Phi_u^Y . X(\Phi_t^X)) . Y(\Phi_t^X(x)) + df_{\hat{q}} . dY_{\Phi_t^X} . X(\Phi_t^X))\}|_{(0,0,0)}(f) =$$

$$= \left\{d(df)_x . (\underbrace{d\Phi_0^Y . X(x))}_{0} . Y(x) + df_x . dY_x . X(x)\right\}(f) = \big(X(Y(f))\big)(x).$$

Analogously, we consider $K(0, u, s) = f(\Phi_s^X \circ \Phi_u^Y(x))$, and so we have

$$\frac{\partial K}{\partial s} = X(f(\phi_u^Y)), \quad \frac{\partial^2 K}{\partial u \partial s}\,|_{(0,0,0)} = Y(x)(X(f)).$$

Hence $\mathcal{L}_X Y = [X, Y]$.   □

Corollary 1 *Let* $X, Y \in \Gamma(U)$, *so*

$$\frac{d}{dt}\big((\Phi_t^X)^* Y\big) = (\Phi_t^X)^*\big([X, Y]\big)\tag{31}$$

In particular , $\frac{d}{dt}\big((\Phi_t^X)^* Y\big)\,|_{t=0} = [X, Y]$.

Proof To prove the assertion it is enough to note that $(\Phi_t^X)^* Y = d\Phi_{-t}^X \circ Y \circ \Phi_t^X$ is exactly the term in Eq. 30.   □

Exercise

1. Show the Proposition 6 statements.

9 Gradient, Curl and Divergent Differential Operators

In this section, we will define the Gradient, Curl, and Divergent operators that play a prominent role in the theories of Differentiability and Integrability of maps. Let $U \subset \mathbb{R}^n$ be an open subset and let $\beta = \{e_i \mid 1 \leq i \leq n\}$ be an orthonormal basis of $\mathbb{R}^n$. The differential operator "*nabla*" is

$$\nabla = \sum_{i=1}^{n} \frac{\partial}{\partial x_i} . e_i. \tag{32}$$

Definition 15 The gradient operator $\nabla : C^\infty(U) \to C^\infty(U)$ is

$$\nabla(f) = \sum_{i=1}^{n} \frac{\partial f}{\partial x_i} e_i. \tag{33}$$

Proposition 10 (Gradient properties) *Let* $a, b \in \mathbb{R}$ *and* $f, g \in C^\infty(U)$;
(i) $\nabla(af + bg) = a\nabla f + b\nabla g$ (*$\mathbb{R}$-linear*).
(ii) $\nabla(f.g) = \nabla f.g + f.\nabla g$ (*Leibniz's rule*).

Definition 16 Let $U \subset \mathbb{R}^3$ be an open subset and let $\beta = \{e_i \mid 1 \leq i \leq 3\}$ be an orthonormal basis of $\mathbb{R}^3$.

1. The curl operator curl : $\Gamma(U) \to \Gamma(\mathbb{R}^3)$, $V = \sum_{i=1}^{3} v_i e_i \to$ curl(V), is

$$\text{curl}(V) = \sum_{\{1,2,3\}} \left(\frac{\partial v_3}{\partial x_2} - \frac{\partial v_2}{\partial x_3} \right) e_1 + \left(\frac{\partial v_1}{\partial x_3} - \frac{\partial v_3}{\partial x_1} \right) e_2 + \left(\frac{\partial v_2}{\partial x_1} - \frac{\partial v_1}{\partial x_2} \right) e_3.$$

$$\tag{34}$$

Using the cross vector product $\times : \mathbb{R}^3 \times \mathbb{R}^3 \to \mathbb{R}^3$ and ∇, we have curl(V) = $\nabla \times V$.

2. The divergent operator div : $\Gamma(U) \to C^\infty(U)$ is

$$V \to \text{div}(V),$$

$$V = \sum_{i=1}^{n} v_i e_i \to \text{div}(V) = \sum_{i=1}^{n} \frac{\partial v_i}{\partial x_i}. \tag{35}$$

The divergent is also defined by $\mathrm{div}(V) = \nabla.V$ (the dot "." corresponds to the inner product);

$$\nabla.V \;=< \nabla, V > \;=< \sum_{i=1}^{n} \frac{\partial}{\partial x_i} e_i, \sum_{j=1}^{n} v_j e_j > = \sum_{i=1}^{n} \frac{\partial v_i}{\partial x_i}.$$

3. The Laplacian operator on functions is

$$\triangle f = \mathrm{div}(\nabla f) = < \sum_{i=1}^{n} \frac{\partial}{\partial x_i} e_i, \sum_{j=1}^{n} \frac{\partial f}{\partial x_j} e_j > = \sum_{i=1}^{n} \frac{\partial^2 f}{\partial x_i^2}.$$

We have the following fundamental identities in $\mathbb{R}^3$;

$$(i) \; \nabla \times (\nabla f) = 0, \quad (ii) \; \nabla.(\nabla \times V) = 0. \tag{36}$$

Exercises

1. Show that the curl operator satisfies the following identities; let $a, b \in \mathbb{R}$ and $V, W \in \Gamma(U))$;
   (i) If $V = P\vec{i} + Q\vec{j} + R\vec{k}$, show the curl(V) corresponds with taking the skew-symmetric part of the Jacobian matrix dV.
   (ii) $\nabla \times (aV + bW) = a.\nabla \times V + b.\nabla \times W$.
   (iii) $\nabla \times (f.V) = \nabla f \times V + f.\nabla \times V$.
   (iv) $\nabla \times (V \times W) = [\nabla.W + < W, \nabla >](V) - [\nabla.V + < V, \nabla >](W)$.
   (v) $\nabla \times (\nabla V) = 0$, for all $V \in \Gamma(U)$.
   (vi) $\nabla \times (\nabla \times V) = \nabla(\nabla.V) - \triangle V$.
2. Show that the divergent operator satisfies the following identities; let $a, b \in \mathbb{R}$ and $V, W \in \Gamma(U))$;
   (i) $\nabla.(aV + bW) = a.\nabla.V + b.\nabla.W$.
   (ii) $\nabla.(f.V) = < \nabla f, V > + f\nabla.V$.
   (iii) $\nabla.(V \times W) = < \nabla \times V, W > - < V, \nabla \times W >$.
   (iv) $\triangle f = \nabla.(\nabla f)$.
   (v) $\nabla.(\nabla \times V) = 0$, for all $V \in \Gamma(U)$.

The symbols (∇), $(\nabla \times)$ and $(\nabla \cdot)$ are concomitantly used with the symbols grad, curl and div, respectively.

Remark 3 The grad and div operators depend on the inner product in $\mathbb{R}^n$, hence the Laplacian $\triangle$ also depends on the inner product. Consider $\mathbb{R}^n$ endowed with the inner product $< ., . >: \mathbb{R}^n \times \mathbb{R}^n \to \mathbb{R}$ where the matrix relative to the canonical basis $\beta = \{e_1, \ldots, e_n\}$ is $g = (g_{ij})$ and the coefficients g_{ij} are constants.

1. grad.
   Given a function f, the grad(f) is

$$df_x.u = < \text{grad}(f)(x), u > . \tag{37}$$

It follows from Proposition 2, Appendix A, that

$$\text{grad}(f) = \sum_{i=1}^{n} \left(\sum_{j=1}^{n} g^{ij} \frac{\partial f}{\partial x_j} \right) e_i. \tag{38}$$

2. div.

The trace of a matrix $A = (a_{ij})$ with respect to the inner product $G = (g_{ij})$ in $\mathbb{R}^n$ is $\text{tr}_g(A) = \sum_{i,j=1}^{n} g^{ij} a_{ij}$. Then the divergent of a vector field $V = \sum_i v_i e_i$ is

$$\text{div}(V) = \text{tr}_g(dV) = \sum_{i,j=1}^{n} g^{ij} \frac{\partial V_i}{\partial x_j}. \tag{39}$$

3. Laplacian.

The Laplacian of a differentiable function is

$$\triangle f = \text{tr}_g(d\nabla f) = \sum_{i,j=1}^{n} g^{ij} \frac{\partial^2 f}{\partial x_{ij}}. \tag{40}$$

Considering $d^2 f = (\frac{\partial^2 f}{\partial x_{ij}})$ the Hessian matrix of f, we have $\triangle f = \text{tr}_g(d^2 f)$.

4. In Riemannian geometry, the inner product is replaced by a Riemannian metric. A Riemannian metric induces on the tangent plane $T_p U$ an inner product $g_p : T_p U \times T_p U \to \mathbb{R}$. The map $p \to g_p$ is C^∞. We recommend [7] for further reading on Riemannian geometry. We remark that the operator $d(\nabla f)$ depends on the covariant derivative, defined by the Riemannian connection, which depends on Christoffel's symbols. As shown in Sect. 7 in Chap. 3, if the entries g_{ij} are constants, then the Christoffel's symbols are null. In the context of Riemannian geometry, operators have the following form: let $V = \sum_{i=1}^{n} V_i e_i$;

$$\nabla f = \sum_{i=1}^{n} \left(\sum_{j=1}^{n} g^{ij} \frac{\partial f}{\partial x_j} \right) e_i,$$

$$\text{div}(V) = \sum_{i=1}^{n} \frac{1}{\sqrt{\det(g)}} \frac{\partial}{\partial x_i} \left(\sqrt{\det(g)} V_i \right),$$

$$\triangle f = \sum_{i=1}^{n} \frac{1}{\sqrt{\det(g)}} \frac{\partial}{\partial x_i} \left(\sqrt{\det(g)} g^{ij} \frac{\partial}{\partial x_j} \right).$$

In the next chapter, we will give an interpretation of the operators curl and div.

Chapter 5
Vector Integration, Potential Theory

1 Vector Calculus

We will review some operations in vector calculus that will motivate using differential forms when integrating vector fields. The differential forms formalism allows us to generalize the Stokes Theorem to describe the conditions of integrability (Frobenius Theorem), and to write Maxwell's equations succinctly to obtain topological invariants using differentiable tools and many other applications.

1.1 Line Integral

To review the line integral, we use the Physicist's concept of Work. Consider $\vec{F}$ to be a force field acting on the point of mass m. Assume that the magnitude $F = |\vec{F}|$ of the force field is constant and the point undergoes a displacement on a straight line that is s-units from the initial point. In this case, the work W done by the force $\vec{F}$ is

$$W = F \cdot s.$$

When a point of mass m moves on a plane, the displacement describes a C^1-curve $\gamma : [a, b] \to \mathbb{R}^2$ parametrized by $\gamma = (\gamma_1, \gamma_2)$. To determine the Work done, we use a polygonal curve $\gamma_{\mathcal{P}}$ to approximate γ. To define the polygonal, we use a partition $\mathcal{P} = \{a = t_1, \ldots, t_i, \ldots, t_n = b\}$ of an interval $[a, b]$. Considering $P_i = \gamma(t_i)$ the vertices, the polygonal is obtained by joining the vertices with the straight line segments $\overline{P_{i-1}P_i}$. We fix a set of points

$$\{t_i^*; 1 \le i \le n, t_i^* \in [t_{i-1}, t_i]\}.$$

© Springer Nature Switzerland AG 2021
C. M. Doria, *Differentiability in Banach Spaces, Differential Forms and Applications*,
https://doi.org/10.1007/978-3-030-77834-7_5

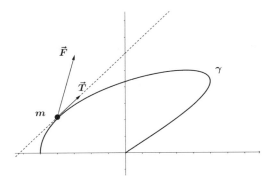

Fig. 1 Work

The Work done by the particle along the straight line segment between the points $\overline{P_{i-1} P_i}$ approximates $W_i = <F(t_i^*), T(t_i^*)> \Delta s_i$, with $T(t_i^*) = \frac{\gamma'(t_i^*)}{|\gamma'(t_i^*)|}$ the tangent vector to γ, at time t_i^*, and

$$\Delta s_i = \sqrt{\left[\frac{\gamma_1(t_{i+1}) - \gamma_1(t_i)}{t_{i+1} - t_i}\right]^2 + \left[\frac{\gamma_2(t_{i+1}) - \gamma_2(t_i)}{t_{i+1} - t_i}\right]^2} \Delta t_i \approx$$

$$\approx \sqrt{[\gamma_1'(t_i^*)]^2 + [\gamma_2'(t_i^*)]^2} . \Delta t_i = | \gamma'(t_i^*) | . \Delta t_i.$$

So the total Work performed along the polygonal line $\gamma_{\mathcal{P}}$ is approximately

$$W_{\mathcal{P}} = \sum_{i=1}^{n} W_i = \sum_{i=1}^{n} <F(t_i^*), T(t_i^*)> \Delta s_i = \sum_{i=1}^{n} <F(t_i^*), T(t_i^*)> | \gamma'(t_i^*) | . \Delta t_i =$$

$$= \sum_{i=1}^{n} <F(t_i^*), \frac{\gamma'(t_i^*)}{| \gamma'(t_i^*)}> . | \gamma'(t_i^*) | . \Delta t_i = \sum_{i=1}^{n} <F(t_i^*), \gamma'(t_i^*)> \Delta t_i.$$

Therefore when the trajectory of a point of mass m is a curve parametrized by γ, the Work W performed along γ is approximately the Work $W_{\mathcal{P}}$ done on $\gamma_{\mathcal{P}}$. Intuitively, when $n \to \infty$, the value of $\| \mathcal{P} \| = \sup_{1 \leq i \leq n} | t_{i+1} - t_i |$ becomes arbitrarily small and the value of $W_{\mathcal{P}}$ becomes close to the value of W. Now we define

$$W = \lim_{\|\mathcal{P}\| \to 0} W_{\mathcal{P}}.$$

If the limit exists, then

$$W = \int_a^b <F(t), \gamma'(t)> dt. \tag{1}$$

Taking $\vec{F} = F_1 e_1 + F_2 e_2$ and $\gamma = (\gamma_1, \gamma_2)$, we have

$$< F(t), \gamma'(t) > dt = F_1(t)\gamma_1'(t)dt + F_2(t)\gamma_2'(t)dt.$$

The starting point for differential forms is to define the linear functionals dx_1, dx_2 as follows: Let $\beta = \{e_1, e_2\}$ be the canonical basis of $\mathbb{R}^2$;

$$dx_1(e_i) = \delta_{1i}, \quad dx_2(e_i) = \delta_{2i}.$$

Then for any vectors $u = u_1 e_1 + u_2 e_2$, $v = v_1 e_1 + v_2 e_2 \in T_p\mathbb{R}^2$, $r, s \in \mathbb{R}$ and $p \in \mathbb{R}^2$, we get

$$dx_1(ru + sv) = ru_1 + sv_1, \quad dx_2(ru + sv) = ru_2 + sv_2.$$

By defining $d\gamma = \frac{d\gamma}{dt}dt = (\gamma_1'dt, \gamma_2'dt)$, we have

$$dx_1(d\gamma) = \gamma_1'(t)dt \quad \text{and} \quad dx_2(d\gamma) = \gamma_2'(t)dt.$$

Therefore

$$W = \int_a^b < F(\gamma(t)), \gamma'(t) > dt = \int_a^b F_1(\gamma(t))\gamma_1'(t)dt + F_2(\gamma(t))\gamma_2'(t)dt =$$

$$= \int_a^b \{F_1(\gamma(t))dx_1(d\gamma) + F_2(\gamma(t))dx_2(d\gamma)\}.$$

$$(2)$$

At every point $\gamma(t) \in U$, consider the linear functional $w_{\gamma(t)} : T_{\gamma(t)}U \to \mathbb{R}$ given by

$$w_\gamma = F_1(\gamma(t))dx_1(d\gamma) + F_2(\gamma(t))dx_2(d\gamma).$$

Using this formalism, Eq. (2) is

$$W = \int_\gamma w. \qquad (3)$$

The Work done along γ is obtained as follows:

$$W = \int_\gamma w = \int_\gamma (F_1 dx_1 + F_2 dx_2) = \int_\gamma \{F_1 dx_1 + F_2 dx_2\}_{\gamma(t)} =$$

$$= \int_a^b (F_1(\gamma(t))dx_1(\gamma'(t)) + F_2(\gamma(t))dx_2(\gamma'(t))) = \int_a^b (F_1(\gamma(t))\gamma_1'(t)dt + F_2(\gamma(t))\gamma_2'(t)dt) =$$

$$= \int_a^b (F_1(\gamma(t))\gamma_1'(t) + F_2(\gamma(t))\gamma_2'(t)) dt.$$

1.2   Surface Integral

We will consider the following physical problem: find the flow throughout of a bounded surface $S \subset \mathbb{R}^3$ of a vector field $\vec{V}$. The vector field could be an electric field, the velocity of a fluid, an electric current and many other examples. Let $\phi : \mathcal{U} \to \mathbb{R}^3$ be a C^∞-parametrization of S given in coordinates as

$$\phi(u, v) = \big(\phi_1(u, v), \phi_2(u, v), \phi_3(u, v)\big), \quad (u, v) \in \mathcal{U}.$$

We assume S is an orientable surface, which means we have a non-null normal C^∞-vector field N given by

$$N = \phi_u \times \phi_v,$$

in which we have $\phi_u = \partial_u \phi = \frac{\partial \phi}{\partial u}$ and

$$\phi_u \times \phi_v = \big(\partial_u\phi_2 \partial_v\phi_3 - \partial_u\phi_3 \partial_v\phi_2 \,,\, \partial_u\phi_3 \partial_v\phi_1 - \partial_u\phi_1 \partial_v\phi_3 \,,\, \partial_u\phi_1 \partial_v\phi_2 - \partial_u\phi_2 \partial_v\phi_1\big).$$

The unitary normal vector is $\hat{n} = \frac{N}{|N|}$.

The idea to define the flow through S is to approximate S by a polyhedron $S_{\mathcal{P}}$ with faces that are images of parallelograms by the parametrization of S. If the vector field $\vec{V}$ is tangent to S, then the flow of $\vec{V}$ across the surface is null. Intuitively, we consider the case when people are walking parallel to a wall (orthogonally to the wall's normal direction) with a door; in this case, there is no flow of people through the door. This example gives good motivation for understanding the flow throughout S.

To define $S_{\mathcal{P}}$, we consider a partition $\mathcal{P}$ of $\mathcal{U}$ by taking a subdivision of $\mathcal{U}$ into small rectangles $\hat{R}_i$, each with area ΔA_i, $i = 1, \ldots, n$. For i, we fix an arbitrary point $P_i^* = (u_i^*, v_i^*) \in \hat{R}_i$. The images of the vertices of rectangle $\hat{R}_i$ define the vertices of the polyhedron $S_{\mathcal{P}}$. Let's consider $\mathcal{U} = [a, b] \times [c, d]$. Taking partitions $\mathcal{P}_u = \{a = u_1, \ldots, u_n = b\}$ and $\mathcal{P}_v = \{c = v_1, \ldots, v_n = d\}$, we define $\mathcal{P} = \mathcal{P}_u \times \mathcal{P}_v = \{(u_i, v_i) \mid 1 \le i \le n\}$, i.e., (see Fig. 2)

$$\mathcal{U} = \bigcup_{k=1}^{n} \bigcup_{l=1}^{m} [u_{k-1}, u_k] \times [v_{l-1}, v_l].$$

Now we fix an order on the set of rectangles $R_{kl} = [u_{i-1}, u_i] \times [v_{j-1}, v_j]$, so that they are indexed by a single index, and we can write $\mathcal{U} = \cup_i R_i$. Each face of the polyhedron $S_{\mathcal{P}}$, denoted by S_i, is determined by the image of the vertices of R_i. For S_i, with area ΔS_i, we approximate the vector field $\vec{V}$ by the constant value $\vec{V}(u_i^*, v_i^*)$, given that $(u_i^*, v_i^*) \in \hat{R}_i$. We define the flow through S_i to be

$$\Phi_i = <\vec{V}(u_i^*, v_i^*), \hat{n}(u_i^*, v_i^*)> .\Delta S_i, \qquad \hat{n} = \frac{N}{|N|}. \tag{4}$$

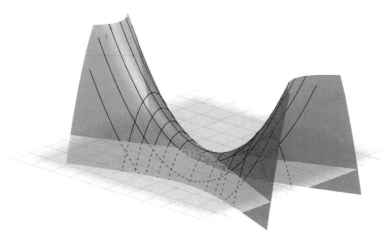

Fig. 2 Polyhedral surface

The area of S_i is approximately $\Delta S_i = | N(u_i^*, v_i^*) | . \Delta A_i$. So now the total flow through $S_{\mathcal{P}}$ is approximately

$$\Phi_{S_{\mathcal{P}}} = \sum_{i=1}^{n} \Phi_i.$$

Let $|| \mathcal{P} || = \sup_{1 \leq i \leq n} \Delta A_i$ be the measure of the largest mesh. Taking the limit $|| \mathcal{P} || \to 0$, the flow through the polyhedron is approximated to the flow throughout S. We define the total flow throughout S as the limit

$$\Phi_S = \lim_{||\mathcal{P}|| \to 0} \Phi_{S_{\mathcal{P}}}. \tag{5}$$

The limit above always exists whenever V is a continuous vector field and S is a C^∞-surface.

Definition 1 Let S be an orientable compact surface embedded in $\mathbb{R}^3$ and parametrized by a C^∞-map $\phi : \mathcal{U} \to \mathbb{R}^3$. The flow of V throughout S is

$$\Phi_S = \int_{\mathcal{U}} < V(u, v), N(u, v) > dA.$$

The integral formula is obtained from Eqs. (4) and (5) since

$$\Phi_S = \int_{\mathcal{U}} < V(u, v), \frac{N(u, v)}{| N(u, v) |} > . | N(u, v) | dA =$$

$$= \int_{\mathcal{U}} < V(u, v), N(u, v) > dA.$$

In coordinates, we have

$$< V(u, v), N(u, v) >= V_1(u, v)N_1(u, v) + V_2(u, v)N_2(u, v) + V_3(u, v)N_3(u, v)$$

and

$$\begin{aligned} V_1 N_1 &= V_1.\left(\partial_u\phi_2\partial_v\phi_3 - \partial_u\phi_3\partial_v\phi_2\right) \\ V_2 N_2 &= V_2.\left(\partial_u\phi_3\partial_v\phi_1 - \partial_u\phi_1\partial_v\phi_3\right) \\ V_3 N_3 &= V_3.\left(\partial_u\phi_1\partial_v\phi_2 - \partial_u\phi_2\partial_v\phi_1\right). \end{aligned} \tag{6}$$

Similar to the line integral, we fix a basis $\beta = \{e_1, e_2, e_3\}$ of $\mathbb{R}^3$ and define the linear functionals $dx_i : T_p\mathbb{R}^3 \to \mathbb{R}, i = 1, 2, 3$. These functionals form the dual basis $\beta^* = \{dx_1, dx_2, dx_3\}$ of β, i.e., $dx_i(e_j) = \delta_{ij}$. We now consider the following skew-symmetric bilinear functions: taking the vectors $\vec{u} = \sum_i u_i e_i$ and $\vec{v} = \sum_j v_j e_j$ in $T_p\mathbb{R}^3$, we consider the product

$$\begin{aligned} dx_i \wedge dx_j :T_p\mathbb{R}^3 &\times T_p\mathbb{R}^3 \to \mathbb{R} \\ dx_i \wedge dx_j(\vec{u}, \vec{v}) &= dx_i(\vec{u}).dx_j(\vec{v}) - dx_i(\vec{v}).dx_j(\vec{u}) = (u_i v_j - v_i u_j). \end{aligned}$$

We note $dx_i \wedge dx_j = -dx_j \wedge dx_i$, in particular $dx_i \wedge dx_i = 0$. Finally, we can write Eq. (6) as

$$\begin{aligned} V_1 N_1 &= V_1.dx_2 \wedge dx_3\left(\partial_u\phi, \partial_v\phi\right) \\ V_2 N_2 &= V_2.dx_3 \wedge dx_1\left(\partial_u\phi, \partial_v\phi\right) \\ V_3 N_3 &= V_3.dx_1 \wedge dx_2\left(\partial_u\phi, \partial_v\phi\right). \end{aligned} \tag{7}$$

Therefore

$$\Phi_\Omega = \int_{\mathcal{U}} \left(V_1 dx_2 \wedge dx_3 + V_2 dx_3 \wedge dx_1 + V_3 dx_1 \wedge dx_2\right)(\partial_u\phi, \partial_v\phi)dudv. \tag{8}$$

Considering

$$w = V_1 dx_2 \wedge dx_3 + V_2 dx_3 \wedge dx_1 + V_3 dx_1 \wedge dx_2,$$

the flow is written as

$$\Phi_S = \int_S w. \tag{9}$$

Exercises

(i) Show that every vector V tangent to S can be written as the linear combination $V = v_1\partial_u\phi + v_2\partial_v\phi$.

(ii) The differential of the map $\phi : \mathcal{U} \to \mathbb{R}^3$, at (u, v), is $d\phi_{(u,v)} : T_{(u,v)}\mathbb{R}^2 \to T_{\phi(u,v)}\mathbb{R}^3$. The Jacobian matrix is

$$d\phi_{(u,v)} = \begin{pmatrix} \partial_u\phi_1 & \partial_v\phi_1 \\ \partial_u\phi_2 & \partial_v\phi_2 \\ \partial_u\phi_3 & \partial_v\phi_3 \end{pmatrix}. \tag{10}$$

Show that the normal vector $N = \phi_u \times \phi_v$ is given in coordinates as

$$N(u, v) = \left(\frac{\partial(\phi_2, \phi_3)}{\partial(u, v)}, \frac{\partial(\phi_3, \phi_1)}{\partial(u, v)}, \frac{\partial(\phi_1, \phi_2)}{\partial(u, v)} \right),$$

and each coordinate is a smaller determinant of the matrix (10).

(iii) Let S be a surface parametrized by a C^∞-map $\phi : \mathcal{U} \to \mathbb{R}^3$. Consider $\vec{w}_1, \vec{w}_2$ tangents to S. Find the products $dx_i \wedge dx_j(\vec{w}_1, \vec{w}_2)$, for $i = 1, 2, 3$. Use the fact that $T_pS = \text{Im}(d\phi)$ and $d\phi = (\partial_u\phi)du + (\partial_v\phi)dv$.

2   Classical Theorems of Integration

We will state the classical integration theorems in a convenient way for our purposes. Let's start with the Fundamental Theorem of Calculus. Let $f : [a, b] \to \mathbb{R}$ be a continuous function and

$$g(x) = \int_a^x f(t)dt.$$

So $g'(x) = f(x)$ and

$$\int_a^b f(t)dt = g(b) - g(a). \tag{11}$$

The boundary of $[a, b]$ is the set $\partial([a, b]) = \{a, b\}$. By orienting the boundary,[1] we have $\partial[a, b] = b - a$. Therefore we can write the formula (11) as

$$\int_{[a,b]} g'(t)dt = \int_{\partial([a,b])} g(t)dt. \tag{12}$$

The theorems about vector field integration generalize (12). To state them, we will use the differential operators gradient, curl or divergence, as defined in Chap. 4. The

---

[1] The orientation will be discussed in Chap. 7.

curl operator can be defined for planar vector fields $\vec{V} = v_1 e_1 + v_2 e_2$, considering that $v_3 = 0$.

Theorem 1 (Green's Theorem) *Let $\Omega \subset \mathbb{R}^2$ be a compact domain with a boundary being a C^1-curve. If $P, Q : \Omega \to \mathbb{R}$ are C^1-functions, then*

$$\int\int_\Omega (\frac{\partial Q}{\partial x} - \frac{\partial P}{\partial y})dxdy = \int_{\partial\Omega} Pdx + Qdy, \qquad (13)$$

that is,

$$\int\int_\Omega \mathrm{curl}(\mathrm{F})dA = \int_{\partial\Omega} \mathrm{F}. \qquad (14)$$

Theorem 2 (Stokes' Theorem) *Let $\mathcal{S} \subset \mathbb{R}^3$ be a compact surface, with a boundary $\partial\mathcal{S}$ that is a C^1-curve. If $P, Q, R : \Omega \to \mathbb{R}$ are C^1-functions, then*

$$\int\int_\mathcal{S} \left[(\frac{\partial Q}{\partial x} - \frac{\partial P}{\partial y})dxdy + (\frac{\partial R}{\partial y} - \frac{\partial Q}{\partial z})dydz + (\frac{\partial P}{\partial z} - \frac{\partial R}{\partial x})dzdx \right] = \int_{\partial\mathcal{S}} Pdx + Qdy + Rdz,$$

that is,

$$\int\int_\mathcal{S} \mathrm{curl}(\mathrm{F})dA = \int_{\partial\mathcal{S}} \mathrm{F}. \qquad (15)$$

Theorem 3 (Gauss' Theorem) *Let $\Omega \subset \mathbb{R}^3$ be a compact domain with a boundary that is a closed C^1-surface. If $P, Q, R : \Omega \to \mathbb{R}$ are C^1-functions, then*

$$\int\int\int_\Omega \left[\frac{\partial P}{\partial x} + \frac{\partial Q}{\partial y} + \frac{\partial R}{\partial z} \right] dxdydz = \int\int_{\partial\Omega} Pdydz + Qdzdx + Rdxdy,$$

that is,

$$\int\int\int_\Omega \mathrm{div}(\mathrm{F})dV = \int\int_{\partial\Omega} \mathrm{F}. \qquad (16)$$

We note the integrals in the identities (12), (14), (15) and (16) are determined by integration on the boundary of the functions depending on the differential operators curl and div. Differential forms will allow us to show that the above theorems are all particular cases of the same theorem, the Stokes Theorem, and it can be extended for any finite dimension n.

2.1 Interpretation of the Curl and Div Operators

The curl and divergence of a vector field in $\mathbb{R}^3$ arises in several mathematical models, so we have to give an interpretation for both. Let $F : \mathbb{R}^3 \to \mathbb{R}^3$ be a differentiable vector field. As maps we have $\mathrm{curl}(F) \in C^\infty(\mathbb{R}^3; \mathbb{R}^3)$ and $\mathrm{div}(F) \in C^\infty(\mathbb{R}^3; \mathbb{R})$.

(1) Curl.

Let $S_t(a)$ be a disc of radius $t > 0$ centered in a, let $A(t)$ be the area of $S_t(a)$ and let $C(t) = \partial S_t(a)$ be the circumference defined by the boundary. If n is the unit normal vector at $S_t(a)$ and $T(t)$ is the unit tangent vector at the point $C(t)$ so that $\{n, T, n \times T\}$ is a positive basis of $\mathbb{R}^3$, then

$$< n, \mathrm{curl}(\mathrm{F})(a) >= \lim_{t \to 0} \frac{1}{A(t)} \int_{\gamma_t(a)} < F, T > ds. \qquad (17)$$

Proof The integral on the right side of the identity (17) measures the circulation of the field F along the curve $C(t)$. The term $< \mathrm{curl}(\mathrm{F})(a), n >$ is the density of the circulation of F at a with respect to a plane perpendicular to n at a. To obtain (17), we define the continuous function $\phi(x) =< \mathrm{curl}(\mathrm{F})(x), n >$. Given $\epsilon > 0$, there is $\delta > 0$ such that if $x \in B_\delta(a)$, then $| \phi(x) - \phi(a) |< \epsilon$. So

$$\phi(a).A(t) = \int_{S_t(a)} \phi(x)dS + \int_{S_t(a)} [\phi(a) - \phi(x)]dS.$$

When we apply the Stokes Theorem, it follows that

$$| \phi(a).A(t) - \int_{\gamma_t(a)} < F, T > d\gamma | \leq \int_{S_t(a)} [\phi(a) - \phi(x)]dS < \epsilon.A(t).$$

In Fig. 3, we have $\mathrm{curl}(\mathrm{F})(A) \neq 0, \mathrm{curl}(\mathrm{F})(B) \neq 0, \mathrm{curl}(\mathrm{F})(C) \neq 0, \mathrm{curl}(\mathrm{F})(D) = 0$ and $\mathrm{curl}(\mathrm{F})(E) = 0$. $\qquad \square$

(2) Divergence.

Let $a \in \mathbb{R}^3$, let $B_t(a)$ be a ball of radius $t > 0$ centered at a, and let $S_t(a) = \partial B_t(a)$ be the sphere defined by the boundary of $B_t(a)$. If $V(t)$ is the volume of $B_t(a)$ and n is the unit normal vector to $S_t(a)$, then

$$\mathrm{div}(\mathrm{F}(a)) = \lim_{t \to 0} \frac{1}{V(t)} \int_{S_t(a)} < F, n > dS. \qquad (18)$$

Proof The integral on the right-hand side of Identity (18) measures the flow of F across the surface $S_t(a)$, which means the divergence of F at a measures the density of the flow of F at a. We recall that the flow measures the amount of F entering through $S_t(a)$ which was diminished by the amount of F exiting through $S_t(a)$. Indeed, the amount entering yields $< F, n >< 0$ while the amount exiting yields $< F, n >< 0$. Therefore if $\mathrm{div}(\mathrm{F})(a) = 0$, then there is no flow; indeed the flow that comes in also goes out. To prove Identity (18), we consider the continuous function $\phi = \mathrm{div}(\mathrm{F})$. Given $\epsilon > 0$, there is $\delta > 0$ such that if $x \in B_\delta(a)$, then $| \phi(x) - \phi(a) |< \epsilon$. So

$$\phi(a).V(t) = \int_{B_t(a)} \phi(x)dV + \int_{B_t(a)} [\phi(a) - \phi(x)]dV.$$

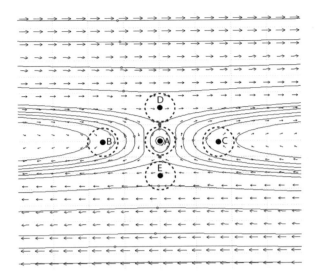

Fig. 3 Curl

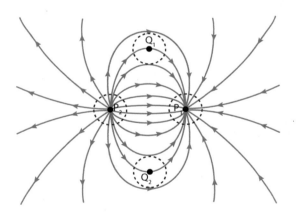

Fig. 4 Divergence

When we apply the Gauss Theorem, we have

$$\mid \phi(a).V(t) - \int_{S_t(a)} < F, n > dS \mid \leq \int_{B_t(a)} [\phi(a) - \phi(x)]dV < \epsilon.V(t).$$

In Fig. 4, we can see div(F) = 0 in the regions where the flow lines enter through the boundary and later leave. For the regions in which there are lines of the flow either only entering or just leaving we have div(F) ≠ 0.   □

Exercise

(1) A vector field F is a solenoidal field if $\mathrm{div}(F) = 0$. Show that the radial field $F(r) = \frac{\vec{r}}{r^3}$ is solenoidal. Show there is no field V such that $F = \mathrm{curl}(V)$.

3 Elementary Aspects of the Theory of Potential

In this section, we will address questions arising from the relationship between the linear differential operators grad, curl and div with integration theorems.

Question 1: Let $f : [a, b] \to \mathbb{R}$ be a continuous function. Is there a function $g : [a, b] \to \mathbb{R}$ such that $g'(x) = f(x)$? If it does, g is called the primitive of f.

Answer 1: Yes, it does exist because of the Fundamental Theorem of Calculus. The function g is

$$g(x) = \int_a^x f(t)dt.$$

When we formulate the same question for maps, the answer is not always positive. Let $\Omega \subset \mathbb{R}^n$ be an open subset and $f : \Omega \to \mathbb{R}$ a C^1-map. We identify the differential df with the gradient vector as follows: at every point $p \in \Omega$, we have $df_p(.) = < \nabla f(p), . >$, and $\nabla f = \sum_{i=1}^n \frac{\partial f}{\partial x_i} e_i$.

Question 2: Let $V = \sum_{i=1}^n v_i e_i : \mathbb{R}^n \to \mathbb{R}^n$ be a C^1-vector field defined on Ω. Is there a differentiable function $f : \Omega \to \mathbb{R}$ such that $V = \nabla f$? Equivalently, does there exist a function f such that

$$v_i = \frac{\partial f}{\partial x_i}, \quad \text{for all } i = 1, \ldots, n \ ? \tag{19}$$

To answer the 2nd question, there is an obstruction to the existence of f due to the identities

$$\frac{\partial^2 f}{\partial x_j \partial x_i} = \frac{\partial^2 f}{\partial x_i \partial x_j}, \quad \text{for all } i, j,$$

because they yield $\frac{\partial v_i}{\partial x_j} = \frac{\partial v_j}{\partial x_i}$ for all i, j. Therefore the 2nd question should be stated as follows;

Question 2': Let $V = \sum_i v_i e_i$ be a C^1-vector field satisfying the identities $\frac{\partial v_i}{\partial x_j} = \frac{\partial v_j}{\partial x_i}$ for all i, j. Is there a function $f : \Omega \to \mathbb{R}$ such that $\nabla f = V$?
    When such an f does exist, it is called the potential function of V. Let's work out a counterexample to question 2'.

Example 1: Consider the vector field $V : \mathbb{R}^2\backslash\{0\} \to \mathbb{R}^2$ given by

$$V(x, y) = \left(\frac{-y}{x^2 + y^2}, \frac{x}{x^2 + y^2}\right).$$

Suppose we have a function $f : \mathbb{R}^2 \to \mathbb{R}$ such that $\nabla f = V$. Let $\gamma : [0, 2\pi] \to \mathbb{R}^2$ be the curve $\gamma(\theta) = (\cos(\theta), \sin(\theta))$ and let $f_\gamma : [0, 2\pi] \to \mathbb{R}$ be the function $f_\gamma(\theta) = f \circ \gamma(\theta)$. By the chain rule,

$$\frac{df_\gamma}{d\theta}(\theta) = \nabla f_{\gamma(\theta)}.\gamma{\cdot}(\theta) = V\big(\cos(\theta), \sin(\theta)\big).\gamma{\cdot}(\theta) =$$
$$=< \big(-\sin(\theta), \cos(\theta)\big), \big(-\sin(\theta), \cos(\theta)\big) >= 1.$$

The line integral $\int_\gamma V$ can be calculated by two different methods:
(i) using the definition;

$$\int_\gamma V = \int_0^{2\pi} < V(\gamma(\theta)), \gamma'(\theta) > d\theta = 2\pi;$$

(ii) using $V = \mathrm{grad}(\mathrm{f})$;

$$\int_\gamma V = \int_0^{2\pi} < V(\gamma(\theta)), \gamma'(\theta) > d\theta = \int_0^{2\pi} < \mathrm{grad}(\mathrm{f}(\theta)), \gamma'(\theta) > d\theta =$$
$$= \int_0^{2\pi} \frac{df_\gamma}{d\theta}(\theta)\, d\theta = f_\gamma(2\pi) - f_\gamma(0) = 0.$$

As the values found are distinct, we conclude that the vector field V does not admit a potential function. V is defined in the region $\mathbb{R}^2\backslash\{0\}$, and we cannot extend it by a C^1-vector field on $\mathbb{R}^2$. So we have learned that the topology of the region matters in order to answer the question as to whether a potential functor exists.

Definition 2 A subset $\Omega \subset \mathbb{R}^n$ is star-shaped with respect to a point $x_0 \in \Omega$ if the segment $\{tx_0 + (1 - t)x \mid t \in [0, 1]\}$ is contained in Ω for all $x \in \Omega$ (see Fig. 5).

Remark 4 In Chap. 3, vector fields that admit a potential function are called conservative fields. They play a fundamental role in physics.

Theorem 4 *Let $\Omega \subset \mathbb{R}^n$ be an open star-shaped subset. If the vector field $V : \Omega \to \mathbb{R}^2$, $V = (v_1, v_2, \ldots, v_n)$ satisfies $\frac{\partial v_i}{\partial x_j} = \frac{\partial v_j}{\partial x_i}$ for all $1 \le i, j \le n$, then we have a potential function f such that $V = \nabla f$.*

Proof We prove the case $n = 2$; the general case can be proved using the same arguments. Without loss of generality, assume $x_0 = 0 \in \Omega$. Consider the function $f : \Omega \to \mathbb{R}$ given by

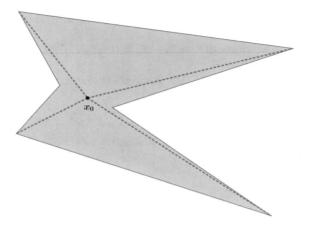

Fig. 5 Star-shaped set

$$f(x_1, x_2) = \int_0^1 \left[x_1.v_1(tx_1, tx_2) + x_2.v_2(tx_1, tx_2) \right] dt.$$

Since Ω is star-shaped with respect to 0, $f(x_1, x_2)$ is well-defined for all $(x_1, x_2) \in \Omega$. By differentiating f, we have

$$\frac{\partial f}{\partial x_1}(x_1, x_2) = \int_0^1 \left[v_1(tx_1, tx_2) + tx_1 \frac{\partial v_1}{\partial x_1}(tx_1, tx_2) + tx_2 \frac{\partial v_2}{\partial x_1}(tx_1, tx_2) \right] dt.$$

It follows from the identity

$$\frac{d}{dt}\left[t.v_1(tx_1, tx_2) \right] = v_1(tx_1, tx_2) + tx_1 \frac{\partial v_1}{\partial x_1}(tx_1, tx_2) + tx_2 \frac{\partial v_1}{\partial x_2}(tx_1, tx_2)$$

that

$$\frac{\partial f}{\partial x_1}(x_1, x_2) = \int_0^1 \frac{d}{dt}\left[t.v_1(tx_1, tx_2) \right] dt + \int_0^1 tx_2 \left(\frac{\partial v_2}{\partial x_1}(tx_1, tx_2) - \frac{\partial v_1}{\partial x_2}(tx_1, tx_2) \right) dt$$

$$= \left[tv_1(tx_1, tx_2) \right]\Big|_0^1 = v_1(x_1, x_2).$$

Analogously, we have $\frac{\partial f}{\partial x_2} = v_2$. $\qquad\square$

The theorem above and the Example 1 reveal that the geometric nature of Ω matters to answer Question 2'.

We will introduce some mathematical language by defining vector spaces associated with a region $\Omega \subset \mathbb{R}^3$ that are useful for detecting whether a potential function exists. We consider $\Gamma(\Omega) = C^\infty(\Omega, \mathbb{R}^k)$ the vector space of C^1-vector fields

$V : \Omega \to \mathbb{R}^k$. The grad, curl and div are linear operators satisfying the identities

$$(i)\ \mathrm{curl}(\mathrm{grad}(f)) = 0, \quad (ii)\ \mathrm{div}(\mathrm{curl}(V)) = 0 \tag{20}$$

for all $f \in C^\infty(\Omega, \mathbb{R})$ and $\forall V \in \Gamma(\Omega)$. For every operator, we define the following vector spaces:

(i) Kernel   (ii) Image

$\mathrm{Ker}(\mathrm{curl}) = \{V \in \Gamma(\Omega) \mid \mathrm{curl}(V) = 0\}, \quad \mathrm{Im}(\mathrm{grad}) = \{V \in \Gamma(\Omega) \mid V = \mathrm{grad}(f)\},$ (21)

$\mathrm{Ker}(\mathrm{div}) = \{V \in \Gamma(\Omega) \mid \mathrm{div}(V) = 0\}. \quad\quad \mathrm{Im}(\mathrm{curl}) = \{V \in \Gamma(\Omega) \mid V = \mathrm{curl}(W)\}.$

So from the Identities (20) (i) and (ii), we have

$$\mathrm{Im}(\mathrm{grad}) \subseteq \mathrm{Ker}(\mathrm{curl}), \quad \mathrm{Im}(\mathrm{curl}) \subseteq \mathrm{Ker}(\mathrm{div}).$$

Definition 3 Let $\Omega \subset \mathbb{R}^3$. We consider the vector spaces associated with Ω:

$$V_0(\Omega) = \mathrm{Im}(\mathrm{grad}), \quad V_1(\Omega) = \frac{\mathrm{Ker}(\mathrm{curl})}{\mathrm{Im}(\mathrm{grad})},$$

$$V_2(\Omega) = \frac{\mathrm{Ker}(\mathrm{div})}{\mathrm{Im}(\mathrm{curl})}, \quad V_3(\Omega) = \frac{C^\infty(\Omega)}{\mathrm{Im}(\mathrm{div})}. \tag{22}$$

The definition extends to $\mathbb{R}^2 = \mathbb{R}^2 \times \{0\} \subset \mathbb{R}^3$.

For any Ω, the spaces $V_i(\Omega)$ can be very intricate. The vector spaces $\mathrm{Ker}(\mathrm{curl})$ and $\mathrm{Im}(\mathrm{grad})$ have infinite dimensions and their quotients may not be vector spaces. In the following chapter, we will use differential forms to define these spaces, which will allow us to obtain a finite dimensional algebra. With these concepts at hand, we have the following statements;

Theorem 5 *Let $\Omega \subset \mathbb{R}^2$ be a star-shaped subset. So*

$$V_0(\Omega) = \mathbb{R}, \ V_1(\Omega) = V_2(\Omega) = 0.$$

Proof Let's check for every dimension.

(1) $V_0(\Omega) = \mathbb{R}$.
    Suppose that $\mathrm{grad}(f) = 0$. It follows that f is constant in Ω. Therefore the map $V_0(\Omega) \to \mathbb{R}$, $f \to c$, in which $c \in \mathbb{R}$ is a constant, defines an isomorphism between vector spaces.
(2) $V_1(\Omega) = 0$.
    It follows from Theorem (4).
(3) $V_2(\Omega) = 0$.
    Let $f \in C^\infty(\Omega, \mathbb{R})$ be an arbitrary function and let $V(x, y) = (0, Q(x, y))$ be a C^∞-vector field with $Q(x, y) = \int_0^1 f(tx, y)x\,dt$. It is straightforward from

$$\frac{\partial Q}{\partial x} = \int_0^1 \frac{\partial}{\partial x} [f(tx, y)x] \, dt = \int_0^1 \frac{d}{dt} [f(tx, y)t] \, dt = f(x, y)$$

that curl(V) = $\frac{\partial Q}{\partial x}$ = f. Hence $H^2(\Omega) = 0$.   □

We call attention to the fact that $V_1(\mathbb{R}^2 \backslash \{0\}) \neq 0$, as shown in Example 1. Next, let's look at the case $\Omega \subset \mathbb{R}^3$.

Theorem 6 *If $\Omega \subset \mathbb{R}^3$ is a star-shaped subset, then*

$$V_0(\Omega) = \mathbb{R}, \quad V_1(\Omega) = V_2(\Omega) = V_3(\Omega) = 0.$$

Proof The reasoning behind proving $V_0(\Omega) = \mathbb{R}$ and $V_1(\Omega) = 0$ is analogous to what was used to prove the last case in $\Omega \subset \mathbb{R}^2$. The proofs below show that $V_2(\Omega) = 0$ and $V_3(\Omega) = 0$.

(1) $V_2(\Omega) = 0$.

Assume Ω is star-shaped with respect to the origin. Let $V : \Omega \to \mathbb{R}^3$, let $V = \sum_i v_i e_i$, be a vector field such that div(V) = 0 and define the vector field $W : \Omega \to \mathbb{R}^3$ as

$$W(x) = \int_0^1 \left[V(tx) \times (tx) \right] dt,$$

with

$$V(tx) \times (tx) = t.(v_2 x_3 - v_3 x_2, \, v_3 x_1 - v_1 x_3, \, v_1 x_2 - v_2 x_1).$$

Since div(V) = 0, we have

$$\text{curl}\left[V(tx) \times (tx) \right] = \frac{d(t^2 V(tx))}{dt}.$$

As a consequence of the identity

$$\text{curl}\left(W(x) \right) = \int_0^1 \frac{d(t^2 V(tx))}{dt} \, dt = V(x),$$

V is the curl of W. Hence $V_2(\Omega) = 0$.

(2) $V_3(\Omega) = 0$.

Let $f \in C^\infty(\Omega, \mathbb{R})$ and consider the vector field $V(x, y, z) = (P(x, y, z), 0, 0)$, with $P(x, y, z)$ as

$$P(x, y, z) = \int_0^1 f(tx, y, z) x \, dt.$$

Since div(V) = f, then $V_3(\Omega, \mathbb{R}) = 0$.   □

Exercises

(1) If $\Omega \subset \mathbb{R}^3$ has k path connected components, show that $V_0(\Omega) \simeq \mathbb{R}^k$.
(2) Let $S = \{(x, y, z) \in \mathbb{R}^3 \mid x^2 + y^2 = 1, z = 0\}$ be the planar unitary circumference in $\mathbb{R}^2 \times \{0\} \subset \mathbb{R}^3$. Consider the function $V : \mathbb{R}^3 - S \to \mathbb{R}^3$ given by

$$V(x, y, z) = \left(\frac{-2xz}{z^2 + (x^2 + y^2 - 1)^2}, \frac{-2yz}{z^2 + (x^2 + y^2 - 1)^2}, \frac{x^2 + y^2 - 1}{z^2 + (x^2 + y^2 - 1)^2} \right).$$

(a) Show that $V \in V_1(\mathbb{R}^3 \backslash S)$.
(b) Find the integral

$$\int_\gamma V$$

given that $\gamma : [-\pi, \pi] \to \mathbb{R}^3$ is

$$\gamma(t) = \left(\sqrt{1 + \cos(t)}, 0, \sin(t) \right).$$

(c) Suppose $V = \nabla f$ and find the line integral $\int_\gamma \nabla f$.
(d) Show that $H^1_{DR}(\mathbb{R}^3 \backslash S^1) \neq 0$.

Chapter 6
Differential Forms, Stokes Theorem

We will introduce the algebra $\Omega^*(U)$ of differential forms on an open subset $U \subset \mathbb{R}^n$, although the formalism to define it on a submanifold of $\mathbb{R}^n$ is the same. The main purpose is to prove the Stokes Theorem and to define De Rham Cohomology. We will start with an introduction to the pointwise concepts which later will be extended to an open set U.

1 Exterior Algebra

Exterior or Grassmannian algebras are the key algebraic structures in studying differential forms. They were introduced in 1844 by Hermann Gunther Grassmann. For a vector space V, the problem addressed by Grassmann was to find an algebra for which the relation $v^2 = 0$ would be satisfied for all $v \in V$. In 1867, Hermann Hankel gave a geometric interpretation of Grassmann's idea which was to use the alternating product of vectors.

Consider $\Lambda(V)$ the Exterior Algebra associated with a vector space V. The construction of $\Lambda(V)$ will be done using the alternating operator $\text{Alt} : T(V) \to T(V)$, and more precisely, we will define $\Lambda(V) = \text{Img}(\text{Alt}(V))$. The Exterior Algebra that interests us is $\Lambda(V^*)$ which is associated with the dual vector space V^*.

Definition 1 Let V be a vector space over $\mathbb{R}$. A p-tensor defined on V is a p-linear function $T : V^{\times p} = V \times \overset{p}{\ldots} \times V \to \mathbb{R}$ such that

$$T(v_1, \ldots \alpha v_j + \beta v_j', \ldots, v_p) = \alpha.T(v_1 \ldots, v_j, v_{j+1}, \ldots, v_p) + \beta.T(v_1, \ldots, v_j', \ldots, v_p),$$

© Springer Nature Switzerland AG 2021
C. M. Doria, *Differentiability in Banach Spaces, Differential Forms and Applications*,
https://doi.org/10.1007/978-3-030-77834-7_6

with $j \in \{1, \ldots, p\}$, for all $v_i, v'_j \in V$ and $\alpha, \beta \in \mathbb{R}$. A p-tensor T is a linear functional $T : V^{\otimes p} \to \mathbb{R}$.[1]

Example 1 Examples.
(i) $p = 1$, linear functionals on V.
(ii) $p = 2$, the inner product of $\mathbb{R}^n$.
(iii) $p = k$, the determinant of a $k \times k$ matrix.

The p-tensors T and S are equal if and only if $T(v_1, \ldots, v_p) = S(v_1, \ldots, v_p)$ for all $\{v_1, \ldots, v_p\} \subset V$.

Let $T^p(V)$ be the set of p-tensors on a vector space V, and $T^1(V) = V^*$. The tensor product between tensors is defined as follows: Let $T \in T^p(V)$ and $S \in T^q(V)$:

$$(T \otimes S)(v_1, \ldots, v_p, v_{p+1}, \ldots, \ldots v_{p+q}) = T(v_1, \ldots, v_p).S(v_{p+1}, \ldots, v_{p+q}). \tag{1}$$

The tensor product over $T(V)$ induces an associative and non-commutative algebraic structure. Let $\beta = \{e_1, \ldots, e_k\}$ be a basis of V and let $\beta^* = \{e_1^*, \ldots, e_k^*\}$ be the corresponding dual basis, i.e., $e_i^*(e_j) = \delta_{ij}$. Consider the sets $C = \{1, 2, \ldots, k\} \subset \mathbb{N}$ and $C^p = C \times .^p. \times C$, $p \in \mathbb{N}$. For pairs of index sequences $I = (i_1, \ldots, i_p)$, $J = (j_1, \ldots, j_p) \in C^p$, we define the elements to be

$$e_I^* = e_{i_1}^* \otimes \cdots \otimes e_{i_p}^* \in (V^*)^{\otimes p}, \quad e_J = (e_{j_1}, \ldots, e_{j_p}) \in V \times .^p. \times V.$$

Therefore we have

$$e_I^*(e_J) = (e_{i_1}^* \otimes \cdots \otimes e_{i_p}^*)(e_{j_1}, \ldots, e_{j_p}) =$$
$$= e_{i_1}^*(e_{j_1}) \ldots e_{i_p}^*(e_{j_p}) = \delta_{IJ}. \tag{2}$$

Proposition 1 *Let V be a k-dimensional vector space, let $\beta = \{e_1, \ldots, e_k\}$ be a basis of V and let $\beta^* = \{e_1^*, \ldots, e_k^*\}$ be the dual basis. Then the set $\{e_{i_1}^* \otimes \cdots \otimes e_{i_p}^* \mid 1 \leq i_1, i_2, \ldots, i_p \leq k\}$ is a basis of $T^p(V)$. Consequently, $\dim(T^p(V)) = k^p$.*

Proof Let $\beta = \{e_1, \ldots, e_k\}$ be a basis of V and let $\beta^* = \{e_1^*, \ldots, e_k^*\}$ be the dual basis.

(i) The set $\{e_{i_1}^* \otimes \cdots \otimes e_{i_p}^* \mid 1 \leq i_1, i_2, \ldots, i_p \leq k\}$ is linearly independent.
Suppose there are coefficients $a_I \in \mathbb{R}$ such that $\sum_I a_I e_I^* = 0$; then

$$\sum_I a_I e_I^*(e_J) = a_J = 0 \quad \Rightarrow \quad a_J = 0 \ \ \forall J.$$

(ii) The set $\{e_{i_1}^* \otimes \cdots \otimes e_{i_p}^* \mid 1 \leq i_1, i_2, \ldots, i_p \leq k\}$ spans $T^p(V)$.
Consider the vectors $w_1, \ldots, w_p$ in V and $T \in T^p(V)$. For $i \in \{1, \ldots, p\}$, we have $w_i = \sum_{l_i} w_i^{l_i} e_{l_i}$ $(i = 1 \ldots, p)$, then

---

[1] The tensor product $V^{\otimes p}$ is defined in Appendix C.

$$T(w_1, \ldots, w_p) = \sum_{l_1} \cdots \sum_{l_p} T(e_{l_1}, \ldots, e_{l_p}) w_1^{l_1} \ldots w_p^{l_p} =$$

$$= \sum_{l_1} \cdots \sum_{l_p} T(e_{l_1}, \ldots, e_{l_p}) e_{l_1}^*(w_1) \ldots e_{l_p}^*(w_p) =$$

$$= \sum_{l_1} \cdots \sum_{l_p} T_{l_1 l_2 \ldots l_p} (e_{l_1}^* \otimes \ldots e_{l_p}^*)(w_1, \ldots, w_p).$$

Therefore we have

$$T = \sum_{l_1} \cdots \sum_{l_p} T_{l_1 l_2 \ldots l_p} e_{l_1}^* \otimes \ldots e_{l_p}^*.$$

Hence $\{e_I^* \mid I \in C^p\}$ spans $T^p(V)$. $\qquad \square$

The space of tensors defined on V is the vector space

$$\mathcal{T}(V) = \bigoplus_{p=0}^{\infty} T^p(V), \quad T^0(V) = \mathbb{R}. \tag{3}$$

Indeed, $\mathcal{T}(V)$ is equal to the tensor algebra $T(V^*)$.

Definition 2 A p-tensor is alternating if

$$T(v_1, \ldots, v_i, \ldots, v_j, \ldots, v_p) = -T(v_1, \ldots, v_j, \ldots, v_i, \ldots, v_p), \quad \forall i, j.$$

By convention, the 1-tensors are alternating.

The classical example of an alternating p-tensor is the determinant of a $p \times p$ matrix.

Let S_p be the symmetric group defined by the set of bijections of the set $\{1, \ldots, p\}$. A permutation $\pi \in S_p$ is either even or odd, depending on whether the number of transpositions of indices are even or odd. Consider $\mid \sigma \mid$ equal to $+1$ if σ is even and -1 if is odd. Then $(-1)^{\mid \pi \mid}$ will be equal to $+1$ or -1. There is a natural action of S_p over $T^p(V)$ given by

$$S_p \times T^p(V) \to T^p(V),$$
$$(\pi, T) \to T^\pi,$$

such that $T^\pi(v_1, v_2, \ldots, v_p) = T(v_{\pi(1)}, v_{\pi(2)}, \ldots, v_{\pi(p)})$ and $(T + S)^\pi = T^\pi + S^\pi$. It follows straightforward that $(T^\pi)^\sigma = T^{\sigma \circ \pi}$. A p-tensor is alternating if

$$T^\pi = (-1)^{\mid \pi \mid} T,$$

for any $\pi \in S_p$. Let

$$\Lambda^P(V) = \{T \in T^p(V) \mid T^\pi = (-1)^{|\pi|}T, \quad \forall \pi \in S_p\} \subset T^p(V),$$

be the subspace of alternating p-tensors. Considering the operator Alt defined as

$$\text{Alt(T)} = \frac{1}{p!} \sum_{\pi \in S_p} (-1)^{|\pi|} T^\pi, \tag{4}$$

we define the alternate linear operator Alt : $T^p(V) \to T^p(V)$ over $T^p(V)$ by extending AlT linearly over $\mathbb{R}$.

Proposition 2 *The operator Alt : $T^p(V) \to T^p(V)$ is a projection over the vector subspace $\Lambda^p(V)$ for all p.*

Proof Let's check that for all $\sigma \in S_p$ and $T \in T^p(V)$, we have $[\text{Alt(T)}]^\sigma = (-1)^{|\sigma|}\text{Alt(T)}$.

$$(\text{Alt(T)})^\sigma = [\frac{1}{p!} \sum_{\pi \in S_p} (-1)^{|\pi|}T^\pi]^\sigma = \frac{1}{p!} \sum_{\pi \in S_p} (-1)^{|\pi|}[T^\pi]^\sigma = \frac{1}{p!}(-1)^{|\sigma|} \sum_{\pi \in S_p} (-1)^{|\sigma|}(-1)^{|\pi|}T^{\sigma \circ \pi} =$$

$$= \frac{1}{p!}(-1)^{|\sigma|} \sum_{\pi \in S_p} (-1)^{|\sigma \circ \pi|}T^{\sigma \circ \pi} = \frac{1}{p!}(-1)^{|\sigma|} \sum_{\pi \in S_p} (-1)^{|\sigma \circ \pi|}T^{\sigma \circ \pi} =$$

$$= \frac{1}{p!}(-1)^{|\sigma|} \sum_{\tau \in S_p} (-1)^{|\tau|}T^\tau = (-1)^{|\sigma|}\text{Alt(T)}.$$

The last equality above uses the fact that S_p is a group, so we have $\pi \in S_p$ such that $\tau = \sigma \circ \pi$ for all $\tau \in S_p$,. Therefore Alt : $T^p(V) \to \Lambda^p(V)$. To verify that Alt is a projection, i.e., Alt $\circ$ Alt $=$ Alt, we note that whenever T is an alternating p-tensor $T^\pi = (-1)^{|\pi|}T$, we have

$$\text{Alt(T)} = \frac{1}{p!} \sum_{\pi \in S_p} (-1)^{|\pi|}T^\pi = \frac{1}{p!} \sum_{\pi \in S_p} (-1)^{|\pi|}(-1)^{|\pi|}T =$$

$$= \frac{1}{p!} \sum_{\pi \in S_p} T = T.$$

Hence $(\text{Alt})^2 = \text{Alt}$.   □

The vector space $\Lambda(V)$ carries an algebraic structure, which we show next.

Definition 3 Let $T \in T^p(V)$ and $S \in T^q(V)$ be alternating tensors; the exterior product between T and S is

$$T \wedge S = \text{Alt}(T \otimes S) \in T^{(p+q)}(V).$$

The exterior product has the following properties;

(1) (associativity) For all $T \in \mathcal{T}^p(V)$, $S \in \mathcal{T}^q(V)$ and $R \in \mathcal{T}^l(V)$,

$$(T \wedge S) \wedge R = T \wedge (S \wedge R).$$

(2) Let $k \in \mathbb{R}$ for all $T \in \mathcal{T}^p(V)$,

$$k \wedge T = T \wedge k = kT,$$

in particular, $1 \in \mathbb{R}$ is the identity element.

(3) (distributivity) For all $T \in \mathcal{T}^p(V)$, $S \in \mathcal{T}^q(V)$ and $R \in \mathcal{T}^l(V)$,

$$T \wedge (S + R) = T \wedge S + T \wedge R.$$

The product is not commutative. Associativity requires a proof; for this purpose we use the following lemma.

Lemma 1 *If* $\mathrm{Alt}(T) = 0$, *then* $T \wedge S = S \wedge T = 0$ *for all* S.

Proof If $T \in \mathcal{T}^p(V)$ and $S \in \mathcal{T}^q(V)$, then $T \otimes S \in \mathcal{T}^{(p+q)}(V)$. Let G be the subgroup of $S_{(p+q)}$ with elements that are permutations fixing the indices $p + 1, \ldots, p + q$, so G is isomorphic to S_q. If $\pi \in G$, then

$$\pi(1, \ldots, p, p + 1, \ldots, p + q) = (\pi(1), \ldots, \pi(p), p + 1, \ldots, p + q).$$

For all $\pi \in G$, we have $(T \otimes S)^\pi = T^\pi \otimes S$ since

$$\left(T \otimes S\right)^\pi (v_1, \ldots, v_p, v_{p+1}, \ldots, v_{p+q}) = (-1)^\pi T(v_{\pi(1)}, \ldots, v_{\pi(p)}).S(v_{p+1}, \ldots, v_{p+q}).$$

So

$$\sum_{\pi \in G}(-1)^\pi (T \otimes S)^\pi = [\sum_{\pi \in G}(-1)^\pi T^\pi] \otimes S = \mathrm{Alt}(T) \otimes S = 0.$$

The subgroup G decomposes S_{p+q} into a disjoint union of left lateral classes which means that we have a set of elements $\{\sigma_i \mid 1 \le i \le l\} \subset S_{p+q}$ such that for any $\tau \in S_{p+q}$, there is $\pi \in G$, and $i \in \{1, \ldots, l\}$, showing that $\tau = \sigma_i.\pi$. Consequently,

$$T \wedge S = \sum_{\tau \in S_{p+q}}(-1)^\tau (T \otimes S)^\tau = \sum_{i=1}^{l}\sum_{\pi \in G}(-1)^{\sigma_i \circ \pi}(T \otimes S)^{\sigma_i \circ \pi} =$$

$$= \sum_{i+1}^{l}(-1)^{\sigma_i}[\sum_{\pi \in G}(-1)^\pi (T \otimes S)^\pi]^{\sigma_i} = \sum_{i=1}^{l}(-1)^{\sigma_i}[\sum_{\pi \in G}(-1)^\pi T^\pi] \otimes S]^{\sigma_i} =$$

$$= \sum_{i=1}^{l}(-1)^{\sigma_i}[\mathrm{Alt}(T) \otimes S]^{\sigma_i} = 0.$$

$\square$

Proposition 3 *The exterior product is associative.*

Proof Given the linearity of Alt : $T^p(V) \to \Lambda^p(V)$, we have

$$(T \wedge S) \wedge R - \text{Alt}(T \otimes S \otimes R) = \text{Alt}[(T \wedge S) \otimes R] - \text{Alt}(T \otimes S \otimes R) =$$
$$= \text{Alt}[(T \wedge S) \otimes R - T \otimes S \otimes R] = \text{Alt}[(T \wedge S - T \otimes S) \otimes R].$$

However $\text{Alt}[T \wedge S - T \otimes S] = 0$. From the last lemma, the above equation is null; consequently, $(T \wedge S) \wedge R = \text{Alt}(T \otimes S \otimes R)$. Analogously,

$$T \wedge (S \wedge R) = \text{Alt}(T \otimes S \otimes R).$$

Hence the exterior product is associative.   □

The vector space of alternating tensors $(\Lambda(V), +)$ associated with an $\mathbb{R}$-vector space V is the $\mathbb{R}$-vector space

$$\Lambda(V) = \bigoplus_p \Lambda^p(V).$$

The exterior algebra associated to V is $(\Lambda(V), +, \wedge)$.

Proposition 4 *Let $\phi, \psi \in \Lambda^1(V)$. Then*

(1) $\phi \wedge \psi = -\psi \wedge \phi$;
(2) $\phi \wedge \phi = 0$.

Proof It follows from the definition that

$$(\phi \wedge \psi)(v_1, v_2) = \text{Alt}(\phi \otimes \psi)(v_1, v_2) = \phi \otimes \psi(v_1, v_2) - \phi \otimes \psi(v_2, v_1) =$$
$$= \phi(v_1).\psi(v_2) - \phi(v_2).\psi(v_1).$$

Now swapping the positions of ϕ and ψ, the result follows.   □

Proposition 5 *Let V be an n-dimensional vector space, let $\{\phi_1, \ldots, \phi_k\}$ be a basis of V^*, and let $\mathcal{I}^p = \{I = (i_1, \ldots, i_p) \mid 1 \leq i_1 \leq i_2 \leq \cdots \leq i_p \leq k\}$. Associated to the multi-index $I = (i_1, \ldots, i_p)$, we consider the alternating p-tensor*

$$\phi_I = \phi_{i_1} \wedge \cdots \wedge \phi_{i_p}.$$

Then the set $\{\phi_I; I \in \mathcal{I}^p\}$ is a basis of $\Lambda^p(V)$. The dimension of $\Lambda^p(V)$ is $\binom{n}{p}$.

Proof To check that $\{\phi_I \mid I \in \mathcal{I}\}$ generates $\Lambda^p(V)$, we recall that a p-tensor T can be written in a unique way as

$$T = \sum_{i_1,\dots,i_p} T_{i_1\dots i_p}\phi_{i_1} \otimes \cdots \otimes \phi_{i_p}.$$

Therefore we have

$$\text{Alt}(T) = \sum_{I\in\mathcal{I}^p} T_{i_1\dots i_p}\text{Alt}(\phi_{i_1} \otimes \cdots \otimes \phi_{i_p}) = \sum_{I\in\mathcal{I}^p} T_I\phi_I.$$

The condition of being linearly independent is elementary and to be verified. If $\{v_1,\dots,v_k\}$ is a basis of V such that $\phi_i(v_j) = \delta_{ij}$ and $v_J = (v_{j_1},\dots,v_{j_p})$, then $\phi_J(v_J) = \delta_{IJ}$ and

$$\sum_{I\in\mathcal{I}^p} T_I\phi_I = 0 \Rightarrow \sum_{I\in\mathcal{I}^p} T_I\phi_I(v_J) = T_J = 0.$$

The cardinality of the set $\{\phi_I \mid I \in \mathcal{I}^p\}$ is $\binom{n}{p}$. □

If V is an n-dimensional vector space of dimension n, then $\Lambda(V) = \oplus_{n=0}^k \Lambda^n(V)$ is a $2n$-dimensional vector space. $\Lambda(V)$ endowed with the exterior product "$\wedge$" becomes an algebra such that $x^2 = x \wedge x = 0$ for every $x \in V$. This relation is a universal relation since V generates $\Lambda(V)$. Furthermore, $\Lambda(V)$ is a graduated algebra, that is, $\Lambda(V)^p \wedge \Lambda(V)^q \subset \Lambda(V)^{p+q}$.

The exterior algebra $\Lambda(V)$ can be defined as a factorial algebra. To this end, we consider the ideal $\mathcal{I} \subset T(V)$ generated by the elements of the form $\{v \otimes v \mid v \in V\}$. By definition, the degree of elements in V is 1; therefore they belong to $T^1(V)$ ($\mathcal{I} \cap T^1 = 0$). Given the canonical homomorphism $j : T(V) \to T(V)/\mathcal{I}$, the restriction of j to V is injective. So we can identify V with its image $j(V)$ and consider it a subset of $\Lambda(V) = T(V)/\mathcal{I}$. Since V generates the algebra $T(V)$, V also generates the algebra $\Lambda(V)$. By construction, $j(v)^2 = 0$ for all $v \in V$.

Proposition 6 (Universal Property) *Let $\mathcal{A}$ be an algebra over the field K and let $f : V \to \mathcal{A}$ be a homomorphism such that*

$$f(v).f(v) = 0, \quad \forall v \in V.$$

Then there is a unique homomorphism $\phi : \Lambda(V) \to \mathcal{A}$ extending f.

Proof Consider the homomorphism $f^* : T(V) \to \mathcal{A}$ given by $f^*(v \otimes w) = f^*(v).f^*(w)$. So we have $f^*(v \otimes v) = 0, \forall v \in V$. Therefore $\mathcal{I} \subset \text{Ker}(f^*)$, and consequently, f^* induces a homomorphism $f^* : T(V)/\mathcal{I} \to \mathcal{A}$. Uniqueness arises from the fact that V spans $\Lambda(V)$. □

Clifford algebras generalize the Exterior Algebra. Let V be an n-dimensional real vector space endowed with a non-degenerate bilinear form $q : V \times V \to \mathbb{R}$, and let $q : V \to \mathbb{R}$ be the associated quadratic form. Considering the relation

$$u \bullet v + v \bullet u = -2q(u, v), \quad \forall u, v \in V,$$

we define the vector spaces $C^k = \{u_1 \bullet \cdots \bullet u_k \mid u_j \in V, 1 \le j \le k\}$ and

$$Cl(V, q) = \mathbb{R} \oplus C^1 \oplus \cdots \oplus C^n.$$

The Clifford Algebra $Cl(V, q)$ is associated with the pair (V, q). When the bilinear form is completely degenerate, i.e., $q = 0$, we have $Cl(V, q) = \Lambda(V)$.

Exercises

(1) Consider $\mathbb{R}$ the vector space generated by $e_1 = 1$ with the quadratic form $q(1) = 1$. Consequently we have $e_1 \bullet e_1 = e_1^2 = -1$. Show that the Clifford Algebra $Cl_1 = Cl(\mathbb{R}, q)$ is isomorphic to $\mathbb{C}$.

(2) Let $\beta = \{e_1, e_2\}$ be an orthonormal basis of $\mathbb{R}^2$; let $< ., . >: \mathbb{R}^2 \times \mathbb{R}^2 \to \mathbb{R}$ be an inner product with the quadratic form $q(u) = < u, u >$. Then we have the identity

$$e_i \bullet e_j + e_j \bullet e_i = -2\delta_{ij}.$$

Show that the Clifford Algebra $Cl_2 = Cl(\mathbb{R}^2, q)$ is isomorphic to the quaternion algebra $\mathbb{H}$.

(3) Show that $T^p(V)$ is a vector space over $\mathbb{R}$.

(4) Show that a p-tensor $T : V^{\times p} \to \mathbb{R}$ induces a homomorphism $T : V^{\otimes p} \to \mathbb{R}$.

(5) Let $T \in \Lambda^p(V)$ and $S \in \Lambda^q(V)$; then

$$T \wedge S = (-1)^{pq} S \wedge T \in \Lambda^{p+q}(V).$$

(6) Let $\beta = \{e_1, \ldots, e_n\}$ be an orthonormal basis of $\mathbb{R}^n$ and $< ., . >: \mathbb{R}^n \times \mathbb{R}^n \to \mathbb{R}$ an inner product. Show that the identity

$$e_i \bullet e_j + e_j \bullet e_i = -2\delta_{ij},$$

defines an algebra $(Cl_n = Cl(\mathbb{R}^n, < ., . >))$.

(7) Show that $Cl(V, q)$ is a factorial algebra generated by V.

2 Orientation on V and on the Inner Product on $\Lambda(V)$

In this section, we deal with two technical concepts that are necessary to carry on with our task which is to study differential forms and to prove the Stokes Theorem.

2.1 *Orientation*

Orientation is a misleading concept because it is easy to define on vector spaces and differentiable manifolds, but it is not so elementary to be able to verify if a manifold is orientable. We don't have to bother with necessary conditions to be orientable; we just need the concept.

The bases $\alpha = \{e_1^\alpha, \ldots, e_n^\alpha\}$ and $\beta = \{e_1^\beta, \ldots, e_n^\beta\}$ of $\mathbb{R}^n$ are equivalent if we have a linear transformation $T : \mathbb{R}^n \to \mathbb{R}^n$, such that

$$T(e_i^\beta) = \sum_{j=1}^n t_{ji} e_j^\alpha$$

and the determinant of the matrix $T = (t_{ji})$ is positive. The relation "$\alpha \sim \beta \Leftrightarrow \det(T) > 0$" defines an equivalent relation in the set $\mathcal{B}_{\mathbb{R}^n}$ of bases of $\mathbb{R}^n$. When we fix a basis of $\mathbb{R}^n$, e.g., the canonical basis $c = \{e_1, \ldots, e_n\}$, we note that there are only 2 distinct equivalence classes in $\mathbb{R}^n$: the positive class corresponding to those bases such that $\det(T) > 0$ and negative when $\det(T) < 0$. We consider the equivalent classes $\mathcal{O}_{\mathbb{R}^n} = (\mathcal{B}_{\mathbb{R}^n}/\sim) = \{-1, 1\}$, such that 1 corresponds to the positive class and -1 to the negative class. To fix an orientation on $\mathbb{R}^n$ involves a class in $\mathcal{O}_{\mathbb{R}^n}$, that is, to fix a basis of $\mathbb{R}^n$. For example, once we fix a basis $c = \{e_1, e_2, \ldots, e_n\}$ by swapping positions to obtain a new ordered basis $c' = \{e_2, e_1, \ldots, e_n\}$, we consider $1 = [c]$, and so $-1 = [c']$.

Let's consider $\beta = \{e_1, \ldots, e_n\}$ a set of C^∞-vector fields on $U \subset \mathbb{R}^n$. We say that β is a frame on an open subset $U \subset \mathbb{R}^n$ if the following conditions are verified;
(i) $\beta(p) = \{e_1(p), \ldots, e_n(p)\}$ is a basis of T_pU for all $p \in U$.
(ii) the map $U \to Gl_n(\mathbb{R})$, $p \mapsto \beta(p)$, is C^∞.
We assume the frame is globally defined on U, which is the case when U is simply connected. For all $p \in U$, the ordered basis $\beta(p) = \{e_1(p), \ldots, e_n(p)\}$ is assumed to be the positive orientation on T_pU. An orientation is assigned to U since frame β is fixed.

Let $U \in \mathbb{R}^n$ and $V \in \mathbb{R}^m$ be oriented opens subsets. A differentiable map $f : U \to V$ preserves the orientation if $\det(df_p) > 0$ for all $p \in U$.

2.2 *Inner Product in $\Lambda(V)$*

When we fix an inner product on V, we can induce an inner product on $\Lambda(V)$.

Let $(V, < ., . >)$ be an n-dimensional vector space endowed with an inner product $< ., . >: V \times V \to \mathbb{R}$. We will define an inner product on $\Lambda(V)$. Let's consider the case $p = 2$: the inner product

$$< w_1 \otimes w_2, \eta_1 \otimes \eta_2 > = < w_1, \eta_1 > . < w_2, \eta_2 >$$

on $V \otimes V$ induces on $\Lambda^2(V)$ an inner product as follows;

$$< w_1 \wedge w_2, \eta_1 \wedge \eta_2 > = < w_1 \otimes w_2 - w_2 \otimes w_1, \eta_1 \otimes \eta_2 - \eta_2 \otimes \eta_1 > =$$

$$= < w_1, \eta_1 > . < w_2, \eta_2 > - < w_1, \eta_2 > . < w_2, \eta_1 > = det \begin{pmatrix} < w_1, \eta_1 > & < w_1, \eta_2 > \\ < w_2, \eta_1 > & < w_2, \eta_2 > \end{pmatrix}.$$

Definition 4 The inner product $< ., . >: V \times V \to \mathbb{R}$ induces on each subspace $\Lambda^k(V)$, $1 \leq k \leq n$, the inner product

$$< w_1 \wedge \cdots \wedge w_k, \eta_1 \wedge \cdots \wedge \eta_k > = det(< w_i, \eta_j >), \quad 1 \leq i, j \leq k. \qquad (5)$$

The subspaces $\Lambda^k(V)$ and $\Lambda^l(V)$ are orthogonal whenever $k \neq l$.

If $\beta = \{e_1 \ldots, e_n\}$ is an orthonormal basis of V, then $\beta_k = \{e_I \mid I = (i_1, \ldots, i_k)\}$ is an orthonormal basis of $\Lambda^k(V)$, and consequently, $\hat{\beta} = \cup_k \beta_k$ is an orthonormal basis of $\Lambda(V)$. When $\dim(V) = n$, the space $\Lambda^n(V)$ has dimension 1, and we fix $\beta_n = \{\vartheta_\beta = e_1 \wedge \cdots \wedge e_n\}$ to be the basis. The element $e_1 \wedge \cdots \wedge e_n$ is the volume element of β.

Definition 5 An orientation on V is given by the volume element $e_1 \wedge \cdots \wedge e_n$, associated to an orthonormal basis β as the positive basis of $\Lambda^n(V)$.

Definition 6 Let V be an n-dimensional oriented vector space with an orientation given by the volume element $e_1 \wedge \cdots \wedge e_n$. The Hodge star-operator $* : \Lambda^k(V) \to \Lambda^{n-k}(V)$ is the linear operator defined by the following identity: let $\omega, \eta \in \Lambda^k(V)$

$$\omega \wedge *\eta = < \omega, \eta > e_1 \wedge \cdots \wedge e_n = < \omega, \eta > \vartheta_\beta. \qquad (6)$$

It follows from the definition that $*(1) = \vartheta_\beta$ and $w \wedge *w = \mid w \mid^2 \vartheta_\beta$. Let's consider the case $n = 4$ and $\eta = e_1 \wedge e_3$, which is an element of the basis of $\Lambda^2(V)$. Given $*\eta = \sum_{i,j} a_{ij} e_i \wedge e_j$ and $w = \sum_{k,l} b_{kl} e_k \wedge e_l$, we have

$$w \wedge *\eta = \sum_{k,l} \sum_{i,j} a_{ij} b_{k,l} e_k \wedge e_l \wedge e_i \wedge e_j = \sum_{k,l} b_{kl} < e_k \wedge e_l, e_1 \wedge e_3 > e_1 \wedge e_2 \wedge e_3 \wedge e_4 =$$

$$= b_{13}.e_1 \wedge e_2 \wedge e_3 \wedge e_4.$$

Let S_4 be the symmetric group permuting 4 elements and consider $\mid \sigma \mid$ the signal of an element $\sigma \in S_4$. So we have

$$w \wedge *\eta = \sum_{k,l} \sum_{i,j} a_{ij} b_{k,l} e_k \wedge e_l \wedge e_i \wedge e_j = [\sum_{\sigma \in S_4} (-1)^{|\sigma|} a_{\sigma(1)\sigma(2)} b_{\sigma(3)\sigma(4)}] e_1 \wedge e_2 \wedge e_3 \wedge e_4.$$

Pairing the expressions we obtained, we have

$$[\sum_{\sigma \in S_4} (-1)^{|\sigma|} a_{\sigma(1)\sigma(2)} b_{\sigma(3)\sigma(4)}] e_1 \wedge e_2 \wedge e_3 \wedge e_4 = b_{13} e_1 \wedge e_2 \wedge e_3 \wedge e_4.$$

Therefore $*\eta = -e_2 \wedge e_4$. The example shows that the operator $*$ is determined with respect to the elements of the basis, so it is the unique operator satisfying the identity (32).

Proposition 7 *The operator $* : \Lambda^k(V) \to \Lambda^{n-k}(V)$ satisfies the following properties:*
(i) $*^2 = (-1)^{k(n-k)}$.
(ii) $< *w, *\eta > = < w, \eta >$.

Proof We will prove the identities using the fact that the operator $*$ is determined by the image of the basis.
(i) Let $\vartheta_\beta = e_1 \wedge \cdots \wedge e_n$ be the volume element, $I = (1, \ldots, k)$ a multi-index and $e_I = e_1 \wedge \cdots \wedge e_k$. From the identity (32), we have $e_I \wedge *e_I = \vartheta_\beta$. Note we have $c \in \mathbb{R}$ such that $*e_I = c.e_J$, $J = (k+1, \ldots, n)$ and $e_I \wedge e_J = \vartheta_\beta$. Since $e_I \wedge *e_I = c\vartheta_\beta = \vartheta_\beta$, we have $c = 1$. Due to the identities $e_J \wedge e_I = (-1)^{k(n-k)} e_I \wedge e_J$ and $*e_I = e_J$, we get $*^2 e_I = *(e_J) = (-1)^{k(n-k)} e_I$. Therefore $*^2 = (-1)^{k(n-k)}$.
(ii) $< *w, *\eta > = *w \wedge *(*\eta) = *w \wedge (-1)^{k(n-k)}\eta = (-1)^{2k(n-k)}\eta \wedge *w = < w, \eta >$. □

Exercises

(1) Show that the bilinear form defined in Eq. 5 is symmetric and positive definite, so it is an inner product.
(2) Let $T : V \to V$ be a linear transformation. Show that T preserves orientation if and only if $\det(T) > 0$, and so there are just two possible orientations on a vector space.
(3) Consider $V = \mathbb{R}^4$ and $\beta = \{e_1, e_2, e_3, e_4\}$ an orthonormal basis. Find $*(e_i \wedge e_j)$ for all $1 \le i, j \le 4$.
(4) When $n = 4$ and $k = 2$, we have $*^2 = 1$. Show that there is a decomposition $\Omega^2(\mathbb{R}^4) = \Omega^2_+ \oplus \Omega^2_-$ with $\Omega^{\pm}$ being the eigenspaces associated to the eigenvalues ± 1 of $*$, respectively.

2.3 *Pseudo-Inner Product, the Lorentz Form*

So far we have worked with a real vector space endowed with an inner product $< ., . > : V \times V \to \mathbb{R}$, whose quadratic form has signature $\tau = \dim(V)$. In Restricted Relativity we have $V = \mathbb{R}^4$, and there is the need to work with a quadratic form with signature $\tau = 2$ called the Lorentz form. According to Sylvester's theorem, on

$\mathbb{R}^4$ the quadratic forms satisfying $\tau = 2$ are equivalent up to an orthogonal linear transformation to

$$L = \begin{pmatrix} 1 & 0 & 0 & 0 \\ 0 & 1 & 0 & 0 \\ 0 & 0 & 1 & 0 \\ 0 & 0 & 0 & -1 \end{pmatrix}.$$

A finite dimensional real vector space V is endowed with a pseudo-inner product if, on V, we have a symmetric bilinear form such that $\tau < \dim(V)$. Let $\beta = \{e_1, e_2, e_3, e_4\}$ be a basis of $\mathbb{R}^4$ such that

$$L(e_i, e_j) = \begin{cases} \delta_{ij}, & \text{if } 1 \leq i, j \leq 3, \\ -\delta_{i4}, & \text{if } 1 \leq i \leq 4. \end{cases}$$

Consequently, considering the dual basis $\beta^* = \{e_1^*, e_2^*, e_3^*, e_4^*\}$ $(L^* = L^{-1})$ on $(\mathbb{R}^4)^*$, we have

$$L^*(e_i^*, e_j^*) = \begin{cases} \delta_{ij}, & \text{if } 1 \leq i, j \leq 3, \\ -\delta_{i4}, & \text{if } 1 \leq i \leq 4. \end{cases}$$

Then we get a pseudo-inner product on $\Lambda(\mathbb{R}^4)^*$ given by

$$L^*(e_i^* \wedge e_j^*, e_k^* \wedge e_l^*) = \begin{cases} 1, & \text{if } 1 \leq i, j \leq 3 \text{ and } i = k \text{ and } j = l \\ -1, & \text{if } i = k \text{ and } j = l = 4, \\ 0, & \text{if } \{i, j\} \neq \{k, l\} \end{cases},$$

$$L^*(e_i^* \wedge e_j^* \wedge e_k^*, e_\alpha^* \wedge e_\beta^* \wedge e_\gamma^*) = \begin{cases} 1, & \text{if } 1 \leq i, j, k, \alpha, \beta, \gamma \leq 3 \text{ and } i = \alpha, j = \beta \text{ and } k = \gamma \\ -1, & \text{if } i = \alpha, j = \beta \text{ and } k = \gamma = 4, \\ 0, & \text{if } \{i, j, k\} \neq \{\alpha, \beta, \gamma\} \end{cases},$$

$$L^*(e_1^* \wedge e_2^* \wedge e_3^* \wedge e_4^*, e_1^* \wedge e_2^* \wedge e_3^* \wedge e_4^*) = -1.$$

Hodge's star-operator is defined using the pseudo-inner product L. Then

$$\omega \wedge (*\eta) = L^*(\omega, \eta) e_1 \wedge e_2 \wedge e_3 \wedge e_4, \tag{7}$$

and $*^2 = (-1)^{1+k(4-k)}$. In this case the eigenvalues of $* : \Omega^2(\mathbb{R}^4) \to \Omega^2(\mathbb{R}^4)$ are $\pm i$. The eigenspaces $\Omega_{\pm}^2$ yields the decomposition $\Omega^2(\mathbb{R}^4) = \Omega_+^2 \oplus \Omega_-^2$.

3 Differential Forms

In this section we will introduce differential forms on $\mathbb{R}^n$ and the Exterior Derivative operator. $\mathbb{R}^n$ carries very simple topological and vectorial structures, and every simply connected open subset $U \subset \mathbb{R}^n$ inherits these properties. In $\mathbb{R}^n$, we have a constant orthonormal frame making many concepts and calculations easy to handle. However this comfortable situation has to be extended to frames depending on the point.

Definition 7 Let $U \subset \mathbb{R}^n$ be an open subset.

(1) A frame on U is a C^∞ map $\mathcal{E} : U \to GL_n(\mathbb{R})$

$$\mathcal{E}(x) = \{e_1(x), \ldots, e_n(x)\},$$

such that the set $\mathcal{E}(x)$ is a basis of $T_x\mathbb{R}^n$ for all $x \in U$. A basis can be identified with an invertible matrix through the bijection

$$\mathcal{E}(x) = \{e_1(x), \ldots, e_n(x)\} \overset{1\text{-}1}{\longleftrightarrow} \left(e_1(x) \ldots e_n(x)\right) \in GL_n(\mathbb{R})$$

in which $e_i(x)$ is a column vector.

(2) The co-frame associated to $\mathcal{E}$ on U is

$$\mathcal{E}^*(x) = \{e_1^*(x), \ldots, e_n^*\}, \quad e_i^*(e_j) = \delta_{ij}.$$

Therefore $\mathcal{E}^*(x)$ is a basis of T_x^*U for all $x \in U$.

We can fix a constant orthonormal frame in $U \subset \mathbb{R}^n$ so that the dual co-frame is also constant. Classically, the co-vectors are written as $dx_i = e_i^*$. For $x \in U$, using the vector spaces T_xU and T_x^*U, we obtain the spaces of the alternating p-tensors $\Lambda^p(T_x^*U)$, $0 \le p \le n$.

Definition 8 Let $\Lambda^p(U) = \bigcup_{x \in U} \Lambda^p(T_x^*U)$ be the vector bundle over U with a fiber at x that is the vector space $\Lambda^p(T_x^*U)$. A differential p-form in U is a section $w : U \to \Lambda^p(U)$, i.e., for every $x \in U$ we have an alternating p-tensor $w_x \in \Lambda^p(T_x^*U)$. A p-form ω is smooth or, equivalently, C^∞ if the map $x \to \omega_x$ is C^∞. Let $\Omega^p(U)$ be the space of the p-forms defined on U.

Example 2 Some differential forms are well-known, as shown in the next examples;

(1) The 0-forms are C^∞-functions defined on U, so $\Omega^0(U) = C^\infty(U)$.
(2) For any function $f \in \Omega^0(U)$, the differential $df_x \in T_x^*U$ is a linear functional whose dependence on x is C^∞; therefore $df \in \Omega^1(U)$. Let $(x_1, \ldots, x_n)$ be the coordinates in U, and let $\mathcal{E}(x) = \{e_1, \ldots, e_n\}$ be the constant frame given by the canonical basis. In this case, we have $e_i^* = dx_i$. Since

$$df_x.u = u_1 \frac{\partial f}{\partial x_1} + \cdots + u_n \frac{\partial f}{\partial x_n} = \left(\sum_{i=1}^n \frac{\partial f}{\partial x_i} e_i^*\right)(u),$$

we have $df = \sum_{i=1}^{n} \frac{\partial f}{\partial x_i} dx_i$.

Since $\mathcal{E}^*(x) = \{e_1^*(x), \ldots, e_n^*(x)\}$ is a basis of $T_x^* U$ for all $x \in U$, the set $\{e_I^*(x)\}$, $I = (i_1, \ldots, i_p)$, $1 \le i_1 < \cdots < i_p \le n$ is a basis of $\Lambda^p(T_x^* \mathcal{U})$. Every p-form in U can be written as

$$w = \sum_I w_I e_I^*, \quad w_I \in C^\infty(\mathbb{R}^n; \mathbb{R}).$$

Let $U \subset \mathbb{R}^n$ and $V \subset \mathbb{R}^m$ be open subsets. A differentiable map $\phi : U \to V$ induces the homomorphism $\phi^* : \Omega^p(V) \to \Omega^p(U)$ given by

$$[\phi^* w]_x(u_1, \ldots, u_n) = w_{\phi(x)}(d\phi_x.u_1, \ldots, d\phi_x.u_n). \tag{8}$$

The p-form $\phi^* w \in \Omega^p(U)$ is the pullback of w by ϕ. When w is a 0-form, the pullback of w by ϕ is the 0-form $\phi^* w = w \circ \phi$.

Example 3 This example explains an important case that is useful to simplify calculus with forms. Let $U \subset \mathbb{R}^n$ and $V \subset \mathbb{R}^m$ be open subsets with coordinates $U = \{(x_1, \ldots, x_n) \mid x_i \in \mathbb{R}\}$ and $V = \{(y_1, \ldots, y_m) \mid y_j \in \mathbb{R}\}$. Fix the canonical frame $\beta = \{e_1, \ldots, e_m\}$ in $\mathbb{R}^m$. Let $\phi : U \to V$ be a C^∞-map defined with coordinates as $\phi = (\phi_1, \ldots, \phi_m)$. We will check the identity

$$\phi^* dy_\alpha = d\phi_\alpha. \tag{9}$$

Therefore we have

$$(\phi^* dy_\alpha)(u) = dy_\alpha(d\phi.u) = dy_\alpha(\sum_{i=1}^{n} u_i \frac{\partial \phi}{\partial x_i}) = \sum_{i=1}^{n} u_i dy_\alpha(\frac{\partial \phi}{\partial x_i}) =$$

$$= \sum_{i=1}^{n} u_i dy_\alpha(\sum_{\beta=1}^{m} \frac{\partial \phi^\beta}{\partial x_i} e_\beta) = \sum_{i=1}^{n} \sum_{\beta=1}^{m} u_i \frac{\partial \phi^\beta}{\partial x_i} dy_\alpha(e_\beta) =$$

$$= \sum_{i=1}^{n} u_i \frac{\partial \phi^\alpha}{\partial x_i} = \left[\sum_{i=1}^{n} \frac{\partial \phi^\alpha}{\partial x_i} dx_i \right](u) = d\phi^\alpha(u).$$

Theorem 1 (Change of Variables) *Let U, V be open subsets of $\mathbb{R}^n$ and let $\phi : U \to V$ be a diffeomorphism preserving the orientation. If w is an integrable n-form in V, then*

$$\int_{\phi(U)} w = \int_U \phi^* w. \tag{10}$$

Proof Let $w = f dy_1 \wedge \cdots \wedge dy_n$; we have

$$\phi^* w = (f \circ \phi) \det (d\phi_x)(dx_1 \wedge \cdots \wedge dx_n).$$

Since $\det (d\phi_x) > 0$, the statement follows from the formula for the change of variables for multiple integrals, Theorem 8 in Appendix A. □

Example 4 Let $U \subset \mathbb{R}^2$ be an open subset, $w = f_1 dx_2 \wedge dx_3 + f_2 dx_3 \wedge dx_1 + f_3 dx_1 \wedge dx_2 \in \Omega^2(\mathbb{R}^3)$ and $\vec{F} = f_1 \vec{i} + f_2 \vec{j} + f_3 \vec{k}$. Let $\phi : U \to \mathbb{R}^3$ be a C^∞-map given by $\phi(u, v) = (u, v, g(u, v))$. The image of ϕ is the surface $S \subset \mathbb{R}^3$ given by the graph of g. We would like to check the identity

$$\int_S w = \int_S \vec{F}.$$

By the Change of Variables Theorem, we have $\int_S w = \int_U \phi^* w$. Therefore

$$\int_S w = \int_U \phi^* w = \int_U \sum_{\{ijk\}} (f_i \circ \phi)\phi^*(dx_j \wedge dx_k) = \sum_{\{ijk\}} \int_U (f_i \circ \phi)(\phi^* dx_j) \wedge (\phi^* dx_k).$$

Since $\phi^*(dx_1) = d\phi_1 = du, \phi^*(dx_2) = d\phi_2 = dv$ and $\phi^*(dx_3) = dg = \frac{\partial g}{\partial u} du + \frac{\partial g}{\partial v} dv$. we get the identities

$$\phi^*(dx_1 \wedge dx_2) = du \wedge dv,$$

$$\phi^*(dx_3 \wedge dx_1) = -\frac{\partial g}{\partial u} du \wedge dv,$$

$$\phi^*(dx_2 \wedge dx_3) = -\frac{\partial g}{\partial v} du \wedge dv.$$

In terms of the coordinate (u, v) on U, the pullback form $\phi^* w$ is written as

$$\phi^* w = [-f_1 \frac{\partial g}{\partial u} - f_2 \frac{\partial g}{\partial v} + f_3] du \wedge dv =$$

$$= < (f_1, f_2, f_3), (-\frac{\partial g}{\partial u}, -\frac{\partial g}{\partial v}, 1) > du \wedge dv.$$

We note the normal vector to S is $\vec{N} = (-\frac{\partial g}{\partial u}, -\frac{\partial g}{\partial v}, 1)$, so

$$\int_U \phi^* w = \int_U < (f_1, f_2, f_3), (-\frac{\partial g}{\partial u}, -\frac{\partial g}{\partial v}, 1) > dudv =$$

$$= \int_U < \vec{F}, \vec{N} > dudv = \int_S \vec{F}.$$

Exercises

(1) Let ϕ and ψ be differentiable maps. Show the following identities;

    (a) $\phi^*(w_1 + w_2) = \phi^* w_1 + \phi^* w_2$;

    (b) $\phi^*(w_1 \wedge w_2) = \phi^* w_1 \wedge \phi^* w_2$;

    (c) $(\phi \circ \psi)^* w = \psi^* \circ \phi^* w$.

(2) Let $\phi : U \to V$ be a diffeomorphism between open sets in $\mathbb{R}^n$. Let $(x_1, \ldots, x_n)$ and $(y_1, \ldots, y_n)$ be coordinate systems in U and V, respectively. Show the identity

$$\left[\phi^*(dy_1 \wedge \cdots \wedge dy_m)\right] = \det[d\phi_x](dx_1 \wedge \cdots \wedge dx_n),$$

for all $y = \phi(x)$.

(3) Find $\int_S w$ given that $w = f_1 dx_2 \wedge dx_3 + f_2 dx_3 \wedge dx_1 + f_3 dx_1 \wedge dx_2 \in \Omega^2(\mathbb{R}^3)$ and S is a surface parametrized by $\phi : U \to \mathbb{R}^3$,

$$\phi(u, v) = \big(\phi_1(u, v), \phi_2(u, v), \phi_3(u, v)\big).$$

3.1 Exterior Derivative

The exterior derivative generalizes the operators grad, curl and div appearing in the classical integration theorems.

Definition 9 Let $U \subset \mathbb{R}^n$ be an open subset. For $p \in \{1, \ldots, n\}$, the exterior derivative operator $d_p : \Omega^p(U) \to \Omega^{p+1}(U)$ is defined as follows:

(1) If $p = 0$ and $w \in \Omega^0(U)$, then

$$d_0 w = dw = \sum_{i=1}^{n} \frac{\partial w_i}{\partial x_i} dx_i. \tag{11}$$

(2) If $p > 0$ and $w = \sum_I w_I dx_I \in \Omega^p(U)$, $w_I \in \Omega^O(U)$, then

$$d_p w = \sum_I dw_I \wedge dx_I. \tag{12}$$

Examples: Let $U \subset \mathbb{R}^n$ be an open subset and F a C^1-vector field defined on U;

(1) If $f \in \Omega^0(U)$, then $df = \sum_{i=1}^{n} \frac{\partial f}{\partial x_i} dx_i$. Therefore $d_0 f$ coincides with the differential df. It follows that $d_0 f = 0$ if and only if $\mathrm{grad}(f) = 0$.

(2) Considering $n = 2$, given $\omega = f_1 dx_1 + f_2 dx_2 \in \Omega^1(U)$ and $F = f_1 \vec{i} + f_2 \vec{j}$, we get

$$d\omega = df_1 \wedge dx_1 + df_2 \wedge dx_2 = \left(\frac{\partial f_1}{\partial x_1}dx_1 + \frac{\partial f_1}{\partial x_2}dx_2\right) \wedge dx_1 + \left(\frac{\partial f_2}{\partial x_1}dx_1 + \frac{\partial f_2}{\partial x_2}dx_2\right) \wedge dx_2$$

$$= \left(\frac{\partial f_2}{\partial x_1} - \frac{\partial f_1}{\partial x_2}\right)dx_1 \wedge dx_2 = \text{curl}(F)dx_1 \wedge dx_2.$$

(3) Considering $n = 3$, given $\omega = f_1 dx_1 + f_2 dx_2 + f_3 dx_3 \in \Omega^1(U)$ and $F = f_1\vec{i} + f_2\vec{j} + f_3\vec{k}$, we get

$$d\omega = \left(\sum_{i=1}^{3}\frac{\partial f_1}{\partial x_i}dx_i\right) \wedge dx_1 + \left(\sum_{i=1}^{3}\frac{\partial f_2}{\partial x_i}dx_i\right) \wedge dx_2 + \left(\sum_{i=1}^{3}\frac{\partial f_3}{\partial x_i}dx_i\right) \wedge dx_3$$

$$= \frac{\partial f_1}{\partial x_2}dx_2 \wedge dx_1 + \frac{\partial f_1}{\partial x_3}dx_3 \wedge dx_1 + \frac{\partial f_2}{\partial x_1}dx_1 \wedge dx_2 + \frac{\partial f_2}{\partial x_3}dx_3 \wedge dx_2 + \frac{\partial f_3}{\partial x_1}dx_1 \wedge dx_3$$

$$+ \frac{\partial f_3}{\partial x_2}dx_2 \wedge dx_3 = \left(\frac{\partial f_2}{\partial x_1} - \frac{\partial f_1}{\partial x_2}\right)dx_1 \wedge dx_2 + \left(\frac{\partial f_3}{\partial x_1} - \frac{\partial f_1}{\partial x_3}\right)dx_1 \wedge dx_3 + \left(\frac{\partial f_3}{\partial x_2} - \frac{\partial f_2}{\partial x_3}\right)dx_2 \wedge dx_3.$$

Therefore $d\omega = 0$ if and only if $\text{curl}(F) = 0$.

(4) Considering $n = 3$, given $w = f_3 dx_1 \wedge dx_2 + f_2 dx_3 \wedge dx_1 + f_1 dx_2 \wedge dx_3 \in \Omega^2(U)$ and $F = f_1\vec{i} + f_2\vec{j} + f_3\vec{k}$, we get

$$d\omega = \frac{\partial f_3}{\partial x_3}dx_1 \wedge dx_2 \wedge dx_3 + \frac{\partial f_2}{\partial x_2}dx_1 \wedge dx_2 \wedge dx_3 + \frac{\partial f_1}{\partial x_1}dx_1 \wedge dx_2 \wedge dx_3 = \text{div}(F)dx_1 \wedge dx_2 \wedge dx_3.$$

Therefore $d\omega = 0$ if and only if $\text{div}(F) = 0$.

(5) If $w \in \Omega^n(U)$, then $d\omega = 0$.

Theorem 2 *For all $p \in \{1, \ldots, n\}$, the exterior derivative operator $d_p : \Omega^p(U) \to \Omega^{p+1}(U)$ satisfies the following properties;*

(1) (linearity) For any $a_1, a_2 \in \mathbb{R}$ and $\omega_1, \omega_2 \in \Omega^p(U)$, we have

$$d_p(a_1\omega_1 + a_2\omega_2) = a_1 d_p\omega_1 + a_2 d_p\omega_2.$$

(2) (multiplication law) If $\omega \in \Omega^p(U)$, then

$$d_p(\omega \wedge \theta) = d_p\omega \wedge \theta + (-1)^p\omega \wedge d\theta.$$

(3) (cocycle condition)

$$d_{p+1} \circ d_p = 0, \quad (d^2 = 0). \tag{13}$$

In addition, the set of operators $\{d_p : \Omega^p \to \Omega^{p+1} \mid 1 \leq p \leq n\}$ is the only one that satisfies the properties above and $d_0 = d$, i.e., the exterior derivative on 0-forms coincides with the usual derivative on functions.

Proof The above properties follow directly from the definitions. To verify the uniqueness, suppose that there is a family of operators $\{D_p : \Omega^p(U) \to \Omega^{p+1}(U)\}$ such that $D_0 = d$. So $D(dx_I) = 0$ since

$$D(dx_{i_1} \wedge \cdots \wedge dx_{i_n}) = \sum_j (-1)^{i_j - 1} dx_{i_1} \wedge \cdots \wedge D(dx_{i_j}) \wedge \cdots \wedge dx_{i_p}$$

and $D(dx_{i_j}) = D(Dx_{i_j}) = 0$. In this way, if $w = \sum_I a_I dx_I$, then

$$Dw = \sum_I \left(D(a_I) \wedge dx_I + a_I D(dx_I) \right) = \sum_I d(a_I) \wedge dx_I = dw. \qquad \square$$

Corollary 1 *Let $U \subset \mathbb{R}^n$, $V \subset \mathbb{R}^m$ be open subsets and let $\phi : U \to V$ be a C^∞-map. So*

$$d_p(\phi^* \omega) = \phi^* d_p \omega,$$

for all $\omega \in \Omega^p(V)$.

Proof Let's consider $\omega \in \Omega^0(V)$; therefore

$$\left(\phi^* d\omega \right)_x (u) = d\omega_{\phi(x)} . d\phi_x(u) = d(\omega \circ \phi)_x(u) = d(\phi^* \omega)_x(u).$$

For the general case, when $d\omega = \sum_I df_I \wedge dx_I \in \Omega^p(V)$, we have

$$\phi^* d\omega = \phi^* \left(\sum_I df_I \wedge dx_I \right) = \sum_I \phi^* (df_I \wedge dx_I) = \sum_I (\phi^* df_I) \wedge (\phi^*(dx_I))$$

$$d\phi^* \omega = d\left[\phi^* \left(\sum_I f_I \wedge dx_I \right) \right] = \sum_I d\left(\phi^* (f_I \wedge dx_I) \right) = \sum_I d(\phi^* df_I) \wedge (\phi^*(dx_I)).$$

Hence $d_p(\phi^* w) = \phi^* d_p w$. $\qquad \square$

The above corollary means that the following diagram is commutative;

$$\begin{array}{ccc}
\Omega^p(V) & \xrightarrow{d_p} & \Omega^{p+1}(V) \\
\downarrow{\phi^*} & & \downarrow{\phi^*} \\
\Omega^p(U) & \xrightarrow{d_p} & \Omega^{p+1}(U).
\end{array}$$

Exercises

The Frobenius Integrability Theorem has the following version using differential forms (see Ref. [45]);

(1) Let $U \subset \mathbb{R}^n$ be an open subset and let $\mathcal{D}$ be a p-dimensional C^∞-distribution on U. A q-form ω annihilates $\mathcal{D}$ if for each $x \in U$,

$$\omega_x(v_1, \ldots, v_q) = 0 \quad \text{for all } v_1, \ldots, v_q \in \mathcal{D}_x.$$

We let $\mathcal{I}(\mathcal{D}) = \{\omega \in \Omega^*(U) \mid \omega \text{ annihilate } \mathcal{D}\}$. Show that in the following items:

(a) $\mathcal{I}(\mathcal{D})$ is an ideal.
(b) $\mathcal{I}(\mathcal{D})$ is locally generated by $(n - p)$ independent 1-forms $\omega_1, \ldots, \omega_{n-p}$ on U.
(c) If $\mathcal{I}(\mathcal{D}) \subset \Omega^*(U)$ is an ideal locally generated by $(n - p)$ independent 1-forms, then there is a unique C^∞-distribution $\mathcal{D}$ of dimension p on U for which $\mathcal{I} = \mathcal{I}(\mathcal{D})$.

(2) An ideal $\mathcal{I} \subset \Omega^*(U)$ is a differential ideal if it is closed under the exterior derivative D, i.e., $d(\mathcal{I}) \subset \mathcal{I}$. Show that the claims in the following items are true;

(a) A C^∞-distribution on U is involutive if and only if the ideal $\mathcal{I}(\mathcal{D})$ is a differential ideal.
(b) Let $i : M \to U$ be the inclusion of a submanifold. M is an integral manifold of an ideal $\mathcal{I} \subset \Omega^*(U)$ if for every $\omega \in \mathcal{I}$, we have $i^*\omega \equiv 0$ (the pull-back form). M is maximal if its image is not a proper subset of the image of any other connected integral manifold of the ideal. Assume $\mathcal{I} \subset \Omega^*(U)$ is a differential ideal locally generated by $(n - p)$ independent 1-forms.. For any $x \in M$, show that we have a unique maximal, connected, integral manifold of $\mathcal{I}$ through x whose dimension is equal to p.

(3) What can we claim from the results above when $\mathcal{I}$ is generated by a closed 1-form ω?

4 De Rham Cohomology

Differential forms are a powerful tool; they are more than just a convenient language to prove Stokes theorem because they allows us to define and compute topological invariants of a differentiable manifold. In this section, we compute the De Rham groups of the sphere S^n and the closed surfaces Σ_g of genus g. The formalism we are developing can be used to define the De Rham cohomology of any differentiable manifold and not just for an open subset $U \subset \mathbb{R}^n$. We recommend reading the section on differential forms on manifolds at the end of this chapter. The exterior derivative defines the sequence

$$\Omega^0(V) \xrightarrow{d} \Omega^1(V) \xrightarrow{d} \cdots \xrightarrow{d} \Omega^p(V) \xrightarrow{d} \Omega^{p+1}(V) \xrightarrow{d} \cdots \xrightarrow{d} \Omega^n(V). \qquad (14)$$

The cocycle condition $d^2 = 0$ defines the differentiable complex $(\Omega^*(U), d)$ defined as

$$\Omega^*(U) = \bigoplus_{p=0}^{n} \Omega^p(U), \quad d : \Omega^p(U) \to \Omega^{p+1}(U).$$

Since $d^2 = 0$, we have the required condition to be a complex, i.e., $\mathrm{Im}(d) \subset \mathrm{Ker}(d)$.

Definition 10 Let $(\Omega^*(U), d)$ be a differentiable complex;

(1) $\omega \in \Omega^p(U)$ is a closed p-form if $d\omega = 0$. The space of closed p-forms is

$$Z^p(U) = \{\omega \in \Omega^p(U) \mid d\omega = 0\}.$$

(2) $\omega \in \Omega^p(U)$ is an exact p-form if we have a $(p-1)$-form $\eta \in \Omega^{p-1}(U)$ such that $\omega = d\eta$. The space of exact p-forms is

$$B^p(U) = \{\omega \in \Omega^p(U) \mid \omega = d\eta, \ \eta \in \Omega^{p-1}(U)\}.$$

From the definition, we have $B^p(U) \subset Z^p(U)$, $\forall p = 0, \ldots, n$. Consider in $Z^p(\mathbb{R}^n)$ the following equivalent relation:

$$\omega \sim \omega' \iff \omega' - \omega \in B^p(\mathbb{R}^n), \quad [\omega] = \omega + d\Omega^{p-1}.$$

Definition 11 Let $U \subset \mathbb{R}^n$ be an open subset;
(i) The pth De Rham cohomology group is the quotient vector space

$$H^p_{DR}(U) = \frac{\mathrm{Ker}(d)}{\mathrm{Im}(d)} = \frac{Z^p(\mathbb{R}^n)}{B^p(\mathbb{R}^n)}. \qquad (15)$$

By convention, $H^p_{DR}(U) = 0$ if $p < 0$, and

$$H^0_{DR}(U) = \mathrm{Ker}(d) = \{f \in C^\infty(U, \mathbb{R}) \mid df = 0\}, \quad H^n_{DR}(U) = \frac{\Omega^n(U)}{d\Omega^{n-1}(U)}.$$

(ii) The p-forms ω and ω' are cohomologous if $\omega \sim \omega'$. In this case, they define the same class $[\omega'] = [\omega] \in H^p_{DR}(U)$.

The vector space $\Omega^p(U)$ is infinite dimensional. We cannot guarantee that the quotient spaces $H^p_{DR}(U)$ are closed spaces. They are closed for a large class of sets, e.g., for closed differentiable manifolds because the exterior derivative operator has a closed image.[2] In the following, we will restrict ourselves to subsets $U \subset \mathbb{R}^n$ that are differentiable manifolds. De Rham's theorem ensures that if U is a compact differentiable manifold, then the vector spaces $H^p_{DR}(U)$, $p = 0, \ldots, n$, are isomorphic to the singular cohomology groups $H^p(U, \mathbb{R})$ with real coefficients, and so they are finite dimensional and $\dim(H^p_{DR}(U)) = b_p(U)$ for all $p = 0, \ldots, \dim(U)$. To answer the question about when a vector field is conservative requires a deep insight into the underlying topological space.

Proposition 8 *The vector space*

$$H^*_{DR}(U) = \bigoplus_{i=0}^{n} H^p_{DR}(U)$$

is an algebra endowed with multiplication:

$$H^p_{DR}(U) \times H^q_{DR}(U) \rightarrow H^{p+q}_{DR}(U)$$
$$([w_1], [w_2]) \rightarrow [w_1].[w_2] = [w_1 \wedge w_2]. \tag{16}$$

Proof If w_1 and w_2 are closed forms, i.e., $dw_1 = dw_2 = 0$, then the form $w_1 \wedge w_2$ is also closed, since

$$d(w_1 \wedge w_2) = dw_1 \wedge w_2 + (-1)^{|w_1|} w_1 \wedge dw_2 = 0.$$

It is enough to show that the product is well-defined since the associativity is inherited from the exterior product. Suppose that $[w_1] = [w_1']$ and $[w_2] = [w_2']$; we need to check that

$$[w_1].[w_2] = [w_1'].[w_2'],$$

that is

$$[w_1 \wedge w_2] = [w_1' \wedge w_2'].$$

Given $w_1' = w_1 + d\eta_1$ e $w_2' = w_2 + d\eta_2$, we have

$$w_1' \wedge w_2' = (w_1 + d\eta_1) \wedge (w_2 + d\eta_2) = w_1 \wedge w_2 + w_1 \wedge d\eta_2 + d\eta_1 \wedge w_2 + d\eta_1 \wedge d\eta_2 =$$
$$= w_1 \wedge w_2 + d[(-1)^p w_1 \wedge \eta_2 + \eta_1 \wedge w_2 + \eta_1 \wedge \eta_2].$$

Therefore $[w_1].[w_2] = [w_1'].[w_2']$. $\square$

---

[2]For more details on the De Rham theory, see Ref. [45].

Since $\phi^* d = d\phi^*$, a differentiable map $\phi : U \to V$ between the open subsets $U \subset \mathbb{R}^n$ and $V \subset \mathbb{R}^m$ induces a homomorphism $\phi^* : H^p_{DR}(V) \to H^p_{DR}(U)$. So the composition of maps

$$U \xrightarrow{\ \phi\ } V \xrightarrow{\ \psi\ } W$$

induces the homomorphism

$$H^p_{DR}(W) \xrightarrow{\ \psi^*\ } H^p_{DR}(V) \xrightarrow{\ \phi^*\ } H^p_{DR}(U),$$

$(\psi \circ \phi)^* = \psi^* \circ \phi^*$.

Next, the results obtained in Chap. 4 will be extended to star-shaped sets.

Lemma 2 (Poincaré Lemma)   *If $U \subset \mathbb{R}^n$ is a star-shaped open subset, then $H^p_{DR}(U) = 0$ for all $p > 0$. and $H^0_{DR}(U) \simeq \mathbb{R}$*

Proof The proof relies on the existence of an operator $S_p : \Omega^p(U) \to \Omega^{p-1}(U)$ such that

$$d_{p-1} S_p + S_{p+1} d_p = id_{\Omega^p}, \text{ if } p > 0,$$
$$S_1 d = \text{id} - e, \text{ if } p = 0,$$

and $e(\omega) = \omega(0)$. The existence of this operator immediately implies the lemma, since given a closed p-form ω, we have

$$\omega = d_{p-1} S_p \omega, \text{ if } p > 0,$$
$$\omega = \omega(0), \text{ if } p = 0.$$

(1) First, we construct a homomorphism $\widehat{S}_p : \Omega^p(U \times \mathbb{R}) \to \Omega^{p-1}(U)$. Every form $w \in \Omega^p(U \times \mathbb{R})$ can be written as

$$\omega = \sum_I f_I(x, t) dx_I + \sum_J g_J(x, t) dt \wedge dx_J,$$

given that $I = (i_1, \ldots, i_p)$ and $J = (j_1, \ldots, j_{p-1})$ are multi-index. Consider

$$\widehat{S}_p(\omega) = \sum_J \left[\int_0^1 g_J(x, t) dt \right] dx_J.$$

Once

$$d\omega = \sum_J \frac{\partial f_I}{\partial x_i} dx_i \wedge dx_I + \sum_I \frac{\partial f_I}{\partial t} dt \wedge dx_I + \sum_J \frac{\partial g_J}{\partial x_i} dx_i \wedge dt \wedge dx_J,$$

we get

$$d\widehat{S}_p(w) + \widehat{S}_{p+1}dw = \sum_{J,i}\Big[\int_0^1 \frac{\partial g_J}{\partial x_i}dt\Big]dx_i \wedge dx_J + \sum_I\Big[\int_0^1 \frac{\partial f_I}{\partial t}dt\Big]dx_I - \sum_{J,i}\Big[\int_0^1 \frac{\partial g_J}{\partial x_i}dt\Big]dx_i \wedge dx_J =$$

$$= \sum_I\Big[\int_0^1 \frac{\partial f_I}{\partial t}dt\Big]dx_I = \sum_I\Big[f_I(x,1) - f(x,0)\Big]dx_I.$$

(2) Let $\psi : \mathbb{R} \to \mathbb{R}$ be a C^∞-function, such that $0 \le \psi \le 1$ and

$$\psi(t) = \begin{cases} 0, t \le 0; \\ 1, t \ge 1; \\ 0 \le \psi(t) \le 1, t \in (0,1). \end{cases}$$

Let $\phi : U \times \mathbb{R} \to U$ be the map $\phi(x,t) = \psi(t)x$. Consider $S_p(w) = \widehat{S}^p(\psi^* w)$. Given $w = \sum_I h_I(x)dx_I$, since

$$\phi^*(dx_i) = x_i\psi'(t)dt + \psi(t)dx_i,$$

we have

$$\phi^*(w) = \sum_I h_I\big(\psi(t)x\big)\big(d\psi(t)x_{i_1} + \psi(t)dx_{i_1}\big) \wedge \cdots \wedge \big(d\psi(t)x_{i_p} + \psi(t)dx_{i_p}\big).$$

In the above expression, the first term independent of dt is

$$\sum_I h_I\big(\psi(t)x\big)\big[\psi(t)\big]^p dx_I.$$

So

$$dS_p(w) + S_{p+1}dw = \begin{cases} \sum_I h_I(x)dx_I = w, \ p > 0; \\ w(x) - w(0), \ p = 0. \end{cases}$$

$\square$

Poincaré's Lemma implies that the De Rham cohomology groups of a star-shaped open subset U are trivial;

Proposition 9 *Let $n > 1$. The De Rham cohomology groups of $\mathbb{R}^n$ are*

$$H_{DR}^i(\mathbb{R}^n) = \begin{cases} \mathbb{R}, \ if\, i = 0, \\ 0, \ if\, i \ne 0. \end{cases}$$

The De Rham cohomology H_{DR}^* carries a ring structure, and indeed, it is a topological invariant. We list below some fundamental properties of this ring; the proofs can be read in the Refs. [4, 22, 45].

(1) If X is connected and contractible, then $H^i_{DR}(X) = 0$ for all $i \neq 0$ and $H^0_{DR}(X) = \mathbb{R}$.

(2) From the De Rham theorem, if X is a paracompact space admitting a differentiable structure, then $H^i_{DR}(X)$ is isomorphic to $H^i(X, \mathbb{R})$ (ith-singular cohomology group) for all $i > 0$. If X is a compact differentiable manifold, $H^i_{DR}(X)$ is finitely generated for all $i = 0, \ldots, \dim(X)$. Since $H^i(X, \mathbb{R})$ is a topological invariant, so also is the group $H^i_{DR}(X)$. Indeed, $H^i(X, \mathbb{R})$ is a vector space over $\mathbb{R}$ whose dimension $b_i(X) = \dim(H^i_{DR}(X))$ is called the ith-Betti number of X.

(3) If X and Y are homotopic topological spaces, then $H^*_{DR}(X) = H^*_{DR}(Y)$. In particular, if Y is contractible, then $H^i_{DR}(X \times Y) = H^i_{DR}(X)$.

Exercise

(1) If U has a k-connected component, show that $H^0_{DR}(U) \simeq \mathbb{R}^k$.

4.1 Short Exact Sequence

Let's consider that (Ω_A, d_A) and (Ω_B, d_B) are two differentiable complexes, such that $\Omega_A = \oplus_{p \geq 0} \Omega^p_A$ and $\Omega_B = \oplus_{p \geq 0} \Omega^p_B$ with cohomologies $H^*_{DR}(A)$ and $H^*_{DR}(B)$.

Definition 12 A map $f : \Omega_A \to \Omega_B$ between the differentiable complexes is a chain map if $f = \oplus_{p \geq 0} f_p$ is the direct sum of maps $f_p : \Omega^p_A \to \Omega^p_B$ such that $f_{p+1} \circ d_A = d_B \circ f_p$.

Consequently the diagram below commutes for all $p \geq 0$;

$$
\begin{array}{ccc}
\Omega^p_A & \overset{d_{A,p}}{\to} & \Omega^{p+1}_A \\
\downarrow f_p & & \downarrow f_{p+1} \\
\Omega^p_B & \overset{d_{B,p}}{\to} & \Omega^{p+1}_B.
\end{array}
$$

The differentiable complexes (Ω_A, d_A), (Ω_B, d_B) and (Ω_C, d_C) define a Short Exact Sequence if there are chain maps $f : \Omega_A \to \Omega_B$ and $g : \Omega_B \to \Omega_C$ such that the sequence below is an exact sequence:

$$
0 \longrightarrow \Omega_A \overset{f}{\longrightarrow} \Omega_B \overset{g}{\longrightarrow} \Omega_C \longrightarrow 0. \tag{17}
$$

It means that f is injective, $\text{Ker}(g) = \text{Im}(f)$ and g is surjective.

Proposition 10 *For all $p \geq 0$, the Short Exact Sequence (17) induces homomorphisms $\delta : H^p_{DR}(C) \to H^{p+1}_{DR}(A)$ and a long exact sequence*

$$
\cdots \to H^p_{DR}(A) \overset{f_p}{\to} H^p_{DR}(B) \overset{g_p}{\to} H^p_{DR}(C) \overset{\delta}{\to} H^{p+1}_{DR}(A) \to \cdots .
$$

Proof The homomorphisms $f_p : H_{DR}^p(A) \to H_{DR}^p(B)$ and $g_p : H_{DR}^p(B) \to H_{DR}^p(C)$ are well-defined and $\mathrm{Im}(f_p) = \mathrm{Ker}(g_p)$. To define the homomorphism $\delta : H_{DR}^p(C) \to H_{DR}^{p+1}(A)$, we call upon the diagram below;

$$
\begin{array}{ccccccccc}
 & & \uparrow & & \uparrow & & \uparrow & & \\
0 & \longrightarrow & \Omega_A^{p+1} & \xrightarrow{f_{p+1}} & \Omega_B^{i+1} & \xrightarrow{g_{p+1}} & \Omega_C^{p+1} & \longrightarrow & 0 \\
 & & {\scriptstyle d_{A,p}}\uparrow & & {\scriptstyle d_{B,p}}\uparrow & & {\scriptstyle d_{C,p}}\uparrow & & \\
0 & \longrightarrow & \Omega_A^i & \xrightarrow{f_p} & \Omega_B^p & \xrightarrow{g_p} & \Omega_C^p & \longrightarrow & 0 \\
 & & \uparrow & & \uparrow & & \uparrow & &
\end{array}
\qquad (18)
$$

Let $c \in \Omega_C^p$ be a closed form, i.e., $d_{C,p}(c) = 0$. Since g_p is surjective, there is $b \in \Omega_B^i$ such that $g_p(b) = c$. The map g is a chain map and since the sequence is exact, we have $g_{p+1}(d_{B,p}(b)) = 0$. So there is $a \in \Omega_A^{p+1}$ such that $d_{B,p}(b) = f(a)$. Defining $\delta(c) = a$, we get that $\delta : \Omega_C^p \to \Omega_A^{p+1}$ induces the well-defined homomorphism $\delta : H_{DR}^p(C) \to H_{DR}^{p+1}(A)$, as shown in the following items;

(i) there is a unique $a \in \Omega_A^{p+1}$ such that $d_{B,p}(b) = f_{p+1}(a)$;

Suppose there are $a, a' \in \Omega_A^{p+1}$ such that $d_{B,p}(b) = f_{p+1}(a) = f_{p+1}(a')$, so we get $a' - a$ belonging to $\mathrm{Ker}(f_{p+1}) = \{0\}$. Hence $a' = a$.

(ii) If $[c'] = [c] \in H_{DR}^p(C)$, then we have $c' = c + d\eta$, $\eta \in \Omega_C^{p-1}$, and therefore $\delta(c') = \delta(c)$.

(iii) $d_{A,p+1}(a) = 0$.

The cocycle condition yields

$$f_{p+2}d_{A,p+1}(a) = d_{B,p+1}f_{p+1}(a) = d_{B,p+1}d_{B,p}(b) = 0.$$

Since f_{p+2} is injective, we have $d_{A,p+1}(a) = 0$. Consequently, a defines a class in $H_{DR}^{p+1}(A)$. $\qquad\square$

5 De Rham Cohomology of Spheres and Surfaces

It is a difficult task to compute the De Rham Cohomology groups. However several examples of groups that can be easily found with relatively simple techniques. We will work out two cases: the sphere S^n and the closed surface Σ_g of genus g. To find the groups, we will make use of the partitions of unity and the Mayer-Vietoris exact sequence.

Partitions of Unity

Partitions of unity are useful tools in situations in which we need to define global objects from local objects.

Given a function $f : \Omega \to \mathbb{R}$, let $\text{supp(f)} = \{x \in \Omega \mid f(x) \neq 0\}$ be the support of f.

Definition 13 A family of subsets $\mathcal{C} = \{C_\lambda\}_{\lambda \in \Lambda}$ contained in $\Omega \subset \mathbb{R}^n$ is locally finite if every point $x \in \Omega$ has a neighborhood $V_x \subset \Omega$ intersecting only a finite number of sets C_λ. In this way, if $\Omega \subset \bigcup_{\lambda \in \Lambda} C_\lambda$, then the family $\mathcal{C} = \{C_\lambda \mid \lambda \in \Lambda\}$ defines a locally finite cover of Ω.

Definition 14 A partition of unity defined in Ω is a family of C^k-functions $\{\phi_\lambda\}_{\lambda \in \Lambda} \in C^k(\Omega, \mathbb{R})$ such that:

(1) For all $\lambda \in \Lambda$ and $x \in \Omega$, we have $\phi_\lambda(x) \geq 0$.
(2) The family of sets $\mathcal{F} = \{\text{supp}(\phi_\lambda)\}_{\lambda \in \Lambda}$ is locally finite in Ω.
(3) For all $x \in \Omega$, we have $\sum_{\lambda \in \Lambda} \phi_\lambda(x) = 1$.

The functions ϕ_λ are called cutoff functions. From the definition, we get

(1) $0 \leq \phi_\lambda(x) \leq 1$ for all $x \in \Omega$ and $\lambda \in \Lambda$.
(2) For all $x \in \Omega$, we have $\lambda \in \Lambda$ such that $\phi_\lambda(x) \geq 0$. The supports define the locally finite cover $\Omega \subset \bigcup_{\lambda \in \Lambda} \text{supp}(\phi_\lambda)$.

Definition 15 A partition of unity $\{\phi_\lambda \mid \lambda \in \Lambda\}$ is subordinated to the cover $\mathcal{C}_\Theta = \{U_\theta \mid \theta \in \Theta\}$ of Ω if, for $\lambda \in \Lambda$, there is $\theta \in \Theta$ such that $\text{supp}(\phi_\lambda) \subset U_\theta$. In this case, the cover $\{\text{supp}(\phi_\lambda) \mid \lambda \in \Lambda\}$ is a refinement of the cover $\{U_\theta \mid \theta \in \Theta\}$.

Theorem 3 *Let $\Omega \in \mathbb{R}^n$ be an open subset. For any open cover $\mathcal{C}_\Lambda = \{U_\lambda \mid \lambda \in \Lambda\}$ of Ω, there is a partition of unity $\{\phi_\lambda\}_{\lambda \in \Lambda}$ subordinated to $\mathcal{C}_\Lambda$.*

The proof of the theorem requires a list of results left as Exercises, some are elementary.

Exercises

(1) (Lindelöf's Theorem) Let $\Omega \subset \mathbb{R}^n$ be an[3] arbitrary set. Show that any cover of Ω admits a countable subcover.
(2) Show that any locally finite family of sets contained in an open set $\Omega \subset \mathbb{R}^n$ is countable.
(3) If $\mathcal{C} = \{C_\lambda \mid \lambda \in \Lambda\}$ is a locally finite family contained in an open subset, show that a compact set $K \subset \Omega$ intersects at most a finite number of sets C_λ.
(4) If $\mathcal{C} = \{C_\lambda \mid \lambda \in \Lambda\}$ is a locally finite family of closed sets contained in Ω, show that the set $C = \bigcup_{\lambda \in \Lambda} C_\lambda$ is closed in Ω.
(5) Let $0 < a < b$ and consider the function (Fig. 1),

$$f(t) = \begin{cases} e^{-\frac{1}{(t+a)(t+b)}}, t \in [-b, -a]; \\ 0, t \notin [-b, -a]. \end{cases}$$

---

[3]The proofs can be read in [29].

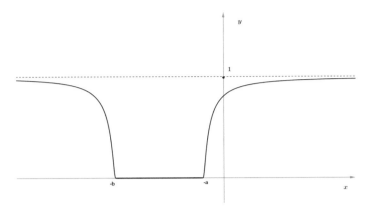

Fig. 1 Graph of $f(t)$

(a) Sketch the graph of $g(t) = \begin{cases} \dfrac{f(t)}{\int_{-\infty}^{\infty} f(t)dt}, & t \le 0, \\ \dfrac{-f(t)}{\int_{-\infty}^{\infty} f(t)dt}, & t > 0. \end{cases}$

(b) Consider the function $h : \mathbb{R} \to \mathbb{R}$ given by $h(s) = \int_{-\infty}^{t} g(s)ds$ and sketch the graph.

(c) Let $\psi : \mathbb{R}^n \to \mathbb{R}$ be the function $\psi(x) = h(|\, x\, |)$. Show that $\psi \in C^\infty(\mathbb{R}^n, \mathbb{R})$ and $\operatorname{supp}(\psi) \subset B_b(0)$.

Lemma 3 *Any open subset $\Omega \subset \mathbb{R}^n$ admits a countable cover $\Omega = \bigcup_{n \in \mathbb{N}} K_n$ by compact sets K_n such that*

$$K_n \subset \operatorname{int}(K_{n+1}), \quad \forall n \in \mathbb{N}.$$

Proof Every point $a \in \Omega$ belongs to a ball $B_\delta(a) \subset \Omega$, so $\Omega \subset \bigcup_{a \in \Omega} B_\delta(a)$. By Lindelöf's theorem, there is a countable subcover such that $\Omega \subset \bigcup_{n \in \mathbb{N}} B_\delta(a_n)$ and $\Omega \subset \bigcup_{n \in \mathbb{N}} \overline{B_\delta(a_n)}$, with $\overline{B_\delta(a_n)}$ being compact sets. The compact sets K_n are defined by induction, as follows; let $K_1 = \overline{B_\delta(a_1)}$ and suppose that

(1) $\Omega \subset K_1 \cup K_2 \cup \cdots \cup K_p \cup B_\delta(a_{p+1}) \cup \ldots$,
(2) $K_i \subset K_{i+1}, 1 \le i \le p$,
(3) $B_\delta(a_1) \cup \cdots \cup B_\delta(a_p) \subset K_p$.

Take the cover of the compact set $K_p \cup \overline{B_\delta(a_{p+1})}$ with a finite number of balls $B_\delta(a)$ and $a \in K_p \cup B_\delta(a_{p+1})$. Consider

$$K_{p+1} = K_p \cup \overline{B_\delta(a_{p+1})}.$$

Repeating this process, we obtain the cover $\{K_n \mid n \in \mathbb{N}\}$.   □

Proof (Theorem 3) Let $\{x_\lambda \mid \lambda \in \Lambda\}$ be a family of points in Ω and let $B_\delta(x_\lambda)$ be open balls such that

$$\Omega \subset \bigcup_{\lambda \in \Lambda} B_\delta(x_\lambda).$$

For $\lambda \in \Lambda$, consider the function $\psi_\lambda(x)$ such that $\text{supp}(\psi_\lambda) \subset B_\delta(x_\lambda)$. By Lemma 3, there is a countable family of compact sets $\{K_k \mid k \in \mathbb{N}\}$ such that $K_k \subset K_{k+1}$ and $\Omega \subset \bigcup_{k \in \mathbb{N}} K_k$. Let $A_1 = \bigcup_{i=1}^{p_1} B_\delta(x_i)$ be a finite cover of K_1. Since $\overline{K_2 - K_1}$ is compact, consider $A_2 = \bigcup_{i=1}^{p_2} B_\delta(x_i)$ a finite cover of $\overline{K_2 - K_1}$. Analogously, $\overline{K_k - K_{k-1}}$ is compact, so

$$A_k = \bigcup_{i=1}^{p_k} B_\delta(x_i)$$

is a finite cover of $\overline{K_k - K_{k-1}}$. Therefore $\Omega \subset \bigcup_{k \in \mathbb{N}} A_k$. After ordering the set of balls, we have $\Omega \subset \bigcup_{n \in \mathbb{N}} B_\delta(x_n)$. For $n \in \mathbb{N}$, consider the function $\phi_n : \Omega \to \mathbb{N}$,

$$\phi_p(x) = \frac{\psi_p(x)}{\sum_{n \in \mathbb{N}} \psi_n(x)}.$$

The collection of functions $\{\phi_p \mid p \in \mathbb{N}\}$ satisfies the following properties;

(1) $0 \leq \phi(_p(x) \leq 1$ for all $x \in \Omega$.
(2) For any $x \in \Omega$, we have $\phi_p(x) = 0$, but for a finite number of $p \in \mathbb{N}$ and $\text{supp}(\psi_p) \subset B_\delta(x_p)$. Consequently, the family $\mathcal{F} = \{\text{supp}(\phi_p) \mid n \in \mathbb{N}\}$ is locally finite in Ω and is also subordinated to the cover $\{B_\delta(x_p) \mid p \in \mathbb{N}\}$.
(3) $\sum_{p \in \mathbb{N}} \phi_p = 1$.

$\square$

The Mayer-Vietoris Sequence

Let M be a set such that $M = U \cup V$, U, V open sets. We have the inclusions

$$M \xleftarrow{\;\;i\;\;} U \sqcup V \xleftarrow{\;i_U \sqcup i_V\;} U \cap V,$$

given that $U \sqcup V$ is the disjoint union. The map $i : U \sqcup V \to M$ is the inclusion, and the maps $i_U : U \cap V \to U$ and $i_V : U \cap V \to V$ are also inclusions. This maps define the homomorphisms

$$\Omega^*(M) \xrightarrow{\;\;i^*\;\;} \Omega^*(U) \oplus \Omega^*(V) \xrightarrow{\;i_U^* \oplus i_V^*\;} \Omega^*(U \cap V). \qquad (19)$$

Consider the homomorphism $j^* : \Omega^*(U) \sqcup \Omega^*(V) \to \Omega^*(U \cap V)$ given by $j^*(\alpha, \beta) = i_U^*(\alpha) - i_V^*(\beta)$.

Lemma 4 (Mayer-Vietoris Sequence) *The sequence*

$$0 \longrightarrow \Omega^*(M) \xrightarrow{\ i^*\ } \Omega^*(U) \oplus \Omega^*(V) \xrightarrow{\ j^*\ } \Omega^*(U \cap V) \longrightarrow 0$$

$$(\alpha, \beta) \longrightarrow i_U^*(\alpha) - i_V^*(\beta)$$

is exact.

Proof Let $w \in \Omega^*(M)$ be a differential form on M. The restrictions $w_U = w \mid_U$, $w_V = w \mid_V$ are also forms on U and V, respectively; hence $j^* \circ i^* = 0$. Let's check that j^* is surjective. Let $\eta \in \Omega^*(U \cap V)$ and let $\{\phi_U, \phi_V\}$ be a partition of unity, subordinate to the cover $\{U, V\}$ of M. The forms $\phi_U.\eta \in \Omega_U^*$ and $\phi_V.\eta \in \Omega_V^*$ satisfy $\eta = \phi_U.\eta - (-\phi_V.\eta)$. Therefore j^* is surjective and $j^*(\phi_U.\eta, -\phi_V.\eta) = \eta$. □

Proposition 10 and Lemma 4 yield the Mayer-Vietoris Exact Sequence (MVS),

$$\cdots \to H_{DR}^p(M) \xrightarrow{i_p} H_{DR}^p(U) \oplus H_{DR}^p(V) \xrightarrow{j_p} H_{DR}^p(U \cap V) \xrightarrow{\delta} H^{p+1}(M) \to \cdots .$$

The MVS is useful to compute De Rham cohomology once the space splits into subsets in which the groups are known. Given the closed form $w \in \Omega^p(U \cap V)$, we can find $\delta(w) \in H_{DR}^{p+1}(M)$ explicitly. Since $d_p w = 0$, we have $0 = d_p j_p \xi = j_{p+1} d_p \xi = 0$, so $d_p \xi = i_{p+1}(a)$, $a = \delta(w)$. We note that $\xi = (\phi_U.w, -\phi_V.w) \in \Omega^*(U) \oplus \Omega^*(V)$ and $d_p \xi = (d_p(\phi_U.w), -d_p(\phi_V.w)) \in \Omega^*(U) \oplus \Omega^*(V)$ satisfy $j_p(\xi) = w$. Since $j_{p+1} d_p \xi = d_p(\phi_U.w) + d_p(\phi_V.w) = 0$, we have $d_p(\phi_U.w) = -d_p(\phi_V.w)$ in $U \cap V$. Then we define

$$\delta(w) = \begin{cases} [d(\phi_U.w)], & \text{in } U, \\ -[d(\phi_V.w)], & \text{in } V. \end{cases} \in H_{DR}^{p+1}(M).$$

Although the groups H_{DR} have been introduced in open subsets of $\mathbb{R}^n$, they can be defined for differentiable submanifolds of $\mathbb{R}^n$ by using the same formalism we already developed. We will ignore this formality to describe their De Rham cohomology groups and we will use the fact that the sphere S^n and closed surface Σ_g of genus g are both compact submanifolds of $\mathbb{R}^n$. The tools so far developed are enough; we just need some topological properties for each space.

1-Sphere S^n

Using the stereographic projection, we have $S^n = U \cup V$, and U, V are open subsets homeomorphic to the unit ball $B^n = \{x \in \mathbb{R}^n; \mid x \mid \leq 1\}$ and $U \cap V$ is homeomorphic to $S^{n-1} \times \mathbb{R}$, as illustrated in Fig. 2. So we have $H_{DR}^p(U) = H_{DR}^p(V) = 0$ and $H_{DR}^p(U \cap V) = H_{DR}^p(S^{n-1})$ for all $p > 0$. From the Mayer-Vietoris exact sequence,

$$\cdots \to H_{DR}^{p-1}(S^n) \xrightarrow{i_{p-1}} 0 \oplus 0 \xrightarrow{j_{p-1}} H_{DR}^{p-1}(S^{n-1}) \xrightarrow{\delta} H_{DR}^p(S^n) \to 0 \cdots$$

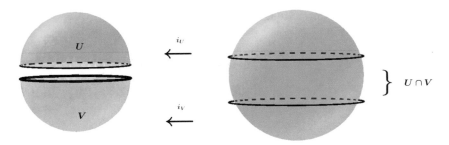

Fig. 2 $S^n = U \cup V$

yields an isomorphism $H_{DR}^{p-1}(S^{n-1}) \xrightarrow{\delta} H_{DR}^p(S^n)$ for all $p > 1$. When $n > 1$ and $p = 1$, j_0 is an isomorphism in

$$0 \to H_{DR}^0(S^n) = \mathbb{R} \xrightarrow{i_0} \mathbb{R} \oplus \mathbb{R} \xrightarrow{j_0} H_{DR}^0(S^{n-1}) = \mathbb{R} \xrightarrow{\delta} H_{DR}^1(S^n) \to 0 \ldots$$

breaking into three exact sequences:

$$(i)\ 0 \to \mathbb{R} \xrightarrow{i_0} \mathbb{R} \oplus \mathbb{R} \xrightarrow{j_0} \mathbb{R} \xrightarrow{\delta} 0,$$

$$(ii)\ 0 \xrightarrow{\delta} H_{DR}^1(S^n) \to 0,$$

$$(iii)\ 0 \to H_{DR}^{p-1}(S^{n-1}) \to H_{DR}^p(S^n) \to 0.$$

Consequently, $H_{DR}^1(S^n) = 0$ for all $n > 1$. When $n = 1$, the sequence becomes

$$0 \to \mathbb{R} \xrightarrow{i_0} \mathbb{R} \oplus \mathbb{R} \xrightarrow{j_0} \mathbb{R} \oplus \mathbb{R} \xrightarrow{\delta} H_{DR}^1(S^1) \to 0 \ldots \tag{20}$$

Proposition 11 *Let $n > 1$. The isomorphisms* $H_{DR}^{p-1}(S^{n-1}) \simeq H_{DR}^p(S^n)$ *and* $H_{DR}^1(S^n) \simeq 0$ *determine* $H_{DR}^*(S^n)$; *indeed,*

$$H_{DR}^i(S^n) = \begin{cases} \mathbb{R}, & \text{if } i = 0, \\ 0, & \text{if } 0 < i < n, \\ H_{DR}^1(S^1), & \text{if } i = n. \end{cases}$$

Proof By recurrence, we get that $H_{DR}^p(S^n) = H_{DR}^{p-k}(S^{n-k}) = 0$. Since $p < n$, if $k = p - 1$, then we have $H_{DR}^p(S^n) = H_{DR}^1(S^{n-p+1}) = 0$. If $p = n$, the same recurrence gives us $H_{DR}^n(S^n) \simeq H_{DR}^1(S^1)$. □

To find $H_{DR}^1(S^1)$, we need to check on the sequence (20),

$$0 \to \mathbb{R} \xrightarrow{i_0} \mathbb{R} \oplus \mathbb{R} \xrightarrow{j_0} \mathbb{R} \oplus \mathbb{R} \xrightarrow{\delta} H_{DR}^1(S^1) \to 0 \ldots \tag{21}$$

The homomorphism i_0 is injective, so we have $\text{Im}(j_0) = \mathbb{R}$. Since δ is surjective, we have $H^1_{DR}(S^1) = (\mathbb{R} \oplus \mathbb{R})/\text{Img}(j_0) = \mathbb{R}$.

Theorem 4 *Let $n > 0$ be finite. The cohomology groups of the sphere S^n are*

$$H^i_{DR}(S^n) = \begin{cases} \mathbb{R}, & \text{if } i = 0, \\ 0, & \text{if } 1 \leq i \leq n - 1, \\ \mathbb{R}, & \text{if } i = n. \end{cases}$$

If $n = 0$, then $H^0_{DR}(S^0) = \mathbb{R} \oplus \mathbb{R}$.

2-Closed Surface Σ_g

The De Rham cohomology of a genus g closed surface Σ_g is

$$H^i_{DR}(\Sigma_g) = \begin{cases} \mathbb{R}, & \text{if } i = 0, \\ \mathbb{R}^{2g}, & \text{if } 1 \leq i \leq n - 1, \\ \mathbb{R}, & \text{if } i = n. \end{cases}$$

Let's find these groups using forms. Since Σ_g is path connected, we get $H^0_{DR}(\Sigma_g) = \mathbb{R}$. To find the 1st and 2nd cohomology groups, we need some lemmas. Let $B = \{x \in \mathbb{R}^2; \mid x \mid < 1\}$ be the open unit disc in $\mathbb{R}^2$.

Lemma 5 *Let $\rho \in \Omega^2(B)$ be a 2-form such that $\int_B \rho = 0$. Then we have $\alpha \in \Omega^1(B)$ such that $\rho = d\alpha$.*

Proof Let $\rho = R(u, v) du dv$ and consider $\psi : \mathbb{R} \to \mathbb{R}$ a cutoff function with support in B and $\int_{-\infty}^{\infty} \psi(t) dt = 1$. Define the functions

$$r(u) = \int_{-\infty}^{\infty} R(u, t) dt,$$

$$\widetilde{R}(u, v) = R(u, v) - r(u)\psi(v).$$

Then $\widetilde{R}$ has compact support in B and, for each $u \in B$, we have

$$\int_B \widetilde{R}(u, v) du dv = \int_B R(u, v) du dv - \int_{-\infty}^{\infty} r(u) du . \int_{-\infty}^{\infty} \psi(v) dv = 0.$$

The function

$$P(u, v) = \int_{-\infty}^{v} \widetilde{R}(u, t) dt$$

has compact support in B and

$$\frac{\partial P}{\partial v} = \widehat{R}(u, v).$$

The function

$$Q(u, v) = \psi(v) \int_{-\infty}^{u} r(t)dt$$

also has compact support in B and

$$\frac{\partial Q}{\partial v} = \psi(v)r(u) = R(u, v) - \widetilde{R}(u, v).$$

Therefore we have $R = \frac{\partial P}{\partial v} + \frac{\partial Q}{\partial u}$. Considering the 1-form $\alpha = -Pdu + Qdv$, we have $\rho = d\alpha$. □

The lemma extends over Σ_g.

Corollary 2   *Let* $\rho \in \Omega^2(\Sigma_g)$ *be a 2-form such that* $\int_{\Sigma_g} \rho = 0$. *Then there is* $\alpha \in$ $\Omega^1(\Sigma_g)$ *such that* $\rho = d\alpha$.

Proof Consider a cover $\Sigma = \bigcup_{i=1}^{n} U_i$ such that U_i is an open subset diffeomorphic to B. We use induction on n. If $n = 1$, then we are reduced to the last lemma. Suppose $n > 1$, consider $V = U_2 \cup \cdots \cup U_n$, so $K = U \cup V$. If either $\Sigma_g \cap U = \emptyset$ or $\Sigma_g \cap V = \emptyset$, we are done, given the lemma. Since Σ_g is connected, there is a point in $\Sigma \cap U \cap V$. Choose a 2-form τ with compact support in $U \cap V$ and $\int_{\Sigma_g} \tau = 1$. Let $\{f_1, f_2\}$ be a partition of unity that is subordinate to the cover $\{U, V\}$. Then $\rho = f_1\rho + f_2\rho$, supp($f_1\rho$) $\subset U$ and supp($f_2\rho$) $\subset V$. Let

$$I = \int_{\Sigma_g} f_1\rho = -\int_{\Sigma_g} f_2\rho.$$

Both 2-forms $f_1\rho - I\tau$ and $f_2 + I\tau$ have compact support in U and V, respectively, and

$$\int_{\Sigma_g} \left(f_1\rho - I\tau\right) = \int_{\Sigma_g} \left(f_2\rho + I\tau\right) = 0.$$

Using the inductive hypothesis, we can find a 1-form α such that supp(α) $\subset U$ and $d\alpha = f_1\rho - I\tau$, and a 1-form β such that supp(β) $\subset V$ and $d\beta = f_2\rho + I\tau$. Hence $\rho = d(\alpha + \beta)$. □

A by-product of the corollary is the isomorphism $H^2_{\mathrm{DR}}(\Sigma_g) = \mathbb{R}$. To find the group $H^1_{\mathrm{DR}}(\Sigma_g)$, let's consider the following functionals;
(i) let $\gamma : [0, 1] \to \Sigma_g$ be a closed curve, i.e., $\gamma(0) = \gamma(1)$. Integration around γ yields a linear map

$$I_\gamma : H^1_{\mathrm{DR}}(\Sigma_g) \to \mathbb{R},$$

$$I_\gamma(\omega) = \int_\gamma \omega. \tag{22}$$

(ii) Given a 1-form $\theta \in \Omega^1(\Sigma_g)$, we get a linear map

$$J_\theta : H^1_{DR}(\Sigma_g) \to \mathbb{R},$$

$$J_\theta(\omega) = \int_\gamma \theta \wedge \omega. \tag{23}$$

Proposition 12 *For any loop γ, we have a 1-form $\theta \in \Omega^1(\Sigma_g)$ with compact support in a collar neighborhood of γ such that $J_\theta = I_\gamma$.*

Proof Let $\mathcal{P} = \{0 = t_0, t_1, \ldots, t_N = 1\}$ be a partition of the interval $[0, 1]$. Consider the image set $\{\gamma(t_0), \ldots \gamma(t_{i-1}), \gamma(t_i), \ldots, \gamma(t_N)\}$, and $\gamma(t_0) = \gamma(t_N)$. We take a cover $\gamma([0, 1]) \subset \bigcup_{i=0}^N U_i$ of $\gamma([0, 1])$ with open subsets U_i such that;
(i) For all i, U_i is diffeomorphic to B;
(ii) $\gamma([t_{i-1}, t_i]) \subset U_i$;
(iii) for any $i \neq j \neq k$, $U_i \cap U_j \cap U_k = \varnothing$.
  Choose small open discs D_i, $0 \leq i \leq N$, such that:
(i) $\overline{D_i} \subset U_i \cap U_{i+1}$ and $\overline{D_0} = \overline{D_N} \subset U_1 \cap U_N$;
(ii) $\gamma(t_i) \in D_i$.
(ii) $D_i \cap D_j = \varnothing$ unless $\{i, j\} = \{0, N\}$.
Let $\{\rho_i \mid 1 \leq i \leq N\}$ be a family of cutoff functions such that
(i) $\text{supp}(\rho_i) \subset D_i$;
(ii) $\int_{D_i} \rho_i = 1$.
  The 2-form $\rho_i - \rho_{i-1}$ has compact support in U_i and $\int(\rho_i - \rho_{i-1}) = 0$. By Lemma 5, there are 1-forms θ_i with compact support in U_i such that

$$d\theta_1 = \rho_1 - \rho_0, \ldots, d\theta_i = \rho_i - \rho_{i-1}, \ldots, d\theta_N = \rho_N - \rho_{N-1}.$$

Considering the 1-form $\theta = \theta_1 + \cdots + \theta_N$, we get

$$d\theta = (\rho_1 - \rho_0) + (\rho_2 - \rho_1) + \cdots + (\rho_N - \rho_{N-1}) = \rho_N - \rho_0 = 0.$$

Let $\zeta \in H^1_{DR}(\Sigma_g)$ and find a representative ω such that $[\omega] = \zeta$ vanishing when restricted to each disk D_k, $0 \leq k \leq N$. Since $H^1_{DR}(U_i) = 0$, we have a function f_i on U_i such that $\omega = df_i$ in U_i. So we have

$$\int_{\Sigma_g} \theta_i \wedge \omega = \int_{U_i} \theta_i \wedge \omega = \int_{U_i} \theta_i \wedge df_i = \int_{U_i} f_i(\rho_i - \rho_{i-1}).$$

Since ω vanishes on all the discs D_k, the function f_i must be constant on $D_i \supset \text{supp}(\rho_i)$ and $D_{i-1} \supset \text{supp}(\rho_{i-1})$. So

$$\int_{\Sigma_g} \theta_i \wedge \omega = f_i(\gamma(t_i)) - f_i(\gamma(t_{i-1})).$$

Fig. 3 16-polygon

On the other hand, since $\phi = df_i$ over the image of the interval $[t_i, t_{i+1}]$, we have

$$\int_\gamma \omega = \sum_{i=1}^{N} \int_{\gamma([t_{i-1}, t_i])} df_i = f_i(\gamma(t_i)) - f_i(\gamma(t_{i-1})).$$

Hence $J_\theta = J_\gamma$.   □

The result from the proposition above is a bilinear pairing

$$H_{\mathrm{DR}}^1(\Sigma_g) \times H_{\mathrm{DR}}^1(\Sigma_g) \to \mathbb{R},$$

which is skew-symmetric and non-degenerate. Therefore $H_{\mathrm{DR}}^1(\Sigma_g)$ must be even dimensional, say isomorphic to $\mathbb{R}^{2N}$. To find N, let's use the fact that $\Sigma_g = U \cup V$, U is diffeomorphic to B and V is homotopic to a bouquet $\vee_{i=1}^{2g} S_i^1$ of 1-spheres. Indeed, Σ_g is obtained by identifying the sides of a $4g$-polygon as shown in Figs.[4] 3, 4, 5, 6, 7, 8, 9 and 10.

Let V be the closure of the neighborhood of the g closed curves obtained by identifying the sides of the 4g-polygon obtained over Σ_g, as shown in Fig. 10, and let U be the complement of V in Σ_g. Indeed, V is a neighborhood of the sides of the $4g$-polygon and U is the disk inside. Using De Rham cohomology of the 1-sphere, we can compute that the De Rham groups of V and the De Rham cohomology of U are isomorphic to the cohomology of $\mathbb{R}^2$. The Mayer-Vietoris exact sequence associated to the sequence

$$\Sigma_g \to U \cup V \to S^1 \times I$$

---

[4]Figures 3, 4, 5, 6, 7, 8, 9 and 10 were kindly provided by Jos Leys. See Ref. [47] (iv).

Fig. 4 1st-step

Fig. 5 2nd-step

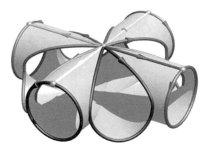

Fig. 6 3rd-step

breaks into two exact sequences: (s1), which is related to the 0-cohomology groups and (s2), which is related to the higher order groups, as follows;

$(s1)\ 0 \to \mathbb{R} \overset{i_0}{\to} \mathbb{R} \oplus \mathbb{R} \overset{j_0}{\to} \mathbb{R} \overset{\delta}{\to} 0$

$(s2)\ 0 \to H^1_{DR}(\Sigma_g) \overset{i_1}{\to} 0 \oplus H^1_{DR}(\vee_{i=1}^{2g} S^1_i) \to H^1_{DR}(S^1 \times I) \overset{\delta}{\to} H^2_{DR}(\Sigma_g) \to 0.$

$$(24)$$

Fig. 7 4th-step

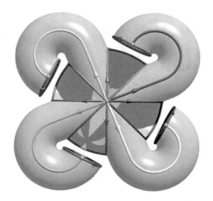

Fig. 8 5th-step

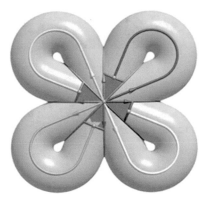

Fig. 9 Σ_4

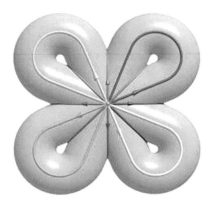

Fig. 10 Σ_4

In the sequence (s2), δ is an isomorphism, so (s2) breaks into (s3) and (s4);

$$(s3)\ 0 \rightarrow H^1_{\text{DR}}(\Sigma_g) \xrightarrow{i_1} 0 \oplus H^1_{\text{DR}}(\vee^{2g}_{i=1} S^1_i) \rightarrow 0,$$

$$(s4)\ 0 \rightarrow H^1_{\text{DR}}(S^1 \times I) \xrightarrow{\delta} H^2_{\text{DR}}(\Sigma_g) \rightarrow 0. \tag{25}$$

Therefore i_1 is an isomorphism, and so $H^1_{\text{DR}}(\Sigma_g) = \mathbb{R}^{2g}$.

To describe the ring structure of H^*_{DR}, we need further concepts beyond the scope of this text. References [22, 34] fully approach the singular homology and cohomology theories.

Exercises

(1) Show that $H^2_{\text{DR}}(\Sigma_g)$ is isomorphic to $\mathbb{R}$.

(2) Show that the linear functional $I_\gamma : H^1_{\text{DR}}(\Sigma_g) \rightarrow \mathbb{R}$ given in Eq. (22) is well-defined.

(3) Show that the linear functional $J_\theta : H^1_{\text{DR}}(\Sigma_g) \rightarrow \mathbb{R}$ given in Eq. (23) is well-defined.

(4) Let $\{D_i \mid 1 \leq i \leq N\}$ be a family of discs in Σ_g such that $D_i \cap D_j = \varnothing$ if $i \neq j$. Given a class $\zeta \in H^1_{\text{DR}}(\Sigma_g$, show that we have a representative $\omega \in Z^1(\Sigma_g)$ such that $\omega|_{D_i} = 0$ (hint: use the cutoff functions $\{\psi_i\}$ associated with the family of discs $\{D_i \mid 1 \leq i \leq N\}$ to obtain $\omega = \omega' - d(\psi_1 f_1 + \cdots + \psi_N f_N)$, and $[\omega'] = \zeta$ is any representative and $\omega'|_{D_i} = df_i$).

(5) Show that the cohomology of the bouquet $\vee^{2g}_{i=1} S^1_i$ is

$$H^0_{\text{DR}}(\vee^{2g}_{i=1} S^1_i) = \mathbb{R}, \quad H^1_{\text{DR}}(\vee^{2g}_{i=1} S^1_i) = \mathbb{R}^{2g}, \quad H^2_{\text{DR}}(\vee^{2g}_{i=1} S^1_i) = 0.$$

(6) Show that $H^1_{\text{DR}}(S^1 \times I) \xrightarrow{\delta} H^2_{\text{DR}}(\Sigma_g)$ is an isomorphism.

6 Stokes Theorem

The Stokes Theorem generalizes the classical integration theorems seen before. The differential forms are the fundamental tool to prove the theorem in full generality.

Throughout the exposition, we consider $I = [0, 1] \subset \mathbb{R}$ to be the closed interval and $I_n = [0, 1] \times \cdots \times [0, 1] = [0, 1]^n$, $[0, 1]^0 = \{0\} = \mathbb{R}^0$, the n-cube.

Definition 16 A singular n-cube in $U \subset \mathbb{R}^n$ is a continuous function $c : [0, 1]^n \to U$.

The simplest cases we have are the following:

(i) $c : [0, 1]^0 \to U$ corresponds to a point $c(0) \in U$;
(ii) $c : [0, 1]^1 \to U$ corresponds to the curve in U;
(iii) the n-cube I_n is also a singular n-cube considering the identity map $I_n :$ $[0, 1]^n \to \mathbb{R}^n$, $I_n(x) = x$

Let $C_n(U)$ be the free group generated by the singular n-cubes such that each element of the group is written as a linear combination of a finite number of singular n-cubes. Any $c \in C_n(U)$ is the sum

$$c = \sum_{i=1}^{k} a_i c_i, \quad a_i \in \mathbb{R}, \quad c_i : I_n \to U \quad \text{singular cube.}$$

The space C_n is a real vector space, its elements are named, each one, an n-chain.

For each n-cube I_n, and for each $i \in \{1, \ldots, n\}$, we define the $(n-1)$-faces of I_n as follows;

$$I_n^{(i,0)}(x) = I_n(x_1, \ldots, x_{i-1}, 0, x_{i+1}, \ldots, x_n),$$
$$I_n^{(i,1)}(x) = I_n(x_1, \ldots, x_{i-1}, 1, x_{i+1}, \ldots, x_n).$$

The n-cube I_n has $2n$-faces of dimension $(n-1)$. The $(n-1)$-cube $I_n^{(i,0)}$ is the $(i, 0)$ face of I_n, and $I_n^{(i,1)}$ is the $(i, 1)$ face of I_n. An n-cube I_n has 0-faces (vertex), 1-faces (edges), 2-faces, ..., $(n-1)$-faces. Now we consider the vector spaces $C_i(U)$ for $0 \leq i \leq n$, in which the elements are the singular i-chains.

Remark 5 Any k-dimensional differentiable submanifold S^k of $\mathbb{R}^n$ admits a decomposition with singular k-cubes.

Definition 17 Let $c : I_n \to U$ be a singular n-cube and $\alpha \in \{0, 1\}$ (see Fig. 11).

(1) A face (i, α) of c is

$$c_{(i,\alpha)} = c \circ I_n^{(i,\alpha)}.$$

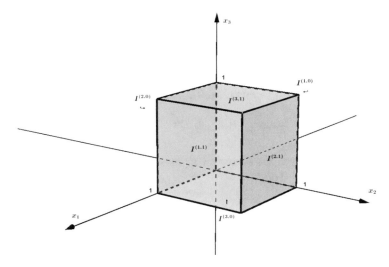

Fig. 11 Boundary of the cube I^3

(2) The boundary of a singular n-cube is the singular $(n-1)$-chain

$$\partial c = \sum_{i=1}^{n} \sum_{\alpha=0}^{1} (-1)^{i+\alpha} c_{(i,\alpha)}.$$

(3) The boundary of an n-chain $c = \sum_{i+1}^{k} n_i c_i$ is

$$\partial c = \sum_{i=1}^{k} n_i \partial c_i.$$

Example 5 The notation is crucial to the exposition.

(1) (i) $n = 2$ and $x = (x_1, x_2)$;

$$I_2^{(1,0)}(x) = I_2(0, x_2), \quad I_2^{(1,1)}(x) = I_2(1, x_2)$$
$$I_2^{(2,0)}(x) = I_2(x_1, 0), \quad I_2^{(2,1)}(x) = I_2(x_1, 1).$$

$$\partial I_2 = -I^{(1,0)} + I^{(2,0)} - I^{(2,1)} + I^{(1,1)}.$$

(2) (ii) $n = 3$ and $x = (x_1, x_2, x_3)$;

$$I_3^{(1,0)}(x) = I_3(0, x_2, x_3), \quad I_3^{(1,1)}(x) = I_3(1, x_2, x_3),$$
$$I_3^{(2,0)}(x) = I_3(x_1, 0, x_3), \quad I_3^{(2,0)}(x) = I_3(x_1, 1, x_3),$$
$$I_3^{(3,0)}(x) = I_3(x_1, x_2, 0), \quad I_3^{(3,1)}(x) = I_3(x_1, x_2, 1).$$

$$\partial I_3 = -I_3^{(1,0)} + I_3^{(1,1)} + I_3^{(2,0)} - I_3^{(2,1)} - I_3^{(3,0)} + I_3^{(3,1)}.$$

Lemma 6 *For $0 \leq i \leq n$, the boundary defines the homomorphism $\partial_i : C_i(U) \to C_{i-1}(U)$, such that $\partial_0 = 0$ and*

$$\partial_{i-1} \circ \partial_i = 0 \quad (\partial^2 = 0).$$

Proof See Ref. [22].

Definition 18 Let $S \subset \mathbb{R}^n$ be a k-dimensional submanifold. For $0 \leq i \leq k$, a singular i-chain $\sigma \in Cj(S)$:
(i) is an i-cycle if $\partial\sigma = 0$. The subspace of i-cycles is

$$\mathcal{Z}_i(S) = \{\sigma \in C_i(S) \mid \partial\sigma = 0\}.$$

(ii) is an i-boundary if we have $\tau \in C_{i+1}(U)$ such that $\sigma = \partial\tau$. The subspace of boundaries is

$$\mathcal{B}_i(S) = \{\sigma \in C_i(S) \mid \sigma = \partial\tau\}.$$

(iii) The ith singular homology group with real coefficients of S and

$$H_i(S; \mathbb{R}) = \frac{\mathcal{Z}_i(S)}{\mathcal{B}_i(S)}. \tag{26}$$

The operators d and ∂ share the similar properties $d^2 = 0$ and $\partial^2 = 0$, respectively; this suggests the existence of some relation between chains and forms, or between singular homology groups and De Rham cohomology groups. This relationship is established when integrating the forms on chains. For our purposes, we will consider only the singular i-cubes whose map $c : I_i \to U$ is differentiable.
   A k-form w on I_k is written as $\omega = f\,dx_1 \wedge \cdots \wedge dx_k$. We define

$$\int_{I_k} \omega = \int_{I_k} f,$$

or equivalently,

$$\int_{I_k} f\,dx_1 \wedge \cdots \wedge dx_k = \int_{I_k} f(x_1, \ldots, x_k)dx_1 \ldots dx_k = \int_0^1 \overset{k}{\ldots} \int_0^1 f(x_1, \ldots, x_k)dc_1 \ldots dc_k,$$

$(dc_i = dx_i \circ dc)$. Let ω be a k-form on U and c a singular k-cube in U. We define

$$\int_c \omega = \int_{I_k} c^* \omega.$$

In local coordinates, the integral above is equal to

$$\int_c f \, dx_1 \wedge \cdots \wedge dx_k = \int_{I_k} (c)^* (f \, dx_1 \wedge \cdots \wedge dx_k) = \int_{I_k} (f \circ c) dx_1 \ldots dx_k.$$

When ω is a 0-form, a function, and $c : \{0\} \to U$ is a 0-singular cube in U, we set $\int_c \omega = w(0)$. The integral of a k-form ω along a k-chain $c = \sum_{i=1}^{l} n_i c_i$ is

$$\int_c \omega = \sum_{i=1}^{l} a_i \int_{c_i} \omega. \tag{27}$$

The integral of a 1-form on a 1-chain is a line integral, and the integration of 2-forms on a 2-chain is a surface integral.

Definition 19 A k-chain $c \in U \subset \mathbb{R}^n$ is orientable if the orientation on $\mathbb{R}^n$ induces a non-null k-form on c.

Theorem 5 (Stokes Theorem) *Let ω be a $(k-1)$-form on an open subset $U \subset \mathbb{R}^n$, let c be an orientable k-chain on U and $d : \Omega^{k-1}(U) \to \Omega^k(U)$ the exterior derivative. Then*

$$\int_c d\omega = \int_{\partial c} \omega. \tag{28}$$

Proof Consider $c = \mathrm{id}_{I_k} : I_k \to U$ and ω a $(k-1)$-form on I_k. The boundary of c is

$$\partial c = \sum (-1)^{i+\alpha} c^{i,\alpha} \quad (c_{i,\alpha} = c : I^{(i,\alpha)} \to I_k).$$

Given an orthonormal basis of $\mathbb{R}^n$, so we get an orientation on $\mathbb{R}^n$. A $(k-1)$-form ω is written as

$$\omega = \sum_{i=1}^{n} f_i dx_1 \wedge \cdots \wedge d\widehat{x}_i \wedge \cdots \wedge dx_k.$$

Noting that

$$c^*_{(j,\alpha)}(dx_i) = \begin{cases} 0, & i = j; \\ dx_i, & i \neq j, \end{cases}$$

we get

$$\int_{I_{k-1}} (I_k^{(j,\alpha)})^*(f_i dx_1 \wedge \cdots \wedge d\widehat{x_i} \wedge \cdots \wedge dx_k) = \int_{I_{k-1}} (f_i \circ c_{(j,\alpha)})c_{(j,\alpha)}^*(dx_1) \wedge \cdots \wedge c_{(j,\alpha)}^*(dx_k) =$$

$$= \begin{cases} 0, i = j; \\ \int_{I_k} f_i(x_1, \ldots, \alpha, \ldots, x_k)dx_1 \ldots dx_k, i \neq j. \end{cases}$$

The right-hand side of Eq. (28) is equal to

$$\int_{\partial I_k} f_i dx_1 \wedge \cdots \wedge d\widehat{x_i} \wedge \cdots \wedge dx_k = (-1)^{i+\alpha} \sum_{j=1}^{k} \sum_{\alpha=0}^{1} \int_{I_{k-1}} (I_k^{(j,\alpha)})^*(f_i dx_1 \wedge \cdots \wedge d\widehat{x_i} \wedge \cdots \wedge dx_k) =$$

$$= \sum_{j=1}^{k} [(-1)^{j+1} \int_{I_k} f_i(x_1, \ldots, 1, \ldots, x_k)dx_1 \ldots dx_k + (-1)^j \int_{I_k} f_i(x_1, \ldots, 0, \ldots, x_k)dx_1 \ldots dx_k].$$

The left-hand side of the Eq. (28) gives us

$$\int_{I_k} d(f_i dx_1 \wedge \cdots \wedge d\widehat{x_i} \wedge \cdots \wedge dx_k) = \int_{I_k} \frac{\partial f_i}{\partial x_i} dx_i \wedge dx_1 \wedge \cdots \wedge d\widehat{x_i} \wedge \cdots \wedge dx_k =$$

$$= (-1)^{i-1} \int_{I_k} \frac{\partial f_i}{\partial x_i}.$$

By Fubini's theorem and the Fundamental Theorem of Calculus, we have

$$\int_{I_k} d(f_i dx_1 \wedge \cdots \wedge d\widehat{x_i} \wedge \cdots \wedge dx_k) = (-1)^{i-1} \int_{I_{k-1}} [\int_0^1 \frac{\partial f_i}{\partial x_i} dx_i] dx_1 \ldots d\widehat{x_i} \ldots dx_k =$$

$$= (-1)^{i-1} \int_{I_{k-1}} [f(x_1, \ldots, 1, \ldots, x_k) - f(x_1, \ldots, 0, \ldots, x_k)] dx_1 \ldots d\widehat{x_i} \ldots dx_k = \qquad (29)$$

$$= (-1)^{i-1} \int_{I_k} f(x_1, \ldots, 1, \ldots, x_k) + (-1)^i \int_{I_k} f(x_1, \ldots, 0, \ldots, x_k)dx_1 \ldots dx_k.$$

Matching both sides of Eq. (28), we get

$$\int_{\partial I_k} w = \int_{I_k} dw.$$

Let c_i be any singular k-cube. Using the linear properties of $\int$ and d operators, we get

$$\int_{c_i} d\omega = \int_{I_k} c_i^*(d\omega) = \int_{I_k} d(c_i^*\omega) = \int_{\partial I_k} c_i^*\omega = \int_{\partial c_i} \omega.$$

Therefore if c is an orientable singular k-chain $c = \sum_{i=1}^{l} a_i c_i$, then

$$\int_c d\omega = \sum_{i=1}^{l} a_i \int_{c_i} d\omega = \sum_{i=1}^{l} a_i \int_{\partial c_i} \omega = \int_{\partial c} \omega.$$

Remark 6 The Stokes theorem yields cohomological implications, which we address next.

(1) Let S be a k-dimensional submanifold of $U \subset \mathbb{R}^n$. If $\partial S = \varnothing$, then $\int_S d\eta = 0$ for all $\eta \in \Omega^{k-1}(U)$. Therefore if $[\omega'] = [\omega] \in H_{DR}^k(U)$, that is, $\omega' = \omega + d\eta$, then

$$\int_S \omega' = \int_S \omega.$$

(2) Let S be a k-dimensional submanifold of $U \subset \mathbb{R}^n$ and $\omega \in \Omega^k(U)$ a closed form $(d\omega = 0)$. If c is a $(k-1)$-chain, then

$$\int_{S+\partial c} \omega = \int_S \omega + \int_{\partial c} \omega = \int_S \omega + \int_c d\omega = \int_S \omega.$$

Given a class $[\omega] \in H_{DR}^k(U)$, we get a linear functional

$$\xi : H_k(U; \mathbb{R}) \to \mathbb{R},$$

$$\xi(\omega) = \int_S \omega. \tag{30}$$

(3) De Rham isomorphism.
Gathering the results above, we have a homomorphism $\mathcal{R}_k : H_{DR}^k(U) \to H^k(U, \mathbb{R})$ given by

$$\mathcal{R}_k(\omega)(\sigma) = I_\omega = \int \omega,$$

with $I_\omega : H_k(U, \mathbb{R}) \to \mathbb{R}$ being the linear functional $I_\omega(\sigma) = \int_\sigma \omega$. The De Rham theorem in [45] asserts that $\mathcal{R}_k$ is an isomorphism for all $k > 0$.

7 Orientation, Hodge Star-Operator and Exterior Co-derivative

By defining an inner product on $\Omega^*(U)$, we can define the L^2-dual operator d^* of d. Considering $\omega, \eta \in \Omega^k(U)$, for $x \in U$, we have $\omega_x, \eta_x \in \Lambda(T_x^*U)$. The inner product defined in Definition 4 induces a function $< \omega, \eta >_x: U \to \mathbb{R}$ given as

$< \omega, \eta >_x =< \omega_x, \eta_x >$. Letting $d\vartheta = dx_1 \wedge \cdots \wedge dx_n$ be the volume form, the inner product $< ., . >: \Omega^k(U) \times \Omega^k(U) \to \mathbb{R}$ is given by

$$< \omega, \eta >= \int_U < \omega_x, \eta_x > d\vartheta. \tag{31}$$

Definition 20 An open subset $U \subset \mathbb{R}^n$ is orientable, if in $\Omega^n(U)$, there is an n-form ω such that $\omega(x) \neq 0$ for all $x \in U$. If $\mid \omega \mid= 1$, ω is named the unit volume form.

Definition 21 Let $U \subset \mathbb{R}^n$ be an open subset and $d\vartheta = e_1 \wedge \cdots \wedge e_n$ the volume form. The Hodge star-operator $* : \Omega^k(V) \to \Omega^{n-k}(V)$ is the linear operator defined by the following: let $w, \eta \in \Omega^k(V)$;

$$w \wedge *\eta =< w, \eta > d\vartheta. \tag{32}$$

Hodge's star-operator inherits all properties shown in Proposition 7;
(i) $*^2 = (-1)^{k(n-k)}$.
(ii) $< *w, *\eta >=< w, \eta >$.
(iii) $d* = *d$.
   Property (i) induces an isomorphism $* : H^k_{DR}(U) \to H^{n-k}_{DR}(U)$ related to the well-known Poincaré duality in singular cohomology. In what follows, we will consider the space of the forms $\Omega^*_c(U)$ with compact support in U, so they cancel on the boundary when there is one.

Definition 22 The exterior co-derivative $d^*_k : \Omega^k_c(U) \to \Omega^{k-1}_c$ is the operator given by

$$d^*_k = (-1)^{nk+1} * d_{n-k} * . \tag{33}$$

The co-derivative is the (L^2) dual operator of the exterior derivative operator with respect to the inner product in $\omega \in \Omega^k(U)$, as shown by the following identities: let $\omega \in \Omega^k_c(U)$ and $\eta \in \Omega^{n-k}_c(U)$;
(i) $d(\omega \wedge *\eta) = d\omega \wedge *\eta + (-1)^k \omega \wedge d(*\eta)$

$$d(\omega \wedge *\eta) = d\omega \wedge *\eta + (-1)^k \omega \wedge d(*\eta). \tag{34}$$

(ii) Integrating both identities above, we get

$$\int_U d(\omega \wedge *\eta) = \int_U d\omega \wedge *\eta + (-1)^k \int_U \omega \wedge d(*\eta) =$$
$$= \int_U < d\omega, \eta > d\vartheta + \int_U < \omega, (-1)^{k(n-k)+k} * d * \eta > d\vartheta. \tag{35}$$

Applying the Stokes theorem, the right-hand side in Eq. 35 is equal to 0. Consequently, we get

$$\int_U < d\omega, \eta > d\vartheta = \int_U < \omega, (-1)^{k(n-k)+k+1} * d * \eta > d\vartheta =$$

$$= \int_U < d\omega, \eta > d\vartheta = \int_U < \omega, (-1)^{nk+1} * d * \eta > d\vartheta.$$

Hence $< d\omega, \eta >=< \omega, d^*\eta >$, for any $\omega \in \Omega_c^k(U)$ and $\eta \in \Omega_c^{k+1}(U)$. The following proposition is straightforward from the definition;

Proposition 13 *The exterior co-derivative* $d^* : \Omega_c^k(U) \to \Omega_c^{k-1}$ *satisfies the co-cycle condition*

$$d_{k-1}^* \circ d_k^* = 0. \tag{36}$$

In this way, defining $d_0^* = 0$, we have the co-complex

$$0 \leftarrow \Omega^0(U) \xleftarrow{d^*} \Omega^1(U) \xleftarrow{d^*} \cdots \xleftarrow{d^*} \Omega^p(U) \xleftarrow{d^*} \Omega^{p+1}(U) \xleftarrow{d^*} \cdots \xleftarrow{d^*} \Omega^n(U). \tag{37}$$

Example 6 Considering $U \subset \mathbb{R}^3$, we have $d^* = (-1)^{3k+1} * d*$. Therefore

$$*(dx_1) = dx_2 \wedge dx_3, \quad *(dx_2) = dx_3 \wedge dx_1, \quad *(dx_3) = dx_1 \wedge dx_2.$$

(1) $d^* : \Omega^1(U) \to \Omega^0(U)$, $d^* = *d*$.
   Let $\omega = F_1 dx_1 + F_2 dx_2 + F_3 dx_3$ and $\vec{F} = F_1\vec{i} + F_2\vec{j} + F_3\vec{k}$, so $d^* = \text{div}(F)$.
(2) $d^* : \Omega^2(U) \to \Omega^1(U)$, $d^* = - * d*$, and $\omega = F_1 dx_2 \wedge dx_3 + F_3 dx_1 \wedge dx_2 + F_2 dx_3 \wedge dx_1$. So we get

$$d^*\omega = \left(\frac{\partial F_3}{\partial x_2} - \frac{\partial F_2}{\partial x_3}\right) dx_1 + \left(\frac{\partial F_1}{\partial x_3} - \frac{\partial F_3}{\partial x_1}\right) dx_2 + \left(\frac{\partial F_2}{\partial x_1} - \frac{\partial F_1}{\partial x_2}\right) dx_3 =$$
$$= [\text{curl}\vec{F}]_1 dx_1 + [\text{curl}\vec{F}]_2 dx_2 + [\text{curl}\vec{F}]_3 dx_3.$$

(3) $d^* : \Omega^3(U) \to \Omega^2(U)$, $d^* = *d*$, and $\omega = f dx_1 \wedge dx_2 \wedge dx_3$);

$$d^*\omega = \frac{\partial f}{\partial x_1} dx_2 \wedge dx_3 + \frac{\partial f}{\partial x_2} dx_3 \wedge dx_1 + \frac{\partial f}{\partial x_3} dx_1 \wedge dx_1 =$$
$$= [\nabla f]_1 dx_2 \wedge dx_3 + [\nabla f]_2 dx_3 \wedge dx_1 + [\nabla f]_3 dx_1 \wedge dx_2.$$

What is shown above allows us to identify the operators $d_3^* \rightsquigarrow \text{grad}$, $d_2^* \rightsquigarrow \text{curl}$ and $d_1^* = \text{div}$.

Exercises

(1) Consider the Hodge star-operator induced by the Lorentz form. Show that

$$d^* = (-1)^{nk} * d * . \tag{38}$$

(2) Consider the operator $\triangle_k : \Omega^k(U) \to \Omega^k(U)$ given by

$$\triangle_k = d^*_{k+1} d_k + d_{k-1} d^*_k . \tag{39}$$

    (a) Show that $\triangle_k$ is a Laplacian operator for all $k \in \mathbb{N}$, $\triangle_k$ (hint: write it in coordinates).

    (b) Assuming ω is compactly supported, show that $\triangle_k \omega = 0$ if and only if $d\omega = d^*\omega = 0$.

    (c) Show $\mathrm{Ker}(d) \perp \mathrm{Im}(d^*)$.

    (d) Show $\triangle_k$ is an elliptic operator, and indeed, a Fredholm operator (hard). (hint: consult Ref. [45])

8   Differential Forms on Manifolds, Stokes Theorem

Let M be an n-dimensional differentiable manifold and let $\mathfrak{A}_M = \{(U_\alpha, \phi_\alpha) \mid \alpha \in \Lambda\}$ be a differentiable structure carried by M with U_α being a contractible open subset for all α.

8.1   Orientation

To give M an orientation, the idea will be to extend the orientation on an open $U_\alpha \subset M$. There are differentiable manifolds that do not admit a frame globally defined, so we must work locally. For $\alpha \in \Lambda$, we fix a frame $\beta_\alpha = \{e_1^\alpha, \ldots, e_n^\alpha\}$, or equivalently, an orientation on U_α.

Definition 23   Given an orientation over M, provided each open subset U_α carries an orientation and whenever $U_\alpha \cap U_\beta \neq \varnothing$, the derivative of $\phi_{\alpha\beta} : \phi_\alpha(U_\alpha \cap U_\beta) \to \phi_\beta(U_\alpha \cap U_\beta)$ preserves the orientation. Therefore the differential

$$d\phi_{\alpha\beta}(x) : T_{\phi_\alpha(x)} U_\alpha \to T_{\phi_\beta(x)} U_\beta$$

sends positive frames to positive frames.

    There are manifolds that do not admit an orientation; they are the non-orientable manifolds. Classic examples are the Möebius Band, the Real Projective Plane $\mathbb{R}P^2$, and the Klein Bottle $\mathbb{K}^2$.

The frames β_α and β_β, defined on U_α and U_β respectively, are compatible if $\det(d\phi_{\beta\alpha}(x)) > 0$ for all $x \in U_\alpha \cap U_\beta$. An orientation is given by extending over M a frame β_α defined on U_α under the compatibility condition.

Another way of defining an orientation is through differential forms. Since we have the bijective relation

$$\beta_\alpha = \{e_1, \ldots, e_n\} \quad \leftrightarrow \quad \vartheta_\beta = e_1 \wedge \cdots \wedge e_n,$$

we can set a positive basis for the one dimensional vector space $\Lambda^n(T_pM)$. We define the equivalence relation

$$\vartheta' \sim \vartheta \quad \leftrightarrow \quad \exists\, \lambda > 0 \quad \text{such that} \quad \vartheta' = \lambda.\vartheta.$$

If T is the matrix sending frame β to frame β', then

$$\vartheta_{\beta'} = \det(T).\vartheta_\beta.$$

Consequently, $\beta' \sim \beta$ if and only if $\vartheta_{\beta'} \sim \vartheta_\beta$. An orientation is assigned to an open subset $U \subset \mathbb{R}^n$ by a volume n-form $\vartheta \in \Omega^n(U)$ that never is annulled in U. The same occurs on a manifold M, when we set a volume n-form.

Exercises

(1) show that the Möebius band is non-orientable.
(2) Show that the sphere S^n is orientable.
(3) Show that the the Klein bottle $\mathbb{K}^2$ is non-orientable.
(4) Show that the Projective Plane $\mathbb{R}P^{2n}$ is non-orientable and the Projective Plane $\mathbb{R}P^{2n+1}$ is orientable for all $n \in \mathbb{N}$.
(5) Show that the fundamental group of a non-orientable manifold M has a normal subgroup isomorphic to $\mathbb{Z}_2$.
   (hint: show there is a Möebius band embedded in M non-homotopic to a point).

8.2   Integration on Manifolds

Fort $0 \leq p \leq n = \dim(M)$, consider the vector fibered spaces

$$\Lambda^0(M) = \bigcup_{x \in M} \Lambda^0(T_xM), \quad \Lambda^P(M) = \bigcup_{x \in M} \Lambda^P(T_xM). \tag{40}$$

A smooth p-form on M is a C^∞-section of the vector bundle $\Lambda^P(M)$, that is,

$$\omega : M \to \Lambda^P(M),$$

such that $\omega_x \in \Lambda(T_x^* M)$ for $x \in M$. Let $\Omega^p(M)$ be the space of p-forms and let

$$\Omega^*(M) = \oplus_{p \geq 0} \Omega^p(M)$$

be the Exterior Algebra of the differential forms on M. For $x \in M$ and $\omega \in \Omega^p(M)$, ω_x is an alternating p-tensor on $T_p X$. Let $f : U \to \mathbb{R}$ be a differentiable function. If $\phi : V \to U$ is a diffeomorphism between open sets preserving orientation, then

$$\int_U f \, dx_1 \dots dx_n = \int_V (f \circ \phi) \mid \det(d\phi) \mid dy_1 \dots dy_n.$$

Let $(M, \mathfrak{A}_M)$ be a differentiable manifold in which the differentiable structure is given by the atlas $\mathfrak{A}_M = \{(\widehat{U}_\alpha, \phi_\alpha) \mid \alpha \in \Lambda\}$ with a boundary that may not be empty. Let $\omega \in \Omega^p(M)$ be a p-form with compact support, i.e., $\text{supp}(\omega) = \{x \in M \mid \omega(x) \neq 0\}$ is compact. Initially, we assume the support of ω is contained in a local chart U_α. Once we have fixed the orientation on U_α, we assume the diffeomorphism $\phi_\alpha : U_\alpha \to \widehat{U}_\alpha$ preserves the orientation. The support of $(\phi^{-1})^* \omega$ is a compact set contained in U_α. Define

$$\int_M \omega = \int_{\widehat{U}_\alpha} (\phi^{-1})^* \omega. \tag{41}$$

Let $\beta \in \Lambda$ be such that $U_\alpha \cap U_\beta \neq \varnothing$, and let $\phi_{\beta\alpha} = \phi_\beta \circ \phi_\alpha^{-1} : \phi_\alpha(U_\alpha \cap U_\beta) \to \phi_\beta(U_\alpha \cap U_\beta)$ be an orientation preserving diffeomorphism. The integral defined by (41) is well-defined; indeed, it does not depend on the chart used because of the identity

$$\int_{\widehat{U}_\beta} (\phi_\beta^{-1})^* \omega = \int_{\widehat{U}_\alpha} \phi_{\beta\alpha}^* (\phi_\beta^{-1})^* \omega = \int_{\widehat{U}_\alpha} (\phi_\beta^{-1} \circ \phi_\beta \circ \phi_\alpha^{-1})^* \omega = \int_{\widehat{U}_\alpha} (\phi_\alpha^{-1})^* \omega.$$

Let's now consider the general case when the support set $\text{supp}(\omega)$ is not contained in a local chart. The recipe is to use partitions of unity subordinated to the cover $M = \cup U_\alpha$, so we can decompose a p-form in a summand of p-forms with supports that are contained in a chart. Consider $\{\rho_\alpha\}_{\alpha \in \Lambda}$ a partition of unity subordinated to the cover $\{U_\alpha\}_{\alpha \in \Lambda}$, so

$$\int_X \omega = \sum_i \int_X \rho_i \omega.$$

We need to check that $\int_M \omega$ does not depend on the partition of unity used; let $\{\rho'_\alpha\}_{\alpha \in \Lambda}$ be another partition and

$$\int_M \omega = \sum_j \int_X \rho'_j \omega,$$

and

$$\int_X \rho'_j \omega = \sum_i \int_X \rho_i \rho'_j \omega.$$

Analogously,

$$\int_M \rho_i \omega = \sum_j \int_M \rho'_j \rho_i \omega.$$

Therefore

$$\sum_i \int_m \rho_i \omega = \sum_i \sum_j \int_M \rho'_j \rho_i \omega = \sum_j \sum_i \int_M \rho_i \rho'_j \omega = \sum_j \int_M \rho'_j \omega.$$

For all $a, b \in \mathbb{R}$ and $w_1, w_2 \in \Omega^p(X)$, we have

$$\int_X \{a\omega_1 + b\omega_2\} = a \int_X \omega_1 + b \int_X \omega_2.$$

Theorem 6 *Let M and N be differentiable manifolds and let $f : M \to N$ be a differentiable map preserving orientation; then*

$$\int_N \omega = \int_M f^* \omega.$$

8.3   Exterior Derivative

The differential operator $d : C^\infty(M) \to C^\infty(M)$ induces the operator $d : \Omega^0(M) \to \Omega^1(M)$, $f \rightsquigarrow df$. Locally, we fix a chart (U_α, ϕ_α) and a frame $\{e_1, \ldots, e_n\}$ over $\widehat{U}_\alpha$, so the co-frame $\{e_1^*, \ldots, e_n^*\}$ allows us to write any k-form ω on U_α as

$$\omega = \sum_I \omega_I^\alpha dx_I, \ \omega_I^\alpha \in C^\infty(X).$$

We can apply the exterior derivative on open subsets of $\mathbb{R}^n$ as follows:

$$d^\alpha w = \sum_I d\omega_I^\alpha \wedge dx_I.$$

Let's check that the definition given does not depend on the local chart we have used. Let $\alpha, \beta \in \Lambda$ be such that $U_\alpha \cap U_\beta \neq \varnothing$; the compatibility condition is

$$\phi_{\beta\alpha}^* d^\beta = d^\alpha \phi_{\beta\alpha}^*.$$

For $p \in= \{0, \ldots, n\}$, we get the exterior derivative operator $d_p : \Omega^p(M) \to \Omega^{p+1}(M)$ satisfying the co-cycle condition $d_{p+1} \circ d_p = 0$. Therefore we have on M the complex

$$0 \xrightarrow{0} \Omega^0(M) \xrightarrow{d} \Omega^1(M) \xrightarrow{d} \cdots \xrightarrow{d} \Omega^p(M) \xrightarrow{d} \Omega^{p+1}(M) \xrightarrow{d} \cdots \xrightarrow{d} \Omega^n(M).$$
(42)

For $p \geq 0$, the vector spaces

$$H_{\mathrm{DR}}^p(M) = \frac{\mathrm{Ker}(d_p)}{\mathrm{Im}(d_{p-1})}$$

define the pth De Rham Cohomology group of M.

8.4   Stokes Theorem on Manifolds

The Stokes Theorem extends over a differentiable manifold. The details of the proof are similar to the case in $\mathbb{R}^n$ (see Ref. [42]); they only require partitions of the unit and checking that the concepts are well-defined up to the local chart.

Theorem 7 *Let M be a differentiable manifold and $S \subset M$ a $(p+1)$-dimensional differentiable submanifold of M with a boundary ∂S. For any p-form $\omega \in \Omega^p(M)$, we have the identity*

$$\int_S d\omega = \int_{\partial S} \omega.$$
(43)

For a complete approach regarding cohomology groups of differentiable manifolds using differential forms, we recommend Ref. [4] .

Chapter 7
Applications to the Stokes Theorem

Applications are widespread in many topics of Pure and Applied Mathematics. To apply the formalism of differential forms and the Stokes Theorem, we will discuss the topics on Harmonic Functions and the geometric formulation of Electromagnetism without delving into the contents.

1 Volumes of the $(n + 1)$-Disk and of the n-Sphere

Let $B^{n+1}(R) = \{x \in \mathbb{R}^{n+1}; \mid x \mid < R\}$ be the $(n + 1)$-dimensional open ball with radius R centered at the origin; let $D_R^{n+1} = \overline{B^{n+1}(R)}$ be the closed ball, and let $S^n(R) = \partial B^{n+1}(R)$ be the n-dimensional sphere of radius R. We denote the $(n + 1)$-volume of $D^{n+1}(R)$ by $V_{n+1}(R)$ and the n-volume of $S^n(R)$ by $A_n(R)$. When $R = 1$, we simply denote V_n and A_n, respectively. Let's check that the following identities are true:

$$V_{n+1}(R) = R^{n+1}V_{n+1},$$
$$A_n(R) = R^n A_n. \tag{1}$$

To prove the identities, we consider the diffeomorphism $f : D^{n+1}(1) \to D^{n+1}(R)$ given by $f(x) = R.x$, the differential is $df_x = R.I$, and so $\det(df) = R^{n+1}$. It follows from the change of variables theorem for multiple integrals that the elements of volume and area are $dV_R = \mid \det(df) \mid .dV$ and $dA_R = \mid \det(df_\partial) \mid .dA$; therefore

$$V_{n+1}(R) = \int_{D^{n+1}(R)} dV_R = \int_{D^{n+1}(1)} \mid J(df_x) \mid dV =$$
$$= R^{n+1}. \int_{D^{n+1}(1)} dV = R^{n+1}V_{n+1}.$$

© Springer Nature Switzerland AG 2021
C. M. Doria, *Differentiability in Banach Spaces, Differential Forms and Applications*,
https://doi.org/10.1007/978-3-030-77834-7_7

Similarly, since $f_\partial : S^n(1) \to S^n(R), df_\partial = R^n.I$, we get

$$A_n(R) = \int_{S^n(R)} dA_R = \int_{S^n(1)} |\det(d(f_\partial)_x)| \, dA = R^n . \int_{S^n(1)} dA = R^n A_n.$$

Proposition 1 *For any $n \geq 0$, we have the identity*

$$V_{n+1}(R) = \frac{R}{n+1} A_n(R). \tag{2}$$

Proof First, we fix a partition $\mathcal{P} = \{0 = r_0, r_1, \ldots, r_n = R\}$ for the interval $[0, R]$ and consider $D^{n+1}(R) = \cup_i \left(S^n(r_i^*) \times [t_{i-1}, r_i] \right)$, with $r_i^* = \frac{r_{i-1}+r_i}{2}$ and $\Delta r = r_i - r_{i-1}$. So we have

$$V_{n+1}(R) = \lim_{\Delta r \to 0} \sum_i A_n(r_i^*)\Delta r = \int_0^R A_n(r)dr = \int_0^R r^n A_n dr = \frac{R^{n+1}}{n+1} A_n = \frac{R}{n+1} A_n(R).$$

$\square$

The volume form of $\mathbb{R}^{n+1}$ is $dV = dx_1 \wedge \cdots \wedge dx_{n+1}$, so we have

$$V_{n+1}(R) = \int_{B^{n+1}(R)} dx_1 \wedge \cdots \wedge dx_{n+1}.$$

The external derivative of the differential n-form

$$\omega_n = \sum_{i=1}^{n+1} (-1)^{i-1} x_i dx_1 \wedge \cdots \wedge dx_{i-1} \wedge dx_{i+1} \wedge \cdots \wedge dx_{n+1} \in \Omega^n(D^{n+1}(R)),$$

is $d\omega_n = (n+1)dx_1 \wedge \cdots \wedge dx_{n+1}$. Since the volume of the n-sphere is

$$A_n(R) = \frac{1}{R} \int_{S^n(R)} \omega_n, \tag{3}$$

from the Stokes Theorem we get that

$$V_{n+1}(R) = \int_{D^{n+1}(R)} dx_1 \wedge \cdots \wedge dx_{n+1} = \frac{1}{n+1} \int_{D^{n+1}(R)} d\omega_n = \frac{1}{n+1} \int_{S^n(R)} \omega_n = \frac{R}{n+1} A_n(R).$$

Let's give $S^{n-1}(1)$ a parametrization using spherical coordinates and also using induction on the dimension:

$$\Psi_1(\theta_1) = \left(\cos(\theta_1), \sin(\theta_1) \right),$$
$$\Psi_2(\theta_1, \theta_2) = \left(\Psi_1(\theta_1)\sin(\theta_2), \cos(\theta_2) \right) = \left(\cos(\theta_1)\sin(\theta_2), \sin(\theta_1)\sin(\theta_2), \cos(\theta_2) \right).$$

$$\vdots$$

$$\Psi_n(\theta_1, \ldots, \theta_{n-1}, \theta_n) = \left(\psi_{n-1}(\theta_1, \ldots, \theta_{n-1}).\sin(\theta_n), \cos(\theta_n) \right).$$

A parametrization of $S^n(1)$ induces a parametrization $\Psi_n : [0, 2\pi] \times [0, \pi] \times \overset{n-1}{\cdots} \times [0, \pi] \to S^n(R)$ on $S^n(R)$:

$$\Psi_n(\theta_1, \ldots, \theta_{n-1}, \theta_n) = R.\big(\psi_{n-1}(\theta_1, \ldots, \theta_{n-1}).\sin(\theta_n), \cos(\theta_n)\big), \quad n \geq 2.$$

By induction, the volume element dA_R of $S^n(R)$ given as the function of angles $\theta_1, \ldots, \theta_n$ is

$$dA_R = \frac{1}{R}\omega_n = R^n.\sin^{n-1}(\theta_n).\sin^{n-2}(\theta_{n-1})\ldots\sin(\theta_2)d\theta_1 d\theta_2 \ldots d\theta_n.$$

So the volume of $S^n(R)$ is

$$A_n(R) = \int_0^{2\pi} d\theta_1.\int_0^{\pi}\sin(\theta_2)d\theta_2 \ldots \int_0^{\pi}\sin^{n-1}(\theta_n)d\theta_n.$$

We will not use the above formula to find A_n; instead we will use a smart trick, which is to know the value of the Gaussian integral

$$\int_{-\infty}^{\infty} e^{-x^2}dx = \sqrt{\pi}.$$

By integrating the function $f(x_1, \ldots, x_{n+1}) = e^{-(x_1^2 + \cdots + x_{n+1}^2)}$ in $\mathbb{R}^{n+1}$, we get

$$\int_{\mathbb{R}^{n+1}} e^{-(x_1^2 + \cdots + x_{n+1}^2)}dx_1 \ldots dx_{n+1} = \int_{-\infty}^{\infty} e^{-x_1^2}dx_1 \ldots \int_{-\infty}^{\infty} e^{-x_{n+1}^2}dx_{n+1} = (\pi)^{\frac{n+1}{2}}.$$

Using the spherical coordinates, the volume n-form in $\mathbb{R}^{n+1}$ is $dV = dx_1 \ldots dx_{n+1} = \rho^n d\rho d\omega_n$, so the integral becomes

$$\int_{\mathbb{R}^{n+1}} e^{-(x_1^2 + \cdots + x_{n+1}^2)}dx_1 \ldots dx_{n+1} = \int_0^{\infty} \rho^n e^{-\rho^2}d\rho.\int_{S^n}\omega_n = \frac{1}{2}\Gamma\left(\frac{n+1}{2}\right).A_n.$$

Hence

$$A_n(R) = \frac{2(\pi)^{\frac{n+1}{2}}}{\Gamma\left(\frac{n+1}{2}\right)}R^n, \quad V_{n+1}(R) = \frac{(\pi)^{\frac{n+1}{2}}}{\Gamma\left(\frac{n}{2} + \frac{3}{2}\right)}R^{n+1}. \tag{4}$$

Remark The Gamma function $\Gamma(t) = \int_0^{\infty} x^{t-1}e^{-x}dx$ shares the following properties:
(i) $\Gamma(1) = 1$,
(ii) $\Gamma(t + 1) = t\Gamma(t)$ (hint: integrating by parts),
(iii) $\int_0^{\infty} x^{n-1}e^{-x^2}dx = \frac{1}{2}\Gamma(\frac{n}{2})$. (hint: change the variable to $u = r^2$ and solve the integral).

Exercises

1. Show the following recurrence identities,

$$V_n = \frac{2\pi}{n} V_{n-2}, \quad S_{n-1} = \frac{2\pi}{n-2} S_{n-3}.$$

2. Show that $V_{2n} = \frac{(\pi)^n}{n!}$, $V_{2n+1} = \frac{2n!(4\pi)^n}{(2n+1)!}$ (hint: $\Gamma(n + \frac{1}{2}) = \frac{(2n)!}{4^n n!} \sqrt{\pi}$).
3. Show that $\lim_{n\to\infty} V_n = 0$ and $\lim_{n\to\infty} A_n = 0$. Give an intuitive reason why the limits are 0.

2   Harmonic Functions

Harmonic functions play a key role in physics and in several topics in Applied and Pure Mathematics.

2.1   Laplacian Operator

Let $U \subset \mathbb{R}^n$ be an open subset. The Laplacian operator on U is the linear differential operator $\triangle : C^2(U) \to C^0(U)$ given by

$$\triangle f = \mathrm{div}(\nabla f). \tag{5}$$

In Cartesian coordinates, the Laplacian is given by

$$\triangle f = \frac{\partial^2 f}{\partial x_1^2} + \frac{\partial^2 f}{\partial x_2^2} + \cdots + \frac{\partial^2 f}{\partial x_n^2}. \tag{6}$$

In the case of a vector quantity $\vec{F} = (f_1, \ldots, f_n)$, the Laplacian of $\vec{F}$ is defined as

$$\triangle F = \left(\triangle f_1, \ldots, \triangle f_n \right).$$

Example 1 In physics, the Laplacian is an operator of fundamental importance since it arises in several models, for example, in models that study heat conduction, wave models, electromagnetic models, Quantum Mechanics and many others. Let's look at two examples:

1. Heat conduction.
   Heat or thermal energy is the transfer of energy from a warmer body to a colder body. According to Fourier's law of heat conduction, the amount of heat Q transferred through a closed surface S per unit of time is proportional to the flow through S of the temperature gradient T, that is

$$\frac{dQ}{dt} = -k.\int_S \nabla T,$$

where k is the thermal conductivity of the material, which we assume to be constant. Thermal conductivity is the property that material has to conduct heat. When in a region Ω there is no heat source, or something absorbing heat, the amount of heat remains constant inside Ω. Therefore for every surface $S \subset \Omega$, we have $\int_S \nabla T = 0$. Applying the Stokes Theorem, we conclude that

$$\int_S \text{div}(\nabla T) = 0.$$

Hence $\triangle T = 0$ in Ω.

2. Gauss' Law.

Also known as the Gaussian Flow Theorem, it states that the flow of the electric field $\vec{E}$ through a closed connected surface $S = \partial\Omega$ (boundaryless) is equal to the total charge within S (or Ω). Let ϵ_0 be the electric constant and Q_S the total electric charge within S, then

$$\Phi_E = \frac{1}{\epsilon_0}\int_S \vec{E}dA.$$

Let ρ be the charge density in Ω. Applying the Stokes Theorem, we obtain

$$\Phi_E = \frac{1}{\epsilon_0}\int_\Omega \text{div}(\vec{E})dV = \int_\Omega \rho dV.$$

Therefore it follows that $\text{div}(\vec{E}) = \frac{\rho}{\epsilon_0}$. Assuming there is no electric charge in the region Ω, we have $\text{div}(\vec{E}) = 0$, and if we also assume there is no magnetic field from Faraday's law, we can derive that $\text{curl}(\vec{E}) = 0$. Therefore

$$\text{curl}\big(\text{curl}(\vec{E})\big) = \nabla\big(\text{div}(\vec{E})\big) - \triangle(\vec{E}) = 0 \Rightarrow \triangle(\vec{E}) = 0.$$

Consequently, in the absence of sources in a region Ω, we have $\triangle\vec{E} = 0$ in Ω.

Many mathematical models have some symmetry allowing us to simplify the equations to understand the questions addressed. Once a symmetry is detected, we should use appropriate coordinates. Next, we will address the question of the existence of a solution for the equation $\triangle f = g$ on the disk D_R^n which has spherical symmetry. For our purposes, it will be extremely convenient to use spherical coordinates in $\mathbb{R}^n$. Using the spherical coordinates $(\rho, \theta_1, \dots, \theta_{n-1})$ in $\mathbb{R}^n$, the Laplacian is

$$\triangle f = \frac{\partial^2 f}{\partial\rho^2} + \frac{n-1}{\rho}\frac{\partial f}{\partial\rho} + \frac{1}{\rho^2}\triangle_{S^{n-1}}f, \tag{7}$$

in which $\triangle_{S^{n-1}} f$ is an operator involving only the 1st and 2nd derivatives with respect to the angles $\theta_1, \ldots, \theta_{n-1}$. The operator $\triangle_{S^{n-1}}$ is known as the Laplace-Beltrami operator on S^{n-1}.

Using the exterior derivative, we have $\triangle f = d^* df$.

2.2 Properties of Harmonic Functions

Definition 1 A function $f \in C^2(U)$ is harmonic if it satisfies the equation

$$\triangle f = 0. \tag{8}$$

There are many examples of harmonic functions in $\mathbb{R}^n$; below we show some of them;

(i) every polynomial $p(x_1, \ldots, x_n) = \sum_{i=1}^n a_i x_i^2$ such that $\sum_{i=1}^n a_i = 0$.

(ii) If $n = 2$, then $v_2(x) = \ln(|x|)$ is harmonic in $\mathbb{R}^2 \backslash \{0\}$.

(iii) If $n \geq 3$, then $v_n(x) = |x|^{2-n}$ is harmonic in $\mathbb{R}^n \backslash \{0\}$.

Due to the importance of this example for the text segment, we will prove that $\triangle v = 0$ in $\mathbb{R}^n \backslash \{0\}$. We call attention to the fact that it is not harmonic in $\mathbb{R}^n$. In spherical coordinates, we have $v_n(\rho) = \rho^{2-n}$. Applying the Eq. (7), we get

$$\frac{(2-n)(1-n)}{\rho^n} + \frac{(n-1)}{\rho} \frac{(2-n)}{\rho^{n-1}} = 0, \quad \rho \neq 0.$$

However, $\int_{B_R^n} \triangle(v_n) dV = (2-n) A_{n-1}$, where A_{n-1} is the n-volume of the sphere S^{n-1}.

(iv) If $f(x)$ is harmonic and differentiable, then $\frac{\partial f}{\partial x_i}$ is harmonic for all $1 \leq i \leq n$. As a result there is an infinite number of harmonic functions in $\mathbb{R}^n$.

The harmonic functions in $\mathbb{R}^2$ enjoy many properties also satisfied by the holomorphic functions in $\mathbb{C}$. We recall that the real and imaginary parts of a holomorphic function are real harmonic functions, and every harmonic function defined on an open subset $U \subset \mathbb{R}^2$ is a real or imaginary part of a holomorphic function. In this way, the results we will present next are similar to those known in the function theory of a single complex variable.

We assume the following conditions along with the rest of the exposition;

(i) the region of integration $\Omega \subset U$ is closed, bounded, and has a non-empty interior with $S = \partial \Omega$ such that the Stokes Theorem is valid.

(ii) the functions are in $C^2(U)$.

Let dV be the volume element in Ω and let dS be the volume element of the boundary S.

It follows from the Stokes Theorem that

$$\int_\Omega \triangle f dV = \int_\Omega \text{div}(\nabla f) dV = \int_S <\nabla f, n> dS = \int_S \frac{\partial f}{\partial n} dS,$$

where n is the normal unit vector to S. Consequently, if f is harmonic, then $\int_S \frac{\partial f}{\partial n} dS = 0$. This imposes a strong constraint on f since the flow of the field ∇f must be 0 through all of the surfaces $S' = \partial \Omega'$.

Green's Identities are fundamental for exploring the properties of the harmonic functions.

Proposition 2 *If $f, g \in C^\infty(U)$, then:*
(i) Green's 1st identity.

$$\int_\Omega (g \triangle f + < \nabla f, \nabla g >) dV = \int_S g \frac{\partial f}{\partial n} dS. \tag{9}$$

(ii) Green's 2nd identity:

$$\int_\Omega (f.\triangle g - g.\triangle f) dV = \int_S (f \frac{\partial g}{\partial n} - g.\frac{\partial f}{\partial n}) dS. \tag{10}$$

Proof We recall that $\operatorname{div}(g.\nabla f) = < \nabla f, \nabla g > + g.\triangle f$.
(i) Considering the function $h = g.\nabla f$, we get the identity (9) applying the Stokes Theorem.
(ii) Let $h = f.\nabla g - g.\nabla f$;

$$\operatorname{div}(f.\nabla g - g.\nabla f) = f.\triangle g - g.\triangle f.$$

It follows from the Stokes Theorem that

$$\int_\Omega \operatorname{div}(f.\nabla g - g.\nabla f) dV = \int_\Omega (f.\triangle g - g.\triangle f) dV = \int_S (f. < \nabla g, n > - g. < \nabla f, n >) dS =$$
$$= \int_S (f.\frac{\partial g}{\partial n} - g.\frac{\partial f}{\partial n}) dS.$$

$\square$

If f is harmonic, then it satisfies the following properties;

1. $\int_S \frac{\partial f}{\partial n} dS = 0$.
2. $\int_\Omega |\nabla f|^2 dV = \int_S f.\frac{\partial f}{\partial n} dS$ (take $f = g$ in (9)).

These identities are enough to prove uniqueness.

Theorem 1 *Let $\Omega \subset U$ be a region where the Stokes Theorem is valid. If $f \in C^2(\operatorname{int}(\Omega))$ and $g \in C^0(\partial \Omega)$, then there is a unique function $u \in C^2(\operatorname{int}(\Omega))$ such that*
(ii) $\triangle u = f$ on $\operatorname{int}(\Omega)$,
(ii) $u|_{\partial\Omega} = g$.

Proof Suppose that there are two solutions $u, v \in C^2(\Omega)$, so the function $w = u - v$ solves the equation

$$\triangle w = 0, \quad \text{on } \Omega,$$
$$w \mid_{\partial\Omega} = 0.$$

Since

$$\int_\Omega \mid \nabla w \mid^2 dV = \int_{\partial\Omega} w \cdot \frac{\partial w}{\partial n} dS = 0,$$

we have $w = 0$ in $\text{int}(\Omega)$. Hence $u = v$. $\qquad\square$

Although we have proved the uniqueness for a bounded region Ω, we can't yet claim the existence of solutions. If Ω is unbounded, then it may have no solutions at all, as shown in the example in which $\Omega = \{x = (x_1, \ldots, x_n) \in \mathbb{R}^n \mid x_n \geq 0\}$, $f(x) = 0$ and $g(x) = x_n$.

Similar to the holomorphic functions on $\mathbb{C}$, the harmonic functions satisfy the Mean Value Property.

Theorem 2 (Mean Value Property) *Consider $n \geq 2$. Let $R > 0$ be such that $D_R^n(a) \subset \Omega \subset \mathbb{R}^n$. If $f \in C^2(U)$ is harmonic in U, then*

$$f(a) = \frac{1}{A_{n-1}(R)} \int_{S_R^{n-1}} f, \tag{11}$$

where $A_n(R) = \int_{S_R^{n-1}(a)} dS$.

Proof Given $\epsilon > 0$, we consider $\Omega = \{x \in U; \epsilon \leq \mid x - a \mid \leq R\}$ and $v_{n,a}(x) = \mid x - a \mid^{2-n}$. Using spherical coordinates in $\mathbb{R}^n$ and letting $\rho = \mid x - a \mid$, we get $\Omega = \{x \in U; \epsilon \leq \rho \leq R\}$ and $v_{n,a}(\rho) = \rho^{2-n}$, with the gradient being $\nabla v_{n,a}(\rho) = \frac{2-n}{\rho^{n-1}}\hat{\rho}$. Using the fact that $v_{n,a}(x)$ is harmonic in Ω and $\partial\Omega = S_R^{n-1} \sqcup (-S_\epsilon^{n-1})$ is an oriented boundary, the Green's 2nd identity yields

$$0 = \int_\Omega (f\triangle v_{n,a} - v_{n,a}\triangle f)dV = \int_{S_R^{n-1}(a)} (f\partial_n v_{n,a} - v_{n,a}\partial_n f)dS - \int_{S_\epsilon^{n-1}(a)} (f\partial_n v_{n,a} - v_{n,a}\partial_n f)dS =$$

$$= \frac{(2-n)}{R^{n-1}} \int_{S_R^{n-1}(a)} f dS - R^{2-n} \int_{S_R^{n-1}(a)} \partial_n f dS - \frac{(2-n)}{\epsilon^{n-1}} \int_{S_\epsilon^{n-1}(a)} f dS + \epsilon^{2-n} \int_{S_\epsilon^{n-1}(a)} \partial_n f dS =$$

$$= \frac{(2-n)}{R^{n-1}} \int_{S_R^{n-1}(a)} f dS - \frac{(2-n)}{\epsilon^{n-1}} \int_{S_\epsilon^{n-1}(a)} f dS.$$

So

$$\frac{1}{R^{n-1}} \int_{S_R^{n-1}(a)} f dS = \frac{1}{\epsilon^{n-1}} \int_{S_R^{n-1}(\epsilon)} f dS.$$

Multiplying both sides of the above identity by $\frac{1}{A_{n-1}}$, we get

$$\frac{1}{A_{n-1}(R)} \int_{S_R^{n-1}(a)} f dS = \frac{1}{A_{n-1}(\epsilon)} \int_{S_\epsilon^{n-1}(a)} f dS.$$

Passing the limit $\epsilon \to 0$, we have

$$f(a) = \frac{1}{A_{n-1}(R)} \int_{S_R^{n-1}(a)} f \, dS.$$

For the case $n = 2$, the method of proof goes along the same lines; however, the auxiliary harmonic function is $v_2(x) = \ln(|x|)$. $\square$

The Mean Value Property plays an outstanding and important role in understanding the properties of harmonic functions.

Theorem 3 (Maximum Principle) *Let $\Omega \subset U$ be a connected region. Assume $f \in C^2(\mathrm{int}(\Omega))$ is a harmonic function in $\mathrm{int}(\Omega)$. If f reaches the maximum value or the minimum value in $\mathrm{int}(\Omega)$, then f is constant in Ω.*

Proof Let $M = \max\{f(x) \mid x \in \Omega\}$. Consider the set $C_M = \{x \in \mathrm{int}(\Omega) \mid f(x) = M\}$. Since f is continuous, C_M is closed. Let's check that C_M is also open. Given $a \in C_M \cap \mathrm{int}(\Omega)$, let $\epsilon > 0$ be such that $B_\epsilon(a) \subset \mathrm{int}(\Omega)$. By the Mean Value Property, we have

$$M = f(a) = \frac{1}{V_n(\epsilon)} \int_{B_\epsilon^n(a)} f \, dS.$$

Consequently f must be a constant equal to M on $\partial B_\epsilon(a)$; otherwise, $f(a)$ could not be the average value on $B_\epsilon^n(a)$. Indeed, if we consider smaller balls containing a, we get $B_\epsilon(a) \subset C_M$, and so C_M is open. Therefore $C_M = \Omega$. For the case of the minimum value, we consider the harmonic function $(-f)$, and the above reasoning applies. $\square$

Exercises

1. Using spherical coordinates in $\mathbb{R}^n$, prove that the Laplacian is given by Eq. (7).
2. Show that $f(x) = \ln(|x|)$ is harmonic in $R^2 \backslash \{0\}$.
3. Show that

$$\int_{B_R^n} \Delta\left(|x|^{2-n}\right) dV = A_{n-1}, \quad \int_{B_R^2} \Delta\left(\ln(|x|)\right) dV = 2\pi,$$

   and conclude that $v_2(x) = \ln(|x|)$ and $v_n(x) = |x|^{2-n}$ are not harmonics in B_R^n.
4. If f is harmonic, show that $|f|$ is also harmonic. Indeed, $\Delta(|f|) = \frac{f}{|f|}\Delta f$.
5. Let $D^2 = \{u \in \mathbb{R}^2; |u| \le 1\}$ be the unity disk and let $f : \partial D^2 \to \mathbb{R}$ be a C^2-function.[1] Using polar coordinates show that a solution to the PDE

---

[1] f has a uniformly convergent Fourier series since it is C^2-differentiable.

$$\triangle u = 0 \text{ in } D^2,$$

$$u\big|_{\partial D^2} = f$$

is given by

$$u(r, \theta) = \frac{a_0}{2} + \sum_{i=1}^{\infty} r^n \big[a_n \cos(n\theta) + b_n \sin(n\theta) \big], \quad r \in [0, 1), \; \theta \in [0, 2\pi).$$

The coefficients a_n and b_n are given as

$$a_n = \frac{1}{2\pi} \int_0^{2\pi} f(t) \cos(nt)\, dt, \quad b_n = \frac{1}{2\pi} \int_0^{2\pi} f(t) \sin(nt)\, dt.$$

The solution $u : D^2 \to \mathbb{R}$ is a harmonic extension of f over the disk D^2.

6. Show that the zeros of a continuous harmonic function $f \neq 0$ are not isolated.
7. Consider $R > 0$ such that $D_R^n(a) \subset \Omega$, with the volume being $V(R)$. If f is harmonic, show that the value of f at $a \in \Omega$ is given by

$$f(a) = \frac{1}{V(R)} \int_{D_R^n(a)} f\, dV,$$

where $V(R)$ is the volume of $D_R^n(a)$.

3   Poisson Kernel for the n-Disk D_R^n

For any $n > 2$, let's consider $D_R^n = \{x \in \mathbb{R}^n ; \, | \, x \, | \leq 1\}$ the n-Disk, $B_R^n = \{x \in \mathbb{R}^n ; \, | \, x \, | < 1\}$ the open n-ball, and $S_R^{n-1} = \{x \in \mathbb{R}^n ; \, | \, x \, | = 1\}$ in the $(n-1)$-sphere. Dirichlet Problem in D_R^n: Given a function $\phi \in C^0(S_R^{n-1})$, show the existence of a function $f \in C^2(B_R^n)$ such that

$$\begin{aligned} \triangle f(x) &= 0, \; x \in B_R^n, \\ f(x) &= \phi(x), \; x \in S_R^{n-1}. \end{aligned} \tag{12}$$

The first step is to use Green's 2nd identity:

$$\int_\Omega f. \triangle g\, dV = \int_{B_R^n} g. \triangle f\, dV + \int_S (f.\partial_n g - g.\partial_n f)\, dS, \quad \left(\frac{\partial f}{\partial n} = \partial_n f \right). \tag{13}$$

Let $\xi \in \Omega$ and consider $\Omega_\epsilon = D_R^n \setminus B_\epsilon^n(\xi)$ and[2] $v(\xi, x) = v_\xi(x) = \mid x - \xi \mid^{2-n}$. As in the proof of the 2nd Green identity, we have (all the derivatives taken with respect to the x variable),

$$\int_{\Omega_\epsilon} (f.\triangle v_\xi - v_\xi.\triangle f)\, dV = \int_{\partial\Omega} \left(f.\frac{\partial v_\xi}{\partial n} - v_\xi.\frac{\partial f}{\partial n} \right) dS - \int_{S_\epsilon^{n-1}(p)} \left(f.\frac{\partial v_\xi}{\partial n} - v_\xi.\frac{\partial f}{\partial n} \right) dS =$$
$$= \int_{\partial\Omega} \left(f.\frac{\partial v_\xi}{\partial n} - v_\xi.\frac{\partial f}{\partial n} \right) dS - \frac{(2-n)}{\epsilon^{n-1}} \int_{S_\epsilon^{n-1}(p)} f\, dS + \epsilon^{2-n} \int_{S_\epsilon^{n-1}(p)} \frac{\partial f}{\partial n}\, dS.$$

Since v_ξ is harmonic in $\text{int}(\Omega_\epsilon)$ with respect to x variable, given $\epsilon \to 0$, we get the 3rd Green identity:

$$f(\xi) = \frac{1}{(2-n)A_{n-1}} \int_{\Omega_\epsilon} (v_\xi.\triangle f)\, dV + \frac{1}{(2-n)A_{n-1}} \int_{\partial\Omega} \left(f.\frac{\partial v_\xi}{\partial n} - v_\xi.\frac{\partial f}{\partial n} \right) dS. \tag{14}$$

If f is harmonic, then $f(\xi)$ is determined by the values of f on $\partial\Omega$ since

$$f(\xi) = \frac{1}{(2-n)A_{n-1}} \int_{\partial\Omega} \left(f.\frac{\partial v_\xi}{\partial n} - v_\xi.\frac{\partial f}{\partial n} \right) dS. \tag{15}$$

Once we have proved the existence of a harmonic function $G_\xi(x)$ in $\text{int}(\Omega)$, satisfying the boundary condition $G_\xi \mid_{\partial\Omega} = 0$, the solution to the problem (12) is

$$f(\xi) = \frac{1}{(2-n)A_{n-1}} \int_{\partial\Omega} \frac{\partial G_\xi}{\partial n}.\phi\, dS. \tag{16}$$

The function G_ξ is Green's function for the Dirichlet problem (12) when $\Omega = D_R^n$. In order to find G_ξ, we will consider some special transformations. The inversion map in a sphere S_R^{n-1} is

$$\mathcal{I} : \mathbb{R}^n \to \mathbb{R}^n$$
$$x \to \mathcal{I}(x) = R^2 \frac{x}{\mid x \mid^2}. \tag{17}$$

Consider the image set $\Omega^{\mathcal{I}} = \mathcal{I}(\Omega)$. Once we take into account the decomposition $\mathbb{R}^n = B_R^n \cup S_R^{n-1} \cup (\mathbb{R}^n \setminus B_R^n)$, where B_R^n is the open ball, the map $\mathcal{I} : \mathbb{R}^n \setminus \{0\} \to \mathbb{R}^n \setminus \{0\}$ satisfies the following properties:
(i) $\mathcal{I} \circ \mathcal{I} = \mathcal{I}$,
(ii) $\mid \mathcal{I}(x) \mid . \mid x \mid = R^2$.
Therefore if $x \in S_R^{n-1}$, then $\mathcal{I}(x) = x$, $\mathcal{I}(B_R^n) = \mathbb{R}^n \setminus B_R^n$ and $\mathcal{I}(\mathbb{R}^n \setminus B_R^n) = B_R^n$.
(iii) $\mathcal{I}$ is a conformal map; that is, we have $\lambda : \mathbb{R}^n \setminus \{0\} \to (0, \infty)$ such that

$$< d\mathcal{I}_x.u, \mathcal{I}_x.v > = \lambda(x). < u, v >, \forall u, v \in T_x(\mathbb{R}^n \setminus \{0\}).$$

---

[2] For the case $n = 2$, Green's function is $v_\xi(x) = \ln(\mid x - \xi \mid)$.

Definition 2 Let $f \in C^0(\mathbb{R}^n \setminus \{0\})$. Kelvin's transform of a function f is the function

$$K[f](x) = \frac{|x|^{2-n}}{R^{2-n}} f(\mathcal{I}(x)). \tag{18}$$

Therefore $K : C^0(\mathbb{R}^n \setminus \{0\}) \to C^0(\mathbb{R}^n \setminus \{0\})$.

Kelvin's transform is linear and $K(K(f)) = f$, that is, $K^{-1} = K$. It was Lord Kelvin, famous for his contributions to thermodynamics, who discovered this transformation. The most relevant property for our purposes is in preserving the harmoniousness of a harmonic function. We will need some preliminary results regarding Kelvin's transform.

Lemma 1 *Let* $p : \mathbb{R}^n \to \mathbb{R}$ *be a homogeneous polynomial of degree m. Then,*

$$\triangle(|x|^{2-n-2m} p) = |x|^{2-n-2m} . \triangle p.$$

Proof Since $p(tx) = t^m p(x)$, we have $< x, \nabla p(x) >= mp(x)$. Considering $f(x) = |x|^k . p(x)$, we have

$$\triangle f = |x|^k \triangle p + 2 < \nabla(|x|^k), \nabla p(x) > + p(x . \triangle(|x|^k)).$$

Applying the identities

$$\nabla(|x|^k) = k |x|^{k-2} x,$$
$$\triangle(|x|^k) = \operatorname{div}(\nabla(|x|^k)) = [k(k-2) + nt] |x|^{k-2},$$

we get

$$\triangle f(x) = |x|^k \triangle p + 2k |x|^{k-2} < x, \nabla p(x) > + [k(k-2) + nt] |x|^{k-2} . p(x) =$$
$$= |x|^k \triangle p + 2k |x|^{k-2} mp(x) + [k(k-2) + nt] |x|^{k-2} . p(x) =$$
$$= |x|^k \triangle p + [k(2m + n + k - 2)] |x|^{k-2} . p(x).$$

The statement follows assuming $k = 2 - 2m - n$. □

Lemma 2 *The Kelvin transform preserves uniform convergence on compact sets.*

Proof Let $\{f_n\} \subset C^2(\mathbb{R}^n \setminus \{0\})$ be a uniformly convergent sequence when restricted to a compact set $K \subset (\mathbb{R}^n \setminus \{0\})$ in which $|x| \leq R^{2-n} M$ ($M > 0$). Given $\epsilon > 0$, we have $n_0 \in \mathbb{N}$ such that if $n, m > n_0$, then $\|f_n - f_m\|_2 < \frac{\epsilon}{M^{2-n}}$ in K. The image set $\mathcal{I}(K)$ is also compact. Given that $y = \mathcal{I}(x)$, we get

$$\|K[f_n] - K[f_m]\|_2 = \sup_{y \in K^{\mathcal{I}}} |K[f_n](y) - K[f_m](y)|_2 \leq \sup_{x \in K} \frac{|x|^{2-n}}{R^{2-n}} |f_n(x) - f_m(x)|_2 \leq$$

$$\leq M^{2-n} \|f_n - f_m\|_2 < \epsilon.$$

Therefore $\{K[f_n]\}_{n \in \mathbb{N}}$ uniformly converges in the Banach space $(C^2(\mathbb{R}^n\setminus\{0\}), \|\cdot\|_2)$. $\quad\square$

Corollary 1 *Let* $f \in C^2(\mathbb{R}^n\setminus\{0\})$, *so*

$$\triangle(K[f]) = \left(\frac{|x|}{R}\right)^4 . K[\triangle f]. \tag{19}$$

Proof First, let p be a homogeneous polynomial of degree m; therefore

$$K[p](x) = \frac{|x|^{2-n}}{R^{2-n}} p(\frac{R^2}{|x|^2}x) = R^{2m+n-2} |x|^{2-n-2m} p(x).$$

From Lemma 1, we have $\triangle K[p](x) = R^{2m} |x|^{2-n-2m} \triangle p$. When p is homogeneous of degree m, the function $\triangle p$ is also homogeneous of degree $(m+2)$. Therefore

$$K[\triangle p](x) = \frac{|x|^{2-n}}{R^{2-n}} \triangle p\left(\frac{R^2}{|x|^2}x\right) = \frac{|x|^{2-n}}{R^{2-n}} \left(\frac{R^2}{|x|^2}\right)^{m+2} \triangle p(x) =$$

$$= \left(\frac{R}{|x|}\right)^4 . R^{2m+n-2} . |x|^{2-n-2m} . \triangle p(x) = \left(\frac{R}{|x|}\right)^4 . \triangle K[p](x).$$

Consider $\{p_n\}$ a polynomial sequence converging uniformly to f in $C^2(\mathbb{R}^n\setminus\{0\})$. Writing the polynomial as a finite sum $p_n(x) = \sum_i p_{n,i}(x)$, with $p_{n,i}(x)$ being a homogeneous polynomial of degree i, it follows from the linearity of both operators, Kelvin's transform and Laplacian ($\triangle p_n = \sum_i \triangle p_{n,i}$) that

$$\triangle K[p_n] = \sum_i \triangle K[p_{n,i}] = \left(\frac{|x|}{R}\right)^4 \sum_i K[\triangle p_{n,i}] = \left(\frac{|x|}{R}\right)^4 K[\triangle p_n].$$

Since p_n converges to f in C^2, we have

$$\triangle K[f] = \lim_{n \to \infty} \triangle K[p_n] = \lim_{n \to \infty} \left(\frac{|x|}{R}\right)^4 K[\triangle p_n] = \left(\frac{|x|}{R}\right)^4 K[\triangle f].$$

$\square$

Theorem 4 *A function* $f \in C^2(\mathbb{R}^n\setminus\{0\})$ *is harmonic if and only if* $K[f]$ *is harmonic.*

Proof The statement follows from the identity

$$\triangle K[f] = \left(\frac{|x|}{R}\right)^4 . K[\triangle f].$$

$\square$

Let $D = \{(x, x) \in B_R^n \times B_R^n\}$ be the diagonal and let $G : \left(B_R^n \times B_R^n\right)\backslash D \to \mathbb{R}$ be the function given by $(G_\xi(x) = G(\xi, x))$:

$$G(\xi, x) = |\, x - \xi \,|^{2-n} - K\big[\,|\, x - \xi \,|^{2-n}\,\big] = |\, x - \xi \,|^{2-n} - \frac{|\, x \,|^{2-n}}{R^{2-n}} \cdot \left|\frac{R^2}{|\, x \,|^2}x - \xi\right|^{2-n}, \quad (20)$$

and we have our Green's function.

Definition 3 Green's function for the Dirichlet problem (12) is

$$G(\xi, x) = |\, x - \xi \,|^{2-n} - \frac{|\, x \,|^{2-n}}{R^{2-n}} \cdot \left|\frac{R^2}{|\, x \,|^2}x - \xi\right|^{2-n}. \quad (21)$$

To check $G(\xi, x) = G_\xi(x)$ is the Green function, we need to verify the conditions:
(i) $\triangle G_\xi = 0$ in B_R^n;
(ii) $G_\xi(x) = 0$, if $x \in S_R^{n-1}$.

(1) If ξ or x belong to S_R^{n-1}, then $G(\xi, x) = 0$.
(2) $G(\xi, x) = G(x, \xi)$.
   It is enough to check the second term, so we use the identity (easily checked by squaring both sides):

$$\left|R^2\frac{x}{|\, x \,|} - |\, x \,| . \xi\right| = \left|R^2\frac{\xi}{|\, \xi \,|} - |\, \xi \,| . x\right|.$$

So we have

$$\frac{|\, x \,|^{2-n}}{R^{2-n}} \cdot \left|\frac{R^2}{|\, x \,|^2}x - \xi\right|^{2-n} = \frac{|\, x \,|^{2-n}}{R^{2-n}} \cdot \frac{1}{|\, x \,|^{2-n}}\left|R^2\frac{x}{|\, x \,|} - |\, x \,| . \xi\right|^{2-n} =$$

$$= \frac{1}{R^{2-n}} \cdot \left|R^2\frac{\xi}{|\, \xi \,|} - |\, \xi \,| . x\right|^{2-n} = \frac{|\, \xi \,|^{2-n}}{R^{2-n}} \cdot \left|R^2\frac{\xi}{|\, \xi \,|^2} - x\right|^{2-n}.$$

$G(\xi, x)$ is symmetric with respect to the variables x and ξ.
(3) $G(\xi, x)$ is harmonic in $\left(B_R^n \times B_R^n\right)\backslash D$.
   It is straightforward from the definition of G.
(4) Since the unit normal vector to S_R^{n-1} at x is $n_x = \frac{x}{R}$, the derivative at the direction along the vector n_x is

$$\frac{\partial G}{\partial n_x}(\xi, x) = \frac{1}{R} < \nabla G(\xi, x), x >= \frac{1}{RA_{n-1}}\frac{R^2 - |\, \xi \,|^2}{|\, x - \xi \,|^n}. \quad (22)$$

By swapping[3] the variables $\xi \to x$ and $x \to \xi$ in Eq. (22), the solution (16) is now

---

[3] Just for an aesthetic reason.

$$f(x) = \frac{1}{(2-n)RA_{n-1}} \int_{S_R^{n-1}} \frac{R^2 - |x|^2}{|x - \xi|^n} \phi(\xi) \, dS_\xi. \tag{23}$$

Definition 4 The Poisson kernel for the Dirichlet problem (12) in the n-Disk D_R^n is

$$P(x, \xi) = \frac{1}{(2-n)RA_{n-1}} \frac{R^2 - |x|^2}{|x - \xi|^n}. \tag{24}$$

Let $\mathcal{G} : C^0(S_R^{n-1}) \rightarrow C^0(B_R^n)$ be the map given by

$$\mathcal{H}[g](x) = \frac{1}{(2-n)RA_{n-1}} \int_{S_R^{n-1}} \frac{R^2 - |x|^2}{|x - \xi|^n} g(\xi) \, dS_\xi. \tag{25}$$

The function $\mathcal{H}[g](x)$ is the only solution to the Dirichlet Problem (12). To prove this statement, we need some previous results.

Theorem 5 *If $\xi \in S_R^{n-1}$, then $P(x, \xi)$ is harmonic in $B_R^n \setminus \{\xi\}$.*

Proof The Poisson kernel $P(x, \xi) = \frac{R^2 - |x|^2}{|x - \xi|^n}$ is C^2-differentiable in $R^n \setminus \{\xi\}$. Considering the functions $u = R^2 - |x|^2$ and $v = |x - \xi|^n$, we have

$$\nabla u = -2x, \quad \Delta u = -2n$$

$$\nabla v = -\frac{n}{|x - \xi|^{n+2}}(x - \xi), \quad \Delta v = \frac{2n}{|x - \xi|^{n+2}}.$$

Applying the identity $\Delta(u.v) = u.\Delta v + 2 < \nabla u, \nabla v > + v.\Delta u$, it follows that $P(x, \xi)$ is harmonic in $R^n \setminus \{\xi\}$. □

Proposition 3 *The Poisson kernel satisfies the following properties;*

(1) $P(x, \xi) > 0$ for all $x \in B_R^n$ and $\xi \in S_R^{n-1}$.
(2) $\int_{S_R^{n-1}} P(x, \xi) \, dS_\xi = R$.
(3) For any $\eta \in S_R^{n-1}$ and $\delta > 0$,

$$\lim_{x \to \eta} \left[\int_{|\xi - \eta| > \delta} P(x, \xi) \, dS_\xi \right] = 0.$$

Proof The proof is straightforward from the definition and properties, as shown next.

(1) It is immediate from the definition of $P(x, \xi)$.
(2) Consider a sequence $\{x_k\}_{k \in \mathbb{N}} \subset B_R^n$ such that $\lim x_k = \eta \in S_R^{n-1}$. Assuming the condition $|x_k - \eta| > \delta$, we get

$$\lim_{k \to \infty} \left| \int_{S_R^{n-1}} \frac{R^2 - |x_k|^2}{|x_k - \xi|^n} \, dS_\xi \right| < \lim_{k \to \infty} \int_{S_R^{n-1}} \frac{R^2 - |x_k|^2}{\delta} \, dS_\xi = 0.$$

(3) Consider $R = 1$; since $| \xi | = 1$, we have

$$\int_{S^{n-1}} \frac{1-|x|^2}{|x-\xi|^n} dS_\xi = \int_{S^{n-1}} \frac{1-|\xi|^2|x|^2}{\left| |\xi| x - \frac{\xi}{|\xi|} \right|^n} dS_\xi = \int_{S^{n-1}} \frac{1-|\xi|^2|x|^2}{\left| |x| \xi - \frac{x}{|x|} \right|^n} dS_\xi =$$

$$= \int_{S^{n-1}} P\left(|x| \xi, \frac{x}{|x|} \right) dS_\xi = P\left(0, \frac{x}{|x|} \right) = 1.$$

The last equality follows from the fact that $P(x, \xi)$ is harmonic, so it satisfies the Mean Value Property. Now we apply the coordinate change $\xi' = \frac{\xi}{R}$ to get

$$\int_{S_R^{n-1}} \frac{R^2-|x|^2}{|x-\xi|^n} dS_\xi = R \int_{S^{n-1}} \frac{1-|\frac{x}{R}|^2}{|\frac{x}{R}-\xi'|^n} dS_{\xi'} = R.$$

$\square$

Theorem 6 *Let $g \in C^0(S_R^{n-1})$ and consider*

$$f(x) = \begin{cases} \mathcal{H}_R[\phi](x), & \text{if } x \in B_R^n, \\ \phi(x), & \text{if } x \in S_R^{n-1}. \end{cases} \tag{26}$$

So f is the unique solution to Problem (12).

Proof The Laplacian operator switches with the integral since f is differentiable. In this way, the Laplacian of $f(x)$ in B_R^{n-1} is

$$\Delta f(x) = \int_{S_R^{n-1}} \Delta P(x, \xi) \phi(\xi) \, dS_\xi = 0.$$

To prove the continuity of f in B_R^n, let's fix $\eta \in S_R^{n-1}$ and $\epsilon > 0$. Since ϕ is continuous, there is $\delta > 0$ such that if $| \xi - \eta | < \delta$, then $| \phi(\xi) - \phi(\eta) | < \epsilon$.

$$\lim_{x \to \eta} | f(x) - f(\eta) | = \lim_{x \to \eta} \left| \int_{S_R^{n-1}} P(x, \xi)(\phi(\xi) - \phi(\eta)) \, dS_\xi \right| \leq \lim_{x \to \eta} \int_{S_R^{n-1}} P(x, \xi) | \phi(\xi) - \phi(\eta) | \, dS_\xi \leq$$

$$\leq \lim_{x \to \eta} \int_{|x-\eta|<\delta} P(x, \xi) | \phi(\xi) - \phi(\eta) | \, dS_\xi + \lim_{x \to \eta} \int_{|x-\eta|>\delta} P(x, \xi) | \phi(\xi) - \phi(\eta) | \, dS_\xi \leq$$

$$\leq \epsilon R + 2 \| \phi \|_0 \lim_{x \to \eta} \int_{|x-\eta|>\delta} P(x, \xi) | \phi(\xi) - \phi(\eta) | \, dS_\xi = 0.$$

In the last inequality, we use item (3) of Proposition 3. $\square$

Indeed the function $f(x)$ defined in (26) is $C^\infty(B_R^n)$ with respect to the variable $x \in B_R^n$. We can check this as follows: let $\alpha = (\alpha_1, \dots, \alpha_n)$ be a multi-index in which $0 \leq \alpha_i$ and $| \alpha | = \sum_{i=1}^n \alpha_i$, the partial derivative $\partial^{|\alpha|} f = \frac{\partial^{|\alpha|} f}{\partial^{\alpha_1} x_1 \dots \partial^{\alpha_n} x_n}$ is

$$\partial^{|\alpha|} f(x) = \int_{B_R^n} \partial^{|\alpha|} P(x, \xi) \phi(\xi) \, dS_\xi.$$

Since $P(x, \xi) \in C^\infty(B_R^n)$ for all $\xi \in S_R^{n-1}$, it is straighforward that $f \in C^\infty(B_R^{n-1})$.

Exercises

(1) Prove Eq. (22) can be obtained for the Poisson kernel.
hint: show the identities

$$(i) \quad \frac{\partial \mid x - p \mid^{2-n}}{\partial x_i} = \frac{(2-n)}{\mid x - p \mid^n}(x_i - p_i),$$

$$(ii) \quad \frac{\partial(\mid \frac{|p|}{R} x - \frac{R}{|p|} p \mid^{2-n})}{\partial x_i} = \frac{(2-n)}{\mid \frac{|p|}{R} x - \frac{R}{|p|} p \mid^n} \frac{|p|}{R} \left(\frac{|p|}{R} x_i - \frac{R}{|p|} p_i \right)$$

and prove that if $\mid x \mid = R$, then $\mid \frac{|p|}{R} x - \frac{R}{|p|} p \mid = \mid x - p \mid$.

(2) Consider $\Omega \subset \mathbb{R}^n$. Let $\{f_n\}_{n \in \mathbb{N}} \subset C^\infty(\Omega)$ be a sequence converging uniformly to f in each compact subset $K \subset \Omega$. Show that f is harmonic and for all of multi-index α, the sequence $\{\partial^{|\alpha|} f_n\}_{n \in \mathbb{N}}$ converges uniformly in each compact subset $K \subset \Omega$ to $\partial^{|\alpha|} f$.

(3) Prove the reverse of the Mean Value Property: let $a \in \Omega$ and $f \in C^0(\Omega)$. If

$$f(x) = \frac{1}{A_{n-1}(R)} \int_{S_R^{n-1}} f(\xi) \, dS_\xi,$$

for all $R > 0$ such that $B_R^n(a) \subset \Omega$, then u is harmonic. To prove continuity, it is necessary to assume Ω is bounded; otherwise, assuming $\Omega = \mathbb{R}^n$, the function below is a counterexample;

$$f(x) = \begin{cases} 1, & x_n > 0, \\ 0, & x_n = 0, \\ -1, & x_n < 0. \end{cases}$$

(4) Let f be an integrable function in Ω. Let $a \in \Omega^o$ and $R > 0$. If for all $B_R^n(a) \subset \Omega$, we have

$$f(x) = \frac{1}{V_{n-1}(R)} \int_{B_R^n(a)} f(\xi) \, dV_\xi,$$

then f is harmonic.

(5) If f is harmonic in $\Omega \subset \mathbb{R}^n$, then f is analytic in Ω.

(6) Solve the Dirichlet problem for the case $n = 2$.

(7) Let $\Omega \subset \mathbb{R}^n$ be a region, that is, a bounded set with non-empty interior and with a boundary that is C^0-regular, and let

$$\| df \|^2 = \sum_{i=1}^n \left(\frac{\partial f}{\partial x_i} \right)^2.$$

The Dirichlet energy functional $E : C^2(U) \to \mathbb{R}$ is given by

$$E(f) = \int_\Omega \| df \|^2 \, dV.$$

Given an open subset $U \subset \text{int}(\Omega)$, show the identity $dE_f . u = \int_U < \Delta f, u > dV$. Then conclude that $\nabla E(f) = \Delta f$.

(8) Let M be a compact manifold. If $f \in C^2(M)$ is harmonic, show that it must be constant.

4 Harmonic Differential Forms

Let $U \subset \mathbb{R}^n$ be a bounded open subset endowed with a Riemannian metric g. We consider $\{(x_1, \ldots, x_n) \mid x_i \in \mathbb{R}\}$ a local system of coordinates on U. Let $\beta = \{e_1, \ldots, e_n\}$ be a local frame, in which $e_i = \frac{\partial}{\partial x_i}$ satisfies the conditions;

(i) $\partial_{e_i} e_j = \frac{\partial e_j}{\partial x_i} = 0$ for all i, j;

(ii) $[e_i, e_j] = 0$ for all i, j.

Let Ω^p_{cs} be the space of p-forms in U with compact support, we wish to avoid the boundary terms. The exterior derivative $d_p : \Omega^p_{cs}(U) \to \Omega^{p+1}_{cs}(U)$ is

$$d_p = \sum_j e_j \wedge \partial_{e_j}. \tag{27}$$

Let $f \in \Omega^0_{cs}(U)$, so

$$d_0 f = \sum_j e_j \wedge \partial_{e_j} f = \sum_j e_j \wedge \frac{\partial f}{\partial x_j} = \sum_j \frac{\partial f}{\partial x_j} e_j = \nabla f.$$

Given a p-form $\omega = f e_{i_1} \wedge e_{i_2} \wedge \cdots \wedge e_{i_p}$, we get

$$d_p \omega = \sum_j e_j \wedge \partial_{e_j} \left(f e_{i_1} \wedge e_{i_2} \wedge \cdots \wedge e_{i_p} \right) = \quad = \sum_j \frac{\partial f}{\partial x_j} e_j \wedge e_{i_1} \wedge e_{i_2} \wedge \cdots \wedge e_{i_p}.$$

In $\Omega^p_{cs}(U)$, we have the inner product

$$< \omega, \eta >= \int_U \left(\omega \wedge *\eta \right) \mathrm{dvol}_g$$

and the norm is defined as

$$| \omega |_2 = \left[\int_U | \omega |^2 \, dvol_g \right]^{\frac{1}{2}}.$$

So far we have considered differential forms, i.e., p-forms $\omega : U \to \Lambda^p(U)$ differentiable with respect to the variable $x \in U$ for all $0 \le p \le n$. In this section, we need to consider the space $L^2(\Omega^p_{cs}(U))$ of L^2-forms (square integrable);

$$L^2(\Omega^p_{cs}(U)) = \{\omega \in \Omega^p_{cs}(U); \ \mid \omega \mid_2 < \infty\}.$$

$$L^2(\Omega^*_{cs}(U)) = \bigoplus_{p=0}^{n} L^2(\Omega^p_{cs}(U)). \tag{28}$$

For $0 \le p \le n$, the spaces $L^2(\Omega^p_{cs}(U))$ are Hilbert spaces. The space of C^r-differential forms $C^r(\Omega^p_{cs}(U))$ is a subspace of $L^2(\Omega^p_{cs}(U))$. The space of L^2-forms has a weaker topology than the space of C^r-forms; it is more appropriate to prove the existence of a solution for a PDE. We call the reader's attention to the space of forms we use along this section; we set the notation $L^2(\Omega^p_{cs}(U))$ for the square integrable p-forms, $C^r(\Omega^p_{cs}(U))$ for the C^r-differentiable p-forms and $\Omega^p_{cs}(U)$ when it does not concern the topology defined on the space of p-forms.

The operator $d^* : L^2(\Omega^{p+1}_{cs}(U)) \to L^2(\Omega^p_{cs}(U))$ is the dual operator of d with respect to the inner product in $L^2(\Omega^p_{cs}(U))$, i.e.,

$$< d\omega, \eta > = < \omega, d^*\eta > .$$

The Laplacian operator $\triangle_p : L^2(\Omega^p_{cs}(U)) \to L^2(\Omega^p_{cs}(U))$ is

$$\triangle_p = d^*_{p+1}d_p + d_{p-1}d^*_p. \tag{29}$$

Given $\eta \in C^r(\Omega^p_{cs}(U))$, our intent is to solve the equation

$$\triangle_p\omega = \eta, \tag{30}$$

given that $\omega \in C^r(\Omega^p_{cs}(U))$. A p-form ω is harmonic if $\triangle_p\omega = 0$. Therefore ω is harmonic in $L^2(\Omega^p_{cs}(U))$ if and only if $d_p\omega = d^*_p\omega = 0$. A harmonic p-form ω represents a cohomology class in De Rham Cohomology. Let $\mathcal{H}^p$ be the space of harmonic p-forms. We introduce the operator $\eth_p : \Omega^p_{cs}(U) \to \Omega^{p-1}_{cs}(U) \oplus \Omega^{p+1}_{cs}(U)$,

$$\eth_p = d^*_p + d_p. \tag{31}$$

So the Laplacian is given by $\eth^2_p = \triangle_p$. Furthermore,

$(i) \ \eth^*_p = \eth_p,$

$(ii) \ \mathrm{Ker}(\eth_p) = \mathrm{Ker}(\eth^2_p), \tag{32}$

$(iii) \ < \omega, \triangle_p\omega > = \mid \eth_p \mid^2_{L^2} = \mid d^*_p\omega \mid^2_{L^2} + \mid d_p\omega \mid^2_{L^2} .$

The subspaces $\mathrm{Im}(d_{p-1}) = d_{p-1}(\Omega_{cs}^{p-1})$, $\mathrm{Im}(d_{p+1}^*) = d_{p+1}^*(\Omega_{cs}^{p+1})$ and $\mathcal{H}^p$ are orthogonal since for any $\theta \in \Omega_{cs}^{p-1}$, $\omega \in \mathcal{H}^p$ and $\eta \in \Omega_{cs}^{p+1}$, we have

$$< d\theta, d^*\eta > = < d^2\theta, \eta > = 0$$
$$< d\theta, \omega > = < \theta, d^*\omega > = 0;$$
$$< \omega, d^*\eta > = < d\omega, \eta > = 0.$$

In order to solve the Eq. (30), we stress the following items:
(1) We will prove the existence of a solution $\omega \in L^2(\Omega_{cs}^p(U))$, indeed, a weak solution.
(2) For the purposes of Differential Topology, we need to prove $\omega \in C^\infty(\Omega_{cs}^p(U))$.

To prove the existence of a weak solution, we will open the tool box of Functional Analysis. Assuming $\eta \neq 0$ and the existence of a solution $\omega \in L^2(\Omega_{cs}^p(U))$ for $\Delta\omega = \eta$, we consider the functional $F : L^2(\Omega_{cs}^p(U)) \to \mathbb{R}$ given by

$$F(\Delta\phi) = < \eta, \phi > .$$

The definition yields the following remarks;
(i) $F(\Delta\theta) = < \omega, \Delta\theta >$ for all $\theta \in L^2(\Omega_{cs}^p(U))$, since

$$< \eta, \theta > = < \Delta\omega, \theta > = < \omega, \Delta\theta > .$$

(ii) $\eta \in (\mathcal{H}^p)^\perp \subset L^2(\Omega_{cs}^p(U))$ $(\eta \neq 0)$ since

$$< \eta, h > = < \Delta\omega, h > = < \omega, \Delta h > = 0, \ \forall h \in \mathcal{H}^p.$$

The strategy to obtain a weak solution for Eq. (30) is to prove the orthogonal decomposition

$$L^2(\Omega_{cs}^p(U)) = \mathcal{H}^p \bigoplus (\mathcal{H}^p)^\perp, \tag{33}$$

and the fact that we have a weak solution if and only if $\eta \in (\mathcal{H}^p)^\perp$. The orthogonal decomposition is achieved by proving the space $\mathcal{H}^p$ is finite dimensional. At this point, the ellipticity of the Laplacian operators Δ_p, $0 \leq p \leq n$, is fundamental. Indeed, ellipticity is a very powerful property for which the consequences are the following;
(i) $\mathcal{H}^p$ is finite dimensional.
(ii) Given $\eta \in C^\infty(\Omega_{cs}^p(U))$, if we have an L^2-weak solution ω, then it is in $C^\infty(\Omega_{cs}^p(U))$.

A full treatment would require a more in-depth approach to analytical issues about Sobolev Spaces, which are beyond the scope of this text. For a full treatment we recommend the Refs. [13, 45].

Lemma 3 *Let $c > 0$ and consider $\{\omega_n\} \subset L^2\big(\Omega_{cs}^p(U)\big)$ a sequence of p-forms on U such that*

$$\mid \omega_n \mid_2 \ \leq \ c, \quad and \quad \mid \triangle\omega_n \mid_2 \ \leq \ c, \ \forall n \in \mathbb{N}.$$

Then a subsequence of $\{\omega_n\}_{n\in\mathbb{N}}$ is a Cauchy sequence in $L^2\big(\Omega_{cs}^p(U)\big)$.

Proof See in Ref. [45].   □

Corollary 2 $\mathcal{H}^p$ *is finite dimension. Furthermore, the space $L^2\big(\Omega_{cs}^p(U)\big)$ admits an orthogonal decomposition*

$$\Omega^p(M) = \mathrm{Im}(d_{p-1}) \oplus \mathrm{Im}(d_{p+1}^*) \oplus \mathcal{H}^p. \tag{34}$$

Proof By Lemma 3, $\mathcal{H}^p$ is a locally compact set. Therefore $\mathcal{H}^p$ is finite dimensional by Theorem 15 in Appendix A. In this case, $\mathcal{H}^p$ admits a closed orthogonal complement $(\mathcal{H}^p)^\perp$ as proved in Proposition 5 in Appendix A. Since $\triangle_p = \eth_p^2$ for all $0 \leq p \leq n$, we get the desired decomposition.   □

Corollary 3 *There is a constant $k > 0$ such that*

$$\mid \omega \mid_2 \ \leq \ k \mid \triangle\omega \mid_2, \tag{35}$$

for all $\omega \in (\mathcal{H}^p)^\perp$.

Proof Otherwise, let us assume (35) is false, so there would be a sequence $\{\omega_n\}_{n\in\mathbb{N}} \subset L^2\big(\Omega_{cs}^p(U)\big)$ with $\mid \omega_n \mid_2 \ \leq \ 1$ and $\mid \triangle\omega_n \mid_2 \ \to \ 0$. By Lemma 3, this would imply the existence of a Cauchy sequence which we can assume to be $\{\omega_n\}_{n\in\mathbb{N}}$. Let $\omega = \lim_n \omega_n$, so

$$\lim_{n \to \infty} <\omega_n, \theta> = <\omega, \theta>, \quad \forall\theta.$$

In this case, we have $\mid \triangle\omega \mid_2 = 0$. But this is a contradiction since the orthogonal complement is closed.   □

Theorem 7 (Hodge, L^2-version) *The space $L^2\big(\Omega_{cs}^p(U)\big)$ admits an orthogonal decomposition into*

$$L^2\big(\Omega_{cs}^p(U)\big) = \mathrm{Im}(d_{p-1}) \oplus \mathrm{Im}(d_{p+1}^*) \oplus \mathcal{H}^p \tag{36}$$

for all $0 \leq p \leq n$. Furthermore, the equation $\triangle_p\omega = \eta$ admits a weak solution if and only if $\eta \perp \mathcal{H}^p(U)$.

Proof Since $L^2\big(\Omega_{cs}^p(U)\big) = \mathcal{H}^p \oplus (\mathcal{H}^p)^\perp$, letting $\beta^h = \{\omega_1, \ldots, \omega_{b(p)}\}$ be a basis of $\mathcal{H}^p$, then an arbitrary form $\omega \in L^2\big(\Omega_{cs}^p(U)\big)$ can uniquely be written as

$$\omega = \sum_{i=1}^{b(p)} <\omega, \omega_i> \omega_i + \omega^\perp,$$

where $\omega^\perp \in (\mathcal{H}^p)^\perp$. Define the projection $P^h : L^2(\Omega_{cs}^p(U)) \to \mathcal{H}^p$ by

$$P^h(\omega) = \sum_{i=1}^{b(p)} < \omega, \omega_i > \omega_i.$$

Let's prove the image set $\text{Im}(\triangle) = \triangle(\Omega_{cs}^p(U))$ is equal to $(\mathcal{H}^p)^\perp$. We have already seen that the decomposition is orthogonal and $\triangle(\Omega_{cs}^p(U)) \subseteq (\mathcal{H}^p)^\perp$. Now we need to check the inclusion $(\mathcal{H}^p)^\perp \subseteq \triangle(\Omega_{cs}^p(U))$. Let $\eta \in (\mathcal{H}^p)^\perp$ and define the linear functional $F : (\mathcal{H}^p)^\perp \to \mathbb{R}$,

$$F(\triangle\theta) = < \eta, \theta > .$$

We note the following items;
(i) F is well-defined;
Assume $\triangle\theta_1 = \triangle\theta_2$; so $\triangle(\theta_1 - \theta_2) = 0$ implies $(\theta_1 - \theta_2) \in \mathcal{H}^p$; therefore $< \eta, (\theta_1 - \theta_2) >= 0$ and $F(\triangle\theta_1) = F(\triangle\theta_2)$.
(ii) F, restricted to $\triangle(\Omega_{cs}^p(U))$, is a bounded linear functional;
applying Corollary 3, we have

$$| F(\triangle\theta) | = |< \eta, \theta >| \le | \eta | . | \theta | \le k | \eta | . | \triangle\eta | .$$

Due to the Hahn-Banach Theorem 14 in Appendix A, F extends to a bounded linear functional on $L^2(\Omega_{cs}^p(U))$. By Riesz's Representation Theorem 12 in Appendix A, there is $\omega \in L^2(\Omega_{cs}^p(U))$ such that

$$F(\triangle\theta) = < \omega, \triangle\theta > .$$

So we get
$$F(\triangle\theta) = < \omega, \triangle\theta >=< \triangle\omega, \theta >=< \eta, \theta > .$$

Therefore $\triangle\omega = \eta$ and $\omega \in L^2(\Omega_{cs}^p(U))$ is a weak solution.   □

The differentiable version of the Hodge Theorem requires proving that $\omega \in C^\infty(\Omega_{cs}^p(U))$; nevertheless, this requires using regularity theory which relies on Sobolev's embedding theorem and the ellipticity of the Laplacian operator.

Theorem 8 (Hodge) *The space $\Omega_{cs}^p(U)$ of C^∞-differential forms admits the orthogonal decomposition*

$$\Omega_{cs}^p(U) = \text{Im}(d_{p-1}) \oplus \text{Im}(d_{p+1}^*) \oplus \mathcal{H}^p \tag{37}$$

for all $0 \le p \le n$. Moreover, the equation $\triangle_p\omega = \eta$ admits a solution in $\Omega_{cs}^p(U)$ if and only if $\eta \perp \mathcal{H}^p(U)$.

4.1 Hodge Theorem on Manifolds

The Hodge theorem extends over a closed[4] differentiable manifold M. A byproduct on an orientable manifold is the finiteness of the De Rham cohomology groups. The orientability allows us to define the bilinear map $\mathfrak{h} : H^p_{DR}(M) \times H^{n-p}_{DR}(M) \to \mathbb{R}$ given by

$$\mathfrak{h}\big([\phi], [\psi]\big) = \int_M \phi \wedge \psi. \tag{38}$$

Indeed, we get the following useful theorem;

Theorem 9 (Poincaré Duality) *Let M be an n-dimensional compact orientable manifold. The bilinear pairing* (38) *determines isomorphisms*

$$H^p_{DR}(M) \simeq \big(H^{n-p}_{DR}(M)\big)^*.$$

Proof Assume $H^p_{DR}(M) \neq 0$. Let $\omega \in \mathcal{H}^p$, $\omega \neq 0$. Since $*\Delta = \Delta *$, the $(n-p)$-form $*\omega$ is harmonic, and we have

$$\mathfrak{h}\big([\omega], [\omega]\big) = \int_M \omega \wedge *\omega = \mid \omega \mid^2_2 \neq 0.$$

The pairing being non-singular implies the isomorphism.   □

Exercises

(1) Let $\omega = f e_{i_1} \wedge e_{i_2} \wedge \cdots \wedge e_{i_p}$ be a p-form and let η be a q-form. Using Eq. 27, show that

$$(i)\ d(d\omega) = \sum_{i,j}[e_i, e_j]e_i \wedge e_j \wedge e_1 \wedge \cdots \wedge e_p = 0.$$

$$(ii)\ d(\omega \wedge \eta) = d\omega \wedge \eta + (-1)^p \omega \wedge d\eta.$$

(2) Considering the contraction operator $\llcorner: \Omega^p(U) \to \Omega^{p-1}(U)$ given by

$$v \llcorner \big(v_1 \wedge \cdots \wedge v_p\big) = \sum_{i+1}^{n}(-1)^{i+1} < v, v_i > v_1 \wedge \cdots \wedge \hat{v}_i \wedge \cdots \wedge v_p, \tag{39}$$

where $(\hat{\cdot})$ indicates deletion, show the following identities: let $\omega \in \Omega^p(U)$ and $\eta \in \Omega^q(U)$;

$$(i)\ v \llcorner (\omega \wedge \eta) = (v \llcorner \omega) \wedge \eta + (-1)^p \omega \wedge (v \llcorner \eta). \tag{40}$$

$$(ii)\ v \llcorner (v \llcorner \omega) = 0, \ \forall \omega. \tag{41}$$

---

[4]Compact and $\partial M = \varnothing$.

(3) Let $* : \Omega^p(U) \to \Omega^{n-p}(U)$ be the Hodge star-operator and $d^* = (-1)^{np+n+1} *$ $d*$. Show that

$$d^* = -\sum_{j=1}^{n} e_j \llcorner \partial_{e_j} \tag{42}$$

(4) Let M be an n-dimensional differentiable manifold. Explain the following:
    (i) The Hodge Theorem on M.
    (ii) The bilinear pairing (38) is well-defined.
    (iii) $H_{DR}^n(M) \simeq \mathbb{R}$.
    (iv) If $\omega \in \mathcal{H}^p$ and $\theta \in \mathcal{H}^q$, then not necessarily $\omega \wedge \theta \in \mathcal{H}^{p+q}$.
(5) Show that the Poincaré duality theorem implies

$$H_{DR}^p(M) \simeq H_{DR}^{n-p}(M).$$

5 Geometric Formulation of the Electromagnetic Theory

Electromagnetic Theory is governed by Maxwell's equations named in honor of the Scottish physicist James Clerk Maxwell (1831–1879) who discovered and published them in the early years of the decade of 1860. Maxwell's equations are partial differential equations. Electricity and Magnetism were considered distinct subjects until the Danish physicist Hans Christian Ørsted (1777–1851) discovered around 1819 that electrical currents create magnetic fields. Soon after, in 1821, André-Marie Ampère (1775–1836), a French physicist and mathematician explored Ørsted's discovery and formulated Ampère's Law. This began the unification of Electricity and Magnetism, originating in Electromagnetic Theory. The history of electromagnetism is rich and very important for developing the theory of Vector Calculus and Differential Forms.

The fundamentals of Electromagnetic Theory are described in terms of the concepts for the Electric Field $\vec{E}$ and the Magnetic Field $\vec{B}$. The notion of a "Field" was a revolutionary idea introduced by the English physicist Michael Faraday (1791–1867) in 1831; Faraday was one of the main developmental leaders who was also responsible for important applications of Electromagnetic Theory to the concept of the electric motor. Also Carl Friederich Gauss (1777–1855), Jean-Baptiste Biot (1774–1862) and Félix Savart (1791–1841) contributed to laying the fundamentals of the electromagnetic theory as we know it today.

From the physics point of view, Electromagnetism is governed by four laws (described below using the Gaussian system of units);

(1) **Gauss' Law** (electric charge). *The flow of the electric field through a closed surface is proportional to the charge contained within the surface.*
    Let $S \subset \mathbb{R}^3$ be a surface such that $S = \partial\Omega$ and Ω is a region in which the Stokes Theorem is valid. According to Gaussian Law, we have

$$\int_S < \vec{E}, n > dS = 4\pi Q, \tag{43}$$

where ϵ_0 is the electric constant. If the charge density inside the region Ω is ρ, then

$$\int_\Omega \operatorname{div}(\vec{E}) dV = 4\pi \int_\Omega \rho dV.$$

Considering that the law applies to any closed surface, we have

$$\operatorname{div}(\vec{E}) = 4\pi\rho, \quad \text{(1st Maxwell's eq.).} \tag{44}$$

(2) **Gauss' Law** (magnetic charge). *Since a magnetic monopole has never been observed in nature, what we observe is that the flux of the field $\vec{B}$ through a closed surface S is zero.*
This Law is motivated by the previous one. Consequently,

$$\operatorname{div}(\vec{B}) = 0, \quad \text{(2nd Maxwell's eq.).} \tag{45}$$

(3) **Faraday's Law**. *The induced electromotive force in any closed circuit is equal to the negative of the time rate of change of the magnetic flux enclosed by the circuit.*
The electromotive force is the line integral of the electric field along a closed curve. Let $S \subset \mathbb{R}^3$ be a surface such that $\partial S = \gamma$ and the Stokes Theorem is valid. The Law states that the electric current induced in γ by a magnetic field $\vec{B}$ is proportional to the flow Φ_B of the magnetic field through the surface S per unit of time. That is,

$$\int_\gamma \vec{E} = -\frac{d\Phi_B}{dt} = -\frac{1}{c} \int_S \frac{\partial \vec{B}}{\partial t} dS. \tag{46}$$

Applying the Stokes Theorem, we obtain

$$\nabla \times \vec{E} = -\frac{1}{c} \frac{\partial \vec{B}}{\partial t}, \quad \text{(3rd Maxwell's eq.).} \tag{47}$$

(4) **Ampère's Law**. *The integral around a closed path of the component of the magnetic field tangent to the direction of the path equals μ_0 times the current through the area within the path.*
To apply Ampère's Law, all currents have to be steady (i.e., they do not change with time). The currents have to be taken with their algebraic signs (those going "out" of the surface are positive, those going "in" are negative). Let S be a surface with a boundary that is a closed curve $\gamma = \partial S$ on which the Stokes theorem is valid.

$$\int_\gamma \vec{B} \, d\gamma = \frac{1}{c} 4\pi I, \tag{48}$$

Setting $\vec{J}$ as a current density in S, we get

$$\nabla \times \vec{B} = \frac{1}{c} 4\pi \vec{J}.$$

Maxwell found that there was a need to add the term $\frac{\partial \vec{E}}{\partial t}$ to Ampère's law, which became known as Ampère-Maxwell's Law as follows:

$$\int_{\gamma} \vec{B} d\gamma = \frac{1}{c} \left[4\pi \int_{S} \vec{J} \, dS + \int_{S} \frac{\partial \vec{E}}{\partial t} \right] dS. \qquad (49)$$

Therefore

$$\nabla \times \vec{B} = \frac{1}{c} \left[4\pi \vec{J} + \frac{\partial \vec{E}}{\partial t} \right], \quad \text{(4th Maxwell's eq.)}. \qquad (50)$$

Putting all the equations together, we get Maxwell's equations

$$
\begin{array}{ll}
(1) \ \nabla . \vec{B} = 0, & (3) \ \nabla . \vec{E} = 4\pi\rho, \\
(2) \ \nabla \times \vec{E} = -\frac{1}{c} \frac{\partial \vec{B}}{\partial t}, & (4) \ \nabla \times \vec{B} = \frac{1}{c} \left[4\pi \vec{J} + \frac{\partial \vec{E}}{\partial t} \right].
\end{array}
\qquad (51)
$$

5.1 Electromagnetic Potentials

Since we are considering the equations defined in $\mathbb{R}^3$, noting that De Rham cohomology is trivial here, we consider a region $\Omega \subset \mathbb{R}^3$, diffeomorphic to $\mathbb{R}^3$, and where the Poincaré Lemma and the Stokes Theorem are both valid.

It follows from the equation $\nabla . \vec{B} = 0$ that we have a vector field $\vec{A}_B = A_1 \vec{i} + A_2 \vec{j} + A_3 \vec{k}$, such that $\vec{B} = \nabla \times \vec{A}_B$. Consequently, Eq. (2) in (51) becomes

$$\nabla \times \left(E + \frac{1}{c} \frac{\partial \vec{A}_B}{\partial t} \right) = 0.$$

Applying the Poincaré Lemma, ϕ is such that $E + \frac{\partial \vec{A}_B}{\partial t} = -\nabla \phi$. Consequently

$$E = -\nabla \phi - \frac{1}{c} \frac{\partial \vec{A}_B}{\partial t}.$$

The vector $\vec{A}_B$ is the magnetic potential, the scalar ϕ is the electric potential, and the pair (A_B, ϕ) is the electromagnetic potential. Introducing these relations into Eq. (4) in (51), we get

$$\nabla \times \vec{B} = \nabla \times (\nabla \times \vec{A}_B) = \nabla(\nabla . \vec{A}_B) - \Delta \vec{A}_B =$$

$$= \frac{4\pi}{c} \vec{J} - \frac{1}{c} \nabla \frac{\partial \phi}{\partial t} - \frac{1}{c^2} \frac{\partial^2 \vec{A}_B}{\partial^2 t}.$$

Therefore

$$\nabla \left(\nabla . \vec{A}_B + \frac{1}{c^2} \nabla \frac{\partial \phi}{\partial t} \right) = \frac{4\pi}{c} \vec{J} + \Delta \vec{A}_B - \frac{1}{c^2} \frac{\partial^2 \vec{A}_B}{\partial^2 t}.$$

Definition 5 The D'Alembertian operator $\Box : C^2(\Omega) \rightarrow C^0(\Omega)$ is

$$\Box = -\Delta + \frac{1}{c^2} \frac{\partial^2}{\partial^2 t}. \tag{52}$$

Therefore

$$\Box \vec{A}_B = \frac{4\pi}{c} \vec{J} - \nabla \left(\nabla . \vec{A}_B + \frac{1}{c^2} \nabla \frac{\partial \phi}{\partial t} \right). \tag{53}$$

Let $f = f(x, t) : \mathbb{R}^3 \times \mathbb{R} \rightarrow \mathbb{R}$ be a C^1-function. Given $B = \nabla \times \vec{A}_B$ and $E = -\nabla\phi - \frac{\partial \vec{A}_B}{\partial t}$, we note that the potentials

$$A' = A_B + \nabla f \quad \text{and} \quad \phi' = \phi - \frac{\partial f}{\partial t},$$

satisfy the same equations. This invariance is called gauge invariance. As a result we have some freedom to choose the electromagnetic potential (A_B, ϕ). We select the Lorenz gauge, which will be further explained in the next section, and we have

$$\nabla . \vec{A}_B + \frac{1}{c^2} \nabla \frac{\partial \phi}{\partial t} = 0. \tag{54}$$

The equations for the potentials are now

$$\Box \vec{A}_B = \frac{4\pi}{c} \vec{J}, \quad \Box \phi = 4\pi\rho. \tag{55}$$

5.2 Geometric Formulation

Let's consider Maxwell's equation in the absence of the sources of magnetic ($\vec{J} = 0$) and electric fields ($\rho = 0$). We will not worry about the unit systems or the physical constants. In classical electromagnetism, ruled by Maxwell's equations, we consider $\mathbb{R}^4 = \mathbb{R}^3 \times \mathbb{R}$ (Space$\times$Time) provided with the Lorentz pseudo-product (Hendrik Lorentz, 1853–1928),

$$L = \begin{pmatrix} 1 & 0 & 0 & 0 \\ 0 & 1 & 0 & 0 \\ 0 & 0 & 1 & 0 \\ 0 & 0 & 0 & -1 \end{pmatrix}.$$

We will use the apparatus of differential forms to obtain Maxwell's equations. We start by considering the 1-form

$$A = A_1(p, t)dx + A_2(p, t)dy + A_3(p, t)dz - \phi(p, t)dt,$$

where $p = (x, y, z) \in \mathbb{R}^3$ is the spacial variables and t is the time. Considering the functions

$$B_x = \frac{\partial A_3}{\partial y} - \frac{\partial A_2}{\partial z}, \quad B_y = \frac{\partial A_1}{\partial z} - \frac{\partial A_3}{\partial x}, \quad B_z = \frac{\partial A_2}{\partial x} - \frac{\partial A_1}{\partial y}.$$

$$E_x = -\frac{\partial \phi}{\partial x}, \quad E_y = -\frac{\partial \phi}{\partial y}, \quad E_z = -\frac{\partial \phi}{\partial z},$$

associated to the magnetic field $\vec{B} = \nabla \times \vec{A} = (B_x, B_y, B_z)$, we have the differential 2-form

$$B = B_z dx \wedge dy + B_x dy \wedge dz + B_y dz \wedge dx,$$

and associated to the electric field $\vec{E} = (E_x, E_y, E_z) = -\nabla\phi$, we have the differential 1-form

$$E = E_x dx + E_y dy + E_z dz.$$

Definition 6 The electromagnetic tensor associated to the electromagnetic potential A is the differential 2-form $F_A = dA$.

Now we have

$$F_A = B + E \wedge dt. \tag{56}$$

Since $d^2 = 0$, we get

$$0 = dF_A = dB + dE \wedge dt = \mathrm{div}(\vec{B})dx \wedge dy \wedge dz + \frac{\partial \vec{B}}{\partial t} \wedge dt + \mathrm{curl}(\vec{E}) \wedge dt =$$

$$= \mathrm{div}(\vec{B})dx \wedge dy \wedge dy + \sum_{i<j} \left(\frac{\partial \vec{B}}{\partial t} + \mathrm{curl}(E) \right) \wedge dt.$$

Therefore we have derived Maxwell's equations (1) and (2) in (51):

$$(1)\ \mathrm{div}(\vec{B}) = 0, \quad (2)\ \mathrm{curl}(\vec{E}) + \frac{\partial \vec{B}}{\partial t} = 0.$$

5.3 *Variational Formulation*

To obtain the other pair of equations, a variational formulation is required since at some point nature optimizes its costs. The admissible space for the variational problem is the Banach space of 1-forms $\Omega^1(\Omega)$ provided with the standard $L^{1,2}$ Sobolev norm $|| \cdot ||_{L^{1,2}}$, that is,

$$\mathcal{A} = \{A \in \Omega^1(\Omega); \ || A ||_{L^{1,2}} < \infty\}.$$

Definition 7 The energy functional $E : \mathcal{A} \to \mathbb{R}$ of the electromagnetic field is

$$E(A) = \frac{1}{2} \int_{\Omega} || F_A ||^2 \, dV. \tag{57}$$

To derive the Euler-Lagrange equations, let $\omega \in \Omega^1(\Omega)$ and $F_{A+t\omega} = F_A + t d\omega$, so

$$dE_A.\omega = \frac{1}{2} \lim_{t \to 0} \left(\frac{E(A + t\omega) - E(A)}{t} \right) = \frac{1}{2} \lim_{t \to 0} \int_{\Omega} \left(\frac{< F_{A+t\omega}, F_{A+t\omega}}{t} > \right) =$$

$$= \frac{1}{2} \int_{\Omega} < F_A d\omega > + < d\omega, F_A > = \int_{\Omega} < d^* F_A, \omega > .$$

The Fréchet derivative of the energy functional at $A \in \mathcal{A}$ is

$$dE_A.\omega = \int_{\Omega} < d^* F_A, \omega > dV.$$

Consequently the critical points satisfy the EL-equation

$$d^* F_A = 0. \tag{58}$$

Since $d^* = *d*$ (Eq. 38), the EL-equation is equivalent to $d(*F_A) = 0$. The Hodge star-operator induced by the Lorentz form L satisfies $*^2 = (-1)^{1+n(n-k)}$, so $* : \Omega^2(\mathbb{R}^4) \to \Omega^2(\mathbb{R}^4)$ satisfies $*^2 = -1$. In this case, the eigenvalues are $\pm i$ and the respective eigenspaces are $\Omega^2_+(\mathbb{R}^4)$ (self-dual) and $\Omega^2_-(\mathbb{R}^4)$ (anti-self-dual). The space $\Omega^2(\mathbb{R}^4)$ decomposes as

$$\Omega^2(\mathbb{R}^4) = \Omega^2_+(\mathbb{R}^4) \oplus \Omega^2_-(\mathbb{R}^4). \tag{59}$$

The electromagnetic field F_A decomposes as $F_A = F_A^+ + F_A^-$, where $*F_A = \pm i F_A$. Let's analyze the case when the solutions to Eq. (58) are self-dual or anti-self-dual;

(1) A is self-dual: $*F_A = i F_A$.

   In this case, we get the equation $d F_A = 0$ which is equivalent to Maxwell's 1st and 2nd equations.

(2) A is anti-self-dual: $*F_A = -i F_A$.

   Analogous to the self-dual case.

(3) $d(*F_A) = 0$.

   Let's find each component of $d(*F_A) = d(*B) + d\big(* (E \wedge dt)\big)$. Since

$$*B = B_x dx \wedge dt + B_y dy \wedge dt + B_z dz \wedge dt,$$

the contribution due to the 2-form B is

$$d(*B) = \left[\left(\frac{\partial B_y}{\partial x} - \frac{\partial B_x}{\partial y}\right) dx \wedge dy + \left(\frac{\partial B_z}{\partial y} - \frac{\partial B_y}{\partial z}\right) dy \wedge dz + \left(\frac{\partial B_x}{\partial z} - \frac{\partial B_z}{\partial x}\right) dz \wedge dx\right] \wedge dt.$$

We set $d(*B) = (\nabla \times B) \wedge dt$. The contribution from the Electric Field component is

$$d\big(* (E \wedge dt)\big) = \operatorname{div}(\vec{E}) dx \wedge dy \wedge dz + \left[\frac{\partial E_x}{\partial t} dy \wedge dz + \frac{\partial E_y}{\partial t} dz \wedge dx + \frac{\partial E_z}{\partial t} dx \wedge dy\right] \wedge dt.$$

By setting $d(*(E \wedge dt)) = \operatorname{div}(\vec{E}) dx \wedge dy \wedge dz + \frac{\partial \vec{E}}{\partial t} \wedge dt$, the equation $d^* F_A = 0$ becomes

$$\operatorname{div}(\vec{E}) dx \wedge dy \wedge dz + \left(\operatorname{curl}(\vec{B}) + \frac{\partial \vec{E}}{\partial t}\right) \wedge dt = 0.$$

Hence

$$\operatorname{div}(\vec{E}) = 0, \ \operatorname{curl}(\vec{B}) + \frac{\partial \vec{E}}{\partial t} = 0.$$

In this case, in the absence of electromagnetic sources, Maxwell's equations (51) are written using the formalism of Differential Forms as

$$\begin{aligned} d F_A &= 0, \\ d^* F_A &= 0. \end{aligned} \tag{60}$$

If F_A is a solution to Maxwell's equation above, then F_A is a harmonic 2-form since $\triangle F_A = (dd^* + d^*d)(F_A)$.

As we have mentioned before, electromagnetic theory has a gauge invariance. Given a differentiable function f, the 1-form $A' = A + df$ satisfies $F_{A'} = F_A$. Indeed, $B' = B$ and $E' = E$. The component df belongs to the image of $d : \Omega^0 \to \Omega^1$. Since $\Omega^1(\Omega)$ decomposes (orthogonal) as

$$\Omega^1(\Omega) = \operatorname{Im}\Big(d : \Omega^0(\Omega) \to \Omega^1(\Omega)\Big) \oplus \operatorname{Ker}\Big(d^* : \Omega^1(\Omega) \to \Omega^0(\Omega)\Big),$$

we consider a representative in the $\text{Ker}(d^*)$ of the gauge class of a 1-form A. Given a 1-form $A = A_1(x, t)dx + A_2(x, t)dy + A_3(x, t)dz - \phi(x, t)dt$, we would like to find a 1-form A' gauge equivalent to A, i.e., to find a 0-form f such that

$$A' = A + df, \qquad d^*A' = 0.$$

This is equivalent to solving the equation $\triangle f = -d^*A$. Thanks to the Hodge Theorem there is such an f. Since $d^*(A) = \text{div}(A)$, where the divergent operator is written with respect to the Lorentz form, we get

$$d^*A = 0 \quad \Leftrightarrow \quad \text{div}(A) = \nabla.\vec{A}_B + \frac{\partial \phi}{\partial t} = 0. \tag{61}$$

The Lorenz gauge is defined by the element A satisfying the equation $d^*A = 0$ among all the 1-forms gauge equivalent to A. This explain the choice taken in Eq. (54).

Gauge invariance is a fundamental principle in the physical models of the Elementary Interactions of nature. Electromagnetism is the oldest example of Gauge Theory, and in Mathematics it corresponds to the Differential Geometry of Vector Fiber Bundles. In Geometry, one of the most fundamental concepts is that of Curvature, which measures how much the space moves away, or curves from a plane. The electromagnetic tensor is an example of a Curvature tensor and the electromagnetic potential A is the 1-form connection with a curvature that is the 2-form $F_A = dA$.

After reading this section, we invite the reader to make a comparison with Yang-Mills Theory, which was briefly introduced in Chap. 3, Example (4), when we introduced the Moduli Spaces of Instantons.

Exercises

(1) Show that the matrix representation of the Electromagnetic Field is given in terms of the fields $\vec{B} = (B_x, B_y, B_z)$ and $\vec{E} = (E_x, E_y, E_z)$, by

$$F_A = \begin{pmatrix} 0 & B_z & -B_y & E_x \\ -B_z & 0 & B_x & E_y \\ B_y & -B_x & 0 & E_z \\ -E_x & -E_y & -E_z & 0 \end{pmatrix}.$$

(2) Show that the Laplacian $\triangle_L$ with respect to the Lorentz form L is the D'Alembertian $-\square$, i.e.,

$$\triangle_L f = -\square f.$$

(3) Show that the electric field and the magnetic field satisfy the equations

$$\square E = 0, \quad \square B = 0.$$

(4) Show that the electric potential and the magnetic potential satisfy the equations

$$\Box A_B = 0, \quad \Box\phi = 0.$$

(5) Show that Maxwell's equation in the presence of sources are

$$d F_A = 0,$$
$$d^* F_A = 4\pi J,$$

where J is a 1-form depending on the charge density ρ and the current density $\vec{J}$.

(6) The Biot-Savart law named after the French physicists Jean Baptiste Biot and Felix Savart specifies the magnetic field generated by a steady distributed current $\vec{j}$ in a region $\Omega \subset \mathbb{R}^3$ as

$$B(r) = \frac{1}{4\pi} \int_\Omega \vec{j}(r') \times \frac{(\vec{r} - \vec{r'})}{|r - r'|^3} \, dV. \tag{62}$$

To show the formula above, follow these steps;

(a) Use the 2nd Green identity and the harmonic function $v(r) = \frac{1}{r}$ in $\mathbb{R}^3 \backslash \{0\}$ to prove that the solution to the equation

$$\triangle \vec{A}_B = 4\pi \vec{j}$$

is $\vec{A}(r) = \frac{1}{4\pi} \int_\Omega \frac{\vec{j}}{|r-r'|} \, dV.$

(b) Use the identity $\vec{B} = \nabla \times \vec{A}_B.$

(c) Using the same arguments, show that

$$\vec{E}(r) = \frac{1}{4\pi} \int_\Omega \rho(r') \frac{\vec{r} - \vec{r'}}{|r - r'|^3} \, dV.$$

(7) Consider a current being carried along a straight wire directed along the z-axis. Show the magnetic field generated by the current is

$$\vec{B} = \frac{I}{2\pi}\left(-y\vec{i} + x\vec{j}\right).$$

Show that the vector potential is

$$\vec{A}_B = -\frac{I}{2\pi} \ln(x^2 + y^2)\vec{k}.$$

6 Helmholtz's Decomposition Theorem

In this section, we will ignore analyzing formalisms and techniques that are required to be accurate and rigorous; let's assume jurisprudence in Advanced Calculus that our arguments are acceptable, as they are for physicists.

The Helmholtz theorem is an interesting application, and very useful in electromagnetism. It certainly precedes the Hodge Theorem in time. Let $\vec{F}$ be a vector field and let $\Omega \subset \mathbb{R}^3$ be a domain with $S = \partial\Omega$ in which the Stokes Theorem and Poincaré's Lemma are both valid.

Theorem 10 (Helmholtz) *Let $\vec{F} \in C^2(\Omega; \mathbb{R}^3)$ be a vector field, so we have $\phi \in C^1(\Omega, \mathbb{R})$ and $A \in C^1(\Omega; \mathbb{R}^3)$ such that $\vec{F} = -\nabla(\phi) + \nabla \times A$, where*

$$
\begin{aligned}
\phi(r) &= \frac{1}{4\pi} \int_\Omega \frac{\nabla'.\vec{F}(r')}{|r - r'|} dV' - \frac{1}{4\pi} \int_S \frac{\vec{F}(r)}{|r - r'|}.n' dS', \\
A(r) &= \frac{1}{4\pi} \int_\Omega \frac{\nabla' \times \vec{F}(r')}{|r - r'|} dV' - \frac{1}{4\pi} \int_S \frac{\vec{F}(r)}{|r - r'|}.n' dS'.
\end{aligned}
\tag{63}
$$

That is, $\vec{F}$ is the sum of one component with null curl and another component with null divergence.

The Helmholtz theorem can be understood using differential forms. Let $\vec{F} = (F_x, F_y, F_y)$ and $F = F_x dx + F_y dy + F_z dz$. In light of Functional Analysis, the operators $d : \Omega^0(\mathbb{R}^3) \to \Omega^1(\mathbb{R}^3)$ and $d^* : \Omega^1(\mathbb{R}^3) \to \Omega^0(\mathbb{R}^3)$, the space $\Omega^1(\mathbb{R}^3)$ admits the orthogonal decomposition

$$
\Omega^1(\mathbb{R}^3) = \text{Img}(d) \oplus \text{Ker}(d^*).
$$

As seen in Chap. 6, we have $df = \nabla f$ and $d^* F = \text{div}(\vec{F}).1$. Theorem 10 is a consequence of the Hodge Theorem.

To prove Helmholtz's theorem using simple calculus tools, we will use Dirac's δ-function, which is quite familiar to physicists and unfamiliar to mathematicians.[5] We will leave some exercises to help the reader become familiar with Dirac's δ-function.

Proof Since $\triangle\left(\frac{1}{|r-r'|}\right) = -4\pi\delta^3(r - r')$, we have

$$
\vec{F}(r) = \int_\Omega \vec{F}(r').\delta^3(r - r') dV' = -\frac{1}{4\pi} \int_\Omega \vec{F}(r').\triangle\left(\frac{1}{|r-r'|}\right) dV' = -\frac{1}{4\pi}\triangle\left(\int_\Omega \frac{\vec{F}(r')}{|r - r'|} dV'\right).
$$

Applying the identities

$(i1)\ \triangle(\vec{v}) = \nabla(\nabla.\vec{v}) - \nabla \times (\nabla \times \vec{v}),\ (i2)\ \nabla \times (f\vec{v}) = -\vec{v} \times \nabla f + f\nabla \times \vec{v},$
$(i3)\ \nabla.(f\vec{v}) = <\nabla f, \vec{v}> + f\nabla.\vec{v}$

---

[5]It is an example in the content of Distribution Theory.

we get

$$\vec{F}(r) \overset{(i1)}{=} -\frac{1}{4\pi}\nabla\left[\nabla.\left(\int_\Omega \frac{\vec{F}(r')}{|r-r'|}dV'\right)\right] + \frac{1}{4\pi}\nabla\times\left[\nabla\times\left(\int_\Omega \frac{\vec{F}(r')}{|r-r'|}dV'\right)\right] \overset{(i2)+(i3)}{=}$$

$$= -\frac{1}{4\pi}\nabla\left[\int_\Omega < \vec{F}(r'), \nabla\left(\frac{1}{|r-r'|}\right) > dV'\right] - \frac{1}{4\pi}\nabla\times\left[\int_\Omega \vec{F}(r')\times\nabla\left(\frac{1}{|r-r'|}>\right)dV'\right].$$

Noting that $\nabla'\left(\frac{1}{|r-r'|}\right) = -\nabla\left(\frac{1}{|r-r'|}\right)$, then

$$\vec{F}(r) = \frac{1}{4\pi}\nabla\left[\int_\Omega < \vec{F}(r'), \nabla'\left(\frac{1}{|r-r'|}\right) > dV'\right] + \frac{1}{4\pi}\nabla\times\left[\int_\Omega \vec{F}(r')\times\nabla'\left(\frac{1}{|r-r'|}>\right) > dV'\right].$$

By resorting to identities (i1), (i2) and (i3), we have the equation

$$\vec{F}(r) = -\nabla\left[\frac{1}{4\pi}\int_\Omega \frac{\nabla'.\vec{F}(r')}{|r-r'|}dV' - \frac{1}{4\pi}\int_\Omega \nabla'.\left(\frac{\vec{F}(r')}{|r-r'|}\right)dV'\right] +$$

$$+ \nabla\times\left[\frac{1}{4\pi}\int_\Omega \frac{\nabla'\times\vec{F}(r')}{|r-r'|}dV' - \frac{1}{4\pi}\int_\Omega \nabla'\times\left(\frac{\vec{F}(r')}{|r-r'|}\right)dV'\right].$$

Applying the Stokes Theorem, we obtain

$$\vec{F}(r) = -\nabla\left[\frac{1}{4\pi}\int_\Omega \frac{\nabla'.\vec{F}(r')}{|r-r'|}dV' - \frac{1}{4\pi}\int_S \frac{\vec{F}(r')}{|r-r'|}dV'\right] +$$

$$+ \nabla\times\left[\frac{1}{4\pi}\int_\Omega \frac{\nabla'\times\vec{F}(r')}{|r-r'|}dV' - \frac{1}{4\pi}\int_S \hat{n}\times\left(\frac{\vec{F}(r')}{|r-r'|}\right)dV'\right],$$

where $\hat{n}$ is the unit normal vector to $S = \partial\Omega$.   □

Considering $\Omega = \mathbb{R}^3$ and assuming $\lim_{r\to\infty}\frac{\vec{F}(r')}{|r-r'|^2} = 0$, the decomposition is simplified to

$$\vec{F}(r) = -\nabla\left[\frac{1}{4\pi}\int_\Omega \frac{\nabla'.\vec{F}(r')}{|r-r'|}dV'\right] - \nabla\times\left[\frac{1}{4\pi}\int_\Omega \frac{\nabla'\times\vec{F}(r')}{|r-r'|}dV'\right].$$

Exercises

(1) Extend the Helmholtz Theorem to a region $\Omega \subset \mathbb{R}^n$.
(2) (Dirac's δ-function) Consider the sequence of functions $\{f_k\}_{k\in\mathbb{N}}$, where f_k : $\mathbb{R}^n \to \mathbb{R}$ is given by $f_k(x) = \sqrt{\frac{k}{\pi}}e^{-k|x|^2}$.

(a) Show $\sup_{x\in\mathbb{R}^n} | f_k(x) |= \sqrt{\frac{k}{\pi}}$ and get to the conclusion

$$\lim_{k\to\infty} f_k(x) = \begin{cases} 0, & \text{if } x \neq 0, \\ \infty, & \text{if } x = 0. \end{cases}$$

(b) Fix $x_0 \in \mathbb{R}^n$ and consider $\delta^n(x - x_0) = \lim_{k\to\infty} f_k(x - x_0)$. Show that

$$\int_{\mathbb{R}^n} \delta^n(x)dx = 1, \quad \int_{\mathbb{R}^n} f(x)\delta^n(x - x_0)dx = f(x_0).$$

We denote $\delta(x) = \delta^1(x)$.

(c) In $\mathbb{R}^n$, prove the identity

$$\Delta\left(\frac{1}{| x - x_0 |^{n-2}}\right) = -A_{n-1}.\delta^n(x - x_0),$$

where A_{n-1} is the volume of the sphere S^{n-1}.

(d) Fix a fixed point $a = (a_1, \ldots, a_n)$ in $\mathbb{R}^n$ and let $x = (x_1, \ldots, x_n)$ be an arbitrary point. Show

$$\delta^n(x - a) = \delta(x_1 - a_1) \ldots \delta(x_n - a_n).$$

(e) Prove the following identities: assume $n = 1$ and $a, \lambda \in \mathbb{R}$
   (i) $\delta(-x) = \delta(x)$.
   (ii) $\delta(\lambda x) = \frac{\delta(x)}{|\lambda|}$.
   (iii) $\delta(\lambda x - a) = \frac{\delta(x - \frac{a}{\lambda})}{|\lambda|}$.
   (iv) $\delta(x^2 - a^2) = \frac{\delta(x-a)+\delta(x+a)}{2|a|}$.
   (v) Let $\{x_n \mid f(x_n) = 0\}$ be the set of zeros of $f : \mathbb{R} \to \mathbb{R}$. Assuming $f'(x_n) \neq 0$, then

$$\delta(f(x)) = \sum_n \frac{\delta(x - x_x)}{| f'(x_n) |}.$$

(3) Let $v(r) = \frac{1}{r}$. Show that

$$\Delta v = -4\pi\delta(r) = \begin{cases} 0, & \text{if } r \neq 0, \\ \infty, & \text{if } r = 0. \end{cases}$$

Extend the result when $v(r) = \frac{1}{r^{n-2}}$ is a harmonic function is $\mathbb{R}^n$.

Remark Due to its value at $x = 0$, Dirac's δ-function is not a function, it is a distribution. The distributions are generalizations of functions; they are naturally treated in the theory of distributions or in context of Sobolev spaces.

Appendix A
Basics of Analysis

The contents of Appendix A contains basic definitions and theorems to support reading Chaps. 1–7. Appendix A should be used only as a reference source along with the References.

1 Sets

Let C be a set, and consider $C^k = C \times \cdots \times C$ the Cartesian product of k-copies of C. Consider $C_n = \{1, 2, 3, \ldots, n\}$ and $C_n^k = C_n \times \overset{k}{\ldots} \times C_n$. C_n^k is a multi-index set of length k. For example, an element a_I, $I = (i_1, i_2, \ldots, i_k) \in C_n^k$ means $a_I = a_{i_1 i_2 \ldots i_k}$.

The Kronecker delta is

$$\delta_{IJ} = \begin{cases} 0, & \text{if } I \neq J, \\ 1, & \text{if } I = J. \end{cases}$$

2 Finite-dimensional Linear Algebra: $V = \mathbb{R}^n$

Let V be a real vector space. We say that V is a vector space of dimension n and denote $\dim(V) = n$ whenever V admits a finite basis $\beta = \{e_1, \ldots, e_n\}$. A theorem ensures that any finitely generated vector space admits a finite basis, i.e., for every vector $u \in V$, we have coefficients $u_1, \ldots, u_n$ such that u is the linear combination of $u = \sum_{i=1}^n u_i e_i$, $u_i \in \mathbb{R}$.

© Springer Nature Switzerland AG 2021
C. M. Doria, *Differentiability in Banach Spaces, Differential Forms and Applications*,
https://doi.org/10.1007/978-3-030-77834-7

2.1 Matrix Spaces

The set of matrices $n \times m$ with real entries is denoted by $M_{n,m}(\mathbb{R})$ and $M_n(\mathbb{R})$ whenever $n = m$. The following sets deserve highlights:

(i) The group of invertible matrices $GL_n(\mathbb{R}) = \{A \in M_n(\mathbb{R}) \mid A \text{ is invertible}\}$.
(ii) The group of special matrices $SL_n(\mathbb{R}) = \{A \in GL_n(\mathbb{R}) \mid \det(A) = 1\}$.
(iii) The orthogonal group of matrices is $O_n = \{A \in GL_n(\mathbb{R}) \mid A^t = A^{-1}\}$ and the special orthogonal group is $SO_n = \{A \in O_n \mid \det(A) = 1\}$.
(iv) The vector space $S_n = \{A \in M_n(\mathbb{R}) \mid A^t = A\}$ of symmetric matrices.
(v) The vector space $A_n = \{A \in M_n(\mathbb{R}) \mid A^t = -A\}$ of skew-symmetric matrices. The vector space $M_n(\mathbb{R})$ splits as $M_n(\mathbb{R}) = S_n \oplus A_n$ since any matrix $A \in M_n(\mathbb{R})$ can be decomposed into

$$A = \frac{A + A^t}{2} + \frac{A - A^t}{2}.$$

When the coefficient field is the complex numbers $\mathbb{C}$, we have the following sets of matrices $M_{n,m}(\mathbb{C})$, $M_n(\mathbb{C})$, $GL_n(\mathbb{C})$, $SL_n(\mathbb{C})$. Instead of symmetric or skew-symmetric matrices, now we have the subspace of Hermitian matrices $H_n = \{A \in M_n(\mathbb{C}) \mid A^* = A\}$ and the subspace of anti-Hermitians $AH_n = \{A \in M_n(\mathbb{C}) \mid A^* = -A\}$, $A^* = \bar{A}^t$ $(A^* = \bar{A}^t)$. Similar to the real case, we have the decomposition $M_n(\mathbb{C}) = H_n \oplus AH_n$.

2.2 Linear Transformations

Let V and W be vector spaces over a field $\mathbb{K} = \mathbb{R}$ or $\mathbb{C}$. A linear transformation $T : V \to W$ is a map such that for any $a, b \in \mathbb{R}$ and $u, v \in V$, we have

$$T(au + bv) = aT(u) + bT(v).$$

(1) Matrix Representation
   By fixing a basis $\alpha = \{e_1, \ldots, e_n\}$ of V, we can associate to a linear transformation $T : V \to W$ a matrix $A = [T]_\alpha$, as shown next;

$$T(e_1) = a_{11}e_1 + \cdots + a_{n1}e_n = \sum_{l=1}^{n} a_{l1}e_l,$$

$$T(e_2) = a_{12}e_1 + \cdots + a_{n2}e_n = \sum_{l=1}^{n} a_{l2}e_l, \qquad \Rightarrow T = \begin{pmatrix} a_{11} & a_{12} & \cdots & a_{1n} \\ a_{21} & a_{22} & \cdots & a_{2n} \\ \vdots & \vdots & \vdots & \vdots \\ a_{n1} & a_{n2} & \cdots & a_{nn} \end{pmatrix} = [T]_\alpha,$$

$$\vdots$$

$$T(e_n) = a_{1n}e_1 + \cdots + a_{nn}e_n = \sum_{l=1}^{n} a_{ln}e_l.$$

(2) **Change of Basis**

Let V be a vector space of dimension n and $T : V \to V$ a linear transformation. Given the bases $\alpha = \{e_1, \ldots e_n\}$ and $\beta = \{u_1, \ldots, u_n\}$ of V and a vector $v \in V$, the matrix taking the representation $[v]_\beta$ to the representation $[v]_\alpha$ is the matrix $P = (p_{ij})$ with entries defined as follows;

$$u_1 = p_{11}e_1 + \cdots + p_{n1}e_n = \sum_{l=1}^{n} p_{l1}e_l,$$

$$u_2 = p_{12}e_1 + \cdots + p_{n2}e_n = \sum_{l=1}^{n} p_{l2}e_l, \qquad \Rightarrow P = \begin{pmatrix} p_{11} & p_{12} & \cdots & p_{1n} \\ p_{21} & p_{22} & \cdots & p_{2n} \\ \vdots & \vdots & \vdots & \vdots \\ p_{n1} & p_{n2} & \cdots & p_{nn} \end{pmatrix} = [P]_\beta^\alpha.$$

$$\vdots$$

$$u_n = p_{1n}e_1 + \cdots + p_{nn}e_n = \sum_{l=1}^{n} p_{ln}e_l.$$

In this case, we say that the matrix $[P]_\beta^\alpha$ takes the basis β to the basis α since

$$[u_i]_\alpha = \begin{pmatrix} p_{1i} \\ p_{2i} \\ \vdots \\ p_{ni} \end{pmatrix} = P. \begin{pmatrix} 0 \\ 0 \\ \vdots \\ 1 \\ \vdots \\ 0 \end{pmatrix} = P_\beta^\alpha.[u_i]_\beta.$$

Assume α is the canonical basis and let β be a basis in which the vectors are eigenvectors of T. In this way, the matrix P is the matrix with column vectors exactly as the vectors in β; $[v]_\alpha = P.[v]_\beta$; that is, P takes a vector represented in β to its representation in the canonical basis α. Let's find the relationship between the matrices $A = [T]_\alpha$ and $B = [T]_\beta$ representing T, respectively. So

$$T(e_i) = \sum_{l=1}^{n} a_{li} e_l, \quad T(e_i) = \sum_{l=1}^{n} a_{li} e_l.$$

However,

$$T(u_i) = T\left(\sum_l p_{li} e_l\right) = \sum_l p_{li} T(e_l) = \sum_l p_{li} \sum_k a_{kl} e_k = \tag{1}$$

$$= \sum_k \left(\sum_l a_{kl} p_{li}\right) e_k. \tag{2}$$

Therefore

$$AP = PB \iff B = P^{-1} AP.$$

Consider $\mathcal{L}(\mathbb{R}^n; \mathbb{R}^m)$ the space of linear transformations $T : \mathbb{R}^n \to \mathbb{R}^m$.

2.3 Primary Decomposition Theorem

For simplicity, let's assume V is a $\mathbb{C}$-vector space and let $T : V \to V$ be a linear map. Consider the characteristic polynomial of T to be given by the product

$$p_c(t) = \prod_{i=1}^{k} (t - \lambda_i)^{n_i},$$

with $\lambda_1, \ldots, \lambda_k$ being the distinct eigenvalues of T, and n_i are positive integers such that

$$n_1 + \cdots + n_k = \dim(V).$$

We define the generalized eigenspace of T belonging to λ_i to be the subspace

$$E(T, \lambda_i) = \mathrm{Ker}\big(T - \lambda_i I\big)^{n_i} \subset V.$$

The following theorem[1] shows that the space V can be decomposed into T-invariant subspaces.

Theorem 1 (Primary Decomposition) *Let V be a $\mathbb{C}$-vector space and let $T : V \to V$ be a linear map; otherwise V is real and T has real eigenvalues. Then V admits the direct sum decomposition*

---

[1]If V is an $\mathbb{R}$-vector space, there are more cases to be considered.

$$V = \bigoplus_{i=1}^{r} E(T, \lambda_i).$$

Indeed, $n_i = \dim\big(E(T, \lambda_i)\big).$

2.4 Inner Product and Sesquilinear Forms

An inner product on a vector space V is a bilinear map $< .,. >: V \times V \to \mathbb{R}$ satisfying the following properties: for any $u, v \in V$,

(i) $< u, v > = < v, u >.$
(ii) $< u, u > \leq 0$, otherwise if $< u, u > = o$, then $u = 0$.
(iii) $|< u, v >| \leq |u| . |v|$.

Given a basis $\beta = \{e_1, \dots, e_n\}$ of V, we get the matrix $G = (g_{ij}^{\beta})$ associated to $< .,. >$ with entries $g_{ij}^{\beta} = < e_i, e_j >$, and so we have $< u, v > = u^t . G . v$ for all u, v. Condition (i) implies that G is a symmetric matrix $(G^t = G)$, and so it is diagonalizable. Let $u \in V$ be an eigenvector associated with the eigenvalue λ; it follows that

$$< u, u > = u^t \cdot G \cdot u = \lambda u^t \cdot u.$$

Therefore we have $\lambda > 0$ and consequently $\det(G) > 0$.

An inner product on V induces an inner product on every vector subspace of V. Then, restricting to the subspace V_{ij}, generated by the vectors $\beta_{ij} = \{e_i, e_j\}$, the inner product is defined as

$$< u, v > = u^t \begin{pmatrix} g_{ii}^{\beta} & g_{ij}^{\beta} \\ g_{ji}^{\beta} & g_{jj}^{\beta} \end{pmatrix} = u^t G(i, j).v .$$

Consequently, $\det\big(G^{\beta}(i, j)\big) = g_{ii}^{\beta} g_{jj}^{\beta} - (g_{ij}^{\beta})^2 > 0$. From now on, the index β will be ignored unless it is necessary to convey information.

Definition 1 A basis $\beta = \{e_1, \dots, e_n\}$ on V is orthonormal if the matrix of $< .,. >$ associated to β is the identity matrix, that is, $g_{ij} = \delta_{ij}$.

Theorem 2 *Let* $(V, < .,. >)$ *be a finite dimensional vector space endowed with an inner product. So we have a basis* $\beta = \{e_1, \dots, e_n\}$ *such that* $< e_i, e_j > = \delta_{ij}$.

An orthonormal basis is useful to perform computations. In some cases, as in Differential Geometry or General Relativity, there is a need to use a non-orthogonal basis.

The concept of a vector space with an inner product extends to complex vector spaces as we defined next.

Definition 2 A bilinear form $< ., . >: V \times V \to \mathbb{C}$ is symmetric sesquilinear, or Hermitian, if it satisfies the following conditions; for all $u, v \in V$

(i) $< v, u > = \overline{< u, v >}$.
(ii) $< u, u > \geq 0$; if $< u, u > = 0$, then $u = 0$.
(iii) $|< u, v >| \leq | u | \cdot | v |$.

From (i), for any $a, b \in \mathbb{C}$ and $u, v \in V$, $< au, bv >= a\bar{b} < u, v >$. Let V be a finite dimensional vector space over $\mathbb{C}$ and fix a basis $\beta = \{e_1, \ldots, e_n\}$, so there is a complex matrix H such that

$$< u, v >= u^* \cdot H \cdot v, \quad u^* = (\bar{u})^t.$$

Also from (i), we get $H^* = H$, and $H^* = \overline{H}^t$.

2.5 The Sylvester Theorem

Let V be a real vector space. A bilinear form $B : V \times B \to \mathbb{R}$ is symmetric if $B(u, v) = B(v, u)$, for all $u, v \in V$. In this case, associated with B, there is the quadratic form $Q : V \to \mathbb{R}$, $Q(u) = B(u, u)$. Reciprocally, given a quadratic form Q, we define the bilinear form B as

$$B(u, v) = \frac{1}{2}\left[Q(u + v) - Q(u) - Q(v)\right]. \tag{3}$$

Next, we will be working with the quadratic form instead of the bilinear form. Given a basis $\beta = \{e_1, \ldots, e_n\}$ of V, Q is represented as $Q^\beta = (q_{ij})$, $q_{ij} = B(e_i, e_j)$. We get $B(u, v) = v^t \cdot Q^\beta \cdot u$. The matrix Q^β being symmetric allows us to consider its diagonal form. Assume Q is non-degenerate, that is, $Q(u) \neq 0$ whenever $u \neq 0$. So the eigenvalues of Q are all non-null. The spectrum of Q is, by definition, the set $\sigma(Q) = \{\lambda \in \mathbb{C} \mid Q.u = \lambda.u\}$. Since Q is symmetric, it follows that $\sigma(Q) \subset \mathbb{R}$. Considering the Q-invariant subspaces

$$V_Q^+ = \{\lambda \in \sigma(Q) \mid \lambda > 0\}, \quad V_Q^- = \{\lambda \in \sigma(Q) \mid \lambda < 0\}, \tag{4}$$

we get the decomposition $V = V_Q^+ \oplus V_Q^-$.

Definition 3 Let V be a real vector space of dimension n and let $Q : V \to \mathbb{R}$ be a non-degenerate quadratic form. Let $p_Q = \dim(V_Q^+)$ and $q_Q = \dim(V_Q^-)$. The rank of Q is $r(Q) = p_Q + q_Q = n$ and its signature is $\tau(Q) = p_Q - q_Q$.

If $\tau(Q)$ is known, then the numbers p_Q and q_Q are determined in the terms n and $\tau(Q)$. By performing a change of basis, the matrix Q changes. Considering the orthogonal bases β, β', in which the matrices representing Q are Q_β and $Q_{\beta'}$, respectively, we have a matrix P such that $Q_{\beta'} = P^t.Q_\beta.P$. Since β and β' are orthonormal bases, then P is also an orthonormal matrix, and so $P^t = P^{-1}$. Therefore Q_β and $Q_{\beta'}$ are similar matrices ($Q_\beta \sim Q_{\beta'}$), which means that there is a matrix $P \in O(n)$ such that $Q_{\beta'} = P^{-1} \cdot Q_\beta \cdot P$. Similar matrices have the same characteristic polynomial; therefore their spectrums are equal.

Theorem 3 (Sylvester) *Let V be a real vector space and $Q : V \to \mathbb{R}$ a quadratic form. So Q is similar to a quadratic form with the matrix given as*

$$Q = \begin{pmatrix} I_{p_Q} & 0 \\ 0 & -I_{q_Q} \end{pmatrix},$$

in which I_{p_Q} and I_{q_Q} are identity matrices with rank p_Q and q_Q, respectively.

Sylvester's Theorem 3 generalizes Theorem 2.

2.6 Dual Vector Spaces

Let V be a real vector space of dimension n.

Definition 4 A linear functional defined on V is a function $f : V \to \mathbb{R}$ satisfying the following condition: for any $a, b \in \mathbb{R}$ and $u, v \in V$,

$$f(au + bv) = af(u) + bf(v).$$

The dual space of V is the space $V^* = \{f : V \to \mathbb{R} \mid f \text{ is linear}\}$ for linear functionals on V.

Proposition 1 V^* *is a vector space of dimension n.*

Proof The vector space structure on V^* is induced from the vector space structure on V in the following manner;

(i) if $f_1, f_2 \in V^*$, then $f_1 + f_2 \in V^*$ since $(f_1 + f_2)(u) = f_1(u) + f_2(u)$.
(ii) if $a \in \mathbb{R}$, $f \in V^*$, then $af \in V^*$ since $(af)(u) = a.f(u)$.

To calculate the dimension of V, we fix a basis $\beta = \{e_1, \ldots, e_n\}$ of V and consider the set $\beta^* = \{e_1^*, \ldots, e_n^*\} \subset V^*$, with $e_i^* : V \to \mathbb{R}$ the functional given as the linear extension of the identity $e_i^*(e_j) = \delta_{ij}$;

$$e_i^*(u) = e^*\left(\sum_k u_k e_k\right) = \sum_k u_k \delta_{ik} = u_i.$$

We claim β^* is a basis of V^*.
(i) β^* is linear independent.

Suppose we have $a_i \neq 0 \in \mathbb{R}$, $1 \leq i \leq n$, such that $a_1 e_1^* + \cdots + a_n e_n^* = 0$. Then, for all i,

$$0 = \left(a_1 e_1^* + \cdots + a_n e_n^*\right)(e_i) = a_i.$$

(ii) β^* span V^*.

Let $u \in V$ be the vector $u = u_1 e_1 + \cdots + u_n e_n$. The identity $u_i = e_i^*(u)$ allows us to write $u = \sum_i e_i^*(u) e_i$. Let $f \in V^*$, so we get $f(u) = \sum_i f(e_i) e_i^*(u)$, and so

$$f = \sum_{i=1}^{n} f(e_i) e_i^*. \tag{5}$$

Therefore it follows from (i) and (ii) that β^* is a basis of V^* and $\dim(V^*) = n$. $\square$

Definition 5 Given a basis $\beta = \{e_1, \ldots, e_n\}$ of V, the dual basis of β in V^* is $\beta^* = \{e_1^*, \ldots, e_n^*\}$, $e_i^*(e_j) = \delta_{ij}$. x

Now let's consider $(V, < ., . >)$ a vector space endowed with an inner product and a matrix defined by the basis $\beta = \{e_1, \ldots, e_n\}$ is $G = (g_{ij})$. The inner product induces the linear map $\mathcal{P} : V \to V^*$, $\mathcal{P}(u) =< u, . >$.

Proposition 2 $\mathcal{P} : V \to V^*$ is an isomorphism of vector spaces.

Proof We claim $\mathcal{P}$ is injective and surjective.

(i) Injectivity.

Let $u_1, u_2 \in V$ be such that $\mathcal{P}(u_1) = \mathcal{P}(u_2)$, so $< u_1 - u_2, u >= 0$ for every $u \in V$. Given $u = u_1 - u_2$, it follows that $\mid u_1 - u_2 \mid = 0$. Hence $u_1 = u_2$.

(ii) surjectivity.

Let $\beta^{\#} = \{f_1, \ldots, f_n\}$ be the set of linear functionals given by $f_i(v) =< e_i, v >$ for $i = 1, \ldots, n$. $\beta^{\#}$ is a basis of V^*, and to prove this, take $u = \sum_{j=1}^{n} e_j^*(u) e_j$,

$$f_i(u) = \sum_{i=1}^{n} f_i(e_j) e_j^*(u) = \sum_{i=1}^{n} g_{ij} e_j^*(u) \Rightarrow f_i = \sum_{j=1}^{n} g_{ji} e_j^*, \ (g_{ij} = g_{ji}).$$

In this way, the matrix $G = (g_{ij})$ takes β^* onto $\beta^{\#}$. Considering the inverse matrix $G^{-1} = (g^{ij})$, it follows that

$$e_i^* = \sum_{j=1}^{n} g^{ji} f_j.$$

Any functional $f \in V^*$ can be written as

$$f = \sum_{i=1}^{n} f(e_i) e_i^* = \sum_{i,j=1}^{n} g^{ji} f(e_i) f_j.$$

Consequently

$$f(u) = \sum_{ij=1}^{n} g^{ji} f(e_i) < e_j, u > = < \sum_{i,j=1}^{n} g^{ji} f(e_i)e_j, u > .$$

Defining the vector

$$u_f = \sum_{i,j=1}^{n} g^{ji} f(e_i)e_j, \tag{6}$$

we get $\mathcal{P}(u_f) = f$. Therefore $\mathcal{P}$ is linear and surjective, so it is an isomorphism of vector spaces. □

The last proposition allows us to define an inner product on V^* induced by the one on V as follows; let $f = \mathcal{P}(u_f)$ and $h = \mathcal{P}(u_h) \in V^*$

$$< f, g > = < u_f, u_h > = \sum_{i,j,k,l} g^{ji} g^{lk} f(e_j)h(e_k)g_{jl} = \sum_{i,j,k} g^{ji} \left(\sum_{l} g_{jl}g^{lk} \right) f(e_j)h(e_k) =$$

$$= \sum_{i,j,l} g^{ij} \delta_{jk} f(e_j)h(e_k) = \sum_{i,j} g^{ij} f(e_i)h(e_j).$$

The inner product matrix of $< ., . >: V^* \times V^* \to \mathbb{R}$ is $G^{-1} = (g^{ij})$, $g^{ij} = < e_i^*, e_j^* >$. Therefore it is possible to extend the inner product on V and V^* to a lot of vector spaces that are constructed using the direct sum $\oplus$ and tensor product $\otimes$ operations, as we discussed in Chap. 6.

3   Metric and Banach Spaces

Let X be a topological space. A metric defined on X, also called a distance function, is a function $d : X \times X \to \mathbb{R}$ which has the following properties: for all $x, y, z \in X$,

(1) $d(x, y) \geq 0$, and $d(x, y) = 0$ if, and only if, $x = y$.
(2) $d(x, y) = d(y, x)$.
(3) $d(x, y) \leq d(x, z) + d(z, y)$.

A metric space is a pair (X, d). The notion of distance allows us to define the limit of a sequence $\{x_n\}_{n\in\mathbb{N}}$ in X;

Definition 6 A sequence $\{x_n\}_{n\in\mathbb{N}}$ converges to a point $a \in X$, which is indicated by $\lim_{n\to\infty} x_n = a$, if for any $\epsilon > 0$, we have $n_0 \in \mathbb{N}$ such that $d(x_n, a) < \epsilon$ for all $n > n_0$.

In the space (X, d), the open ball of radius r and centered at $p \in X$ is the set

$$B_r(p) = \{x \in X \mid d(x, p) < r\}.$$

The closed ball is the closure $\overline{B_r(p)} = \{x \in X \mid d(x, p) \leq r\}$. The topology of the space (X, d) is generated by the open balls. In most places, the letter U is used to denote an open set, unless otherwise stated.

Consider (X, d_X) and (Y, d_Y) metric spaces and $f : X \to Y$ a function.

Definition 7 Let $x_0 \in X$ and $y_0 \in Y$. The limit of f is y_0 when x tends to x_0, which is denoted by $\lim_{x \to x_0} f(x) = y_0$, if for a given $\epsilon > 0$ there is $\delta > 0$ such that if $d_X(x, x_0) < \delta$, then $d_Y(f(x), y_0) < \epsilon$.

Definition 8 A function $f : (X, d_X) \to (Y, d_Y)$ is continuous at $x_0 \in X$ if $\lim_{x \to x_0} f(x) = f(x_0)$. That is, f is defined at x_0 and the limit is $f(x_0)$. Considering $U \subset X$, f is continuous on U if it is continuous at every point $x \in U$.

In the definition of continuity, the value of δ may depend on the values of ϵ and x_0. In some cases, the proofs require that there be no such dependence; this motivates the concept of uniform continuity;

Definition 9 A function $f : (X, d_X) \to (Y, d_Y)$ is uniformly continuous in X if, for all $\epsilon > 0$, there is $\delta > 0$ such that $d_Y(f(p), f(q)) < \epsilon$ for every pair of points $p, q \in X$ such that $d_X(p, q) < \delta$ (δ is independent on p and q).

The following theorem has several applications throughout the text;

Theorem 4 *Let (X, d_X) be a compact metric space and consider $f : (X, d_X) \to (Y, d_Y)$ a continuous function. So f is uniformly continuous.*

Proof See Ref. [38].   □

Consider $\mathbb{K} = \mathbb{R}$ or $\mathbb{C}$. A norm on a $\mathbb{K}$-vector space V is a function $|| \cdot ||: X \to \mathbb{R}$ such that for every $x, y, z \in X$, the following properties are satisfied;

(1) $|| x || \geq 0$, and $|| x || = 0$ if, and only if, $x = 0$.
(2) $|| kx || = | k | \cdot || x ||$, for all $k \in \mathbb{K}$.
(3) $|| y - x || \leq || y - z || + || z - x ||$ (triangular inequality).

A vector space endowed with a norm $|| \cdot ||$ is denoted by $(V, || \cdot ||)$, or simply V when it is clear which norm is defined. All normalized vector spaces are a metric space since $d(x, y) = || y - x ||$ defines a metric on V. However, the reverse is false.

A $\mathbb{K}$-vector space has a finite dimension equal to n if it admits a finite basis with n elements; otherwise, we say that it has infinite-dimension. There are infinite dimensional vector spaces that admit an enumerable basis, and there are also examples that do not admit enumerable bases.

The impossibility to know a priori the precise value of $a = \lim_{n \to \infty} x_n$ makes the next concept very useful;

Definition 10 A sequence $\{x_n\}_{n \in \mathbb{N}} \subset X$ is a Cauchy sequence if for any $\epsilon > 0$ there is $n_0 \in \mathbb{N}$ such that if $n, m > n_0$, then we have $d(x_n, x_m) < \epsilon$.

In $\mathbb{R}^n$, a sequence converges if and only if it is a Cauchy sequence. This is not true for all metric spaces. It is impossible to develop the concepts of calculus in a space where there are non-convergent Cauchy sequences, for example over $\mathbb{Q}$. This question is not only a fault of the field, it may be a fault of the metric space.

Definition 11 A metric space (X, d) is complete if every Cauchy sequence in X converges to a point in X. A Banach space is a complete normalized space $(V, || \cdot ||)$.

Among the normed spaces, we have a special category of spaces with a norm induced by an inner product, i.e., in the case of an $\mathbb{R}$-vector space, or induced by a sesquilinear form if it is a $\mathbb{C}$-vector space.

Definition 12 Let $H = (V, || \cdot ||)$ be a Banach space;

(i) H is a complex Hilbert space when V is a $\mathbb{C}$-vector space equipped with a sesquilinear form $< ., . >: V \times V \to \mathbb{C}$, with norm $|| x || = \sqrt{< x, x >}$.
(ii) H is a real Hilbert space when V is an $\mathbb{R}$-vector space endowed with inner product $< ., . >: V \times V \to \mathbb{C}$ and norm $|| x || = \sqrt{< x, x >}$.

The concept of distance is basic to studying issues related to approximation or convergence. As we have seen, there are different types of structures; the techniques to study convergence depend on the structure defined on the space. In fact, we have the following inclusions;

$$\text{Hilbert Spaces} \subsetneq \text{Normed Spaces} \subsetneq \text{Metric Spaces} \subsetneq \text{Topological Spaces.} \quad (7)$$

Example 1 The examples below play an important role in former chapters.

(1) The spaces $(\mathbb{Q}, d)$, $(\mathbb{R}, d)$ and $(\mathbb{C}, d)$ are normed spaces in which the norm is given by the module function $|| x || = | x |$.
(2) The vector spaces $\mathbb{R}^n$ and $\mathbb{C}^n$ have finite dimension. They are Hilbert spaces with a norm $|| x || = \sqrt{< x, x >}$ induced by $< ., . >$; and an inner product if $\mathbb{K} = \mathbb{R}$ and a sesquilinear form if $\mathbb{K} = \mathbb{C}$.
(3) Consider the vector space $C^0([a, b]) = \{f : [a, b] \to \mathbb{R} \mid f \text{ continuous}\}$ endowed with the norm $|| f ||_0 = \sup_{x \in [a,b]} | f(x) |$ and induced metric $d(f, g) = || g - f ||$. Therefore $\left(C^0([a, b]), d\right)$ is a metric space. Indeed, it is a Banach space.
(4) Let $1 \le p < \infty$. The norm $|| f ||_p = \left[\int_a^b | f(x) |^p \, dx \right]^{1/p}$ induces a metric structure on $C^0([a, b])$. For $p \ne q$, the metrics $|| \cdot ||_p$ and $|| \cdot ||_q$ are non-equivalent. So $\left(C^0([a, b]), || \cdot ||_p \right)$ is a normed space but it is not a Banach space.
(5) L^p spaces, $1 < p < \infty$: Let $X \subset \mathbb{R}^n$ be a closed subset, with non-empty interior and finite volume and let E be a Banach space. Define

$$L^p(X; E) = \left\{ f : X \to E \mid \int_X \mid f \mid^p dv < \infty \right\},\qquad (8)$$

and $dv = dx_1 \dots dx_n$ is the volume element of $\mathbb{R}^n$. Consider the norm

$$\| f \|_{L^p} = \left(\int_X \mid f \mid^p dv \right)^{1/p}.\qquad (9)$$

The space $\left(L^p(X; E), \| f \|_{L^p} \right)$ is a Banach space [28]. For $p = 2$, $L^2(X; E)$ is a Hilbert space.

(6) Consider the space of sequences $l^p(\mathbb{Z}) = \{\{a_n\}_{n\in\mathbb{N}} \mid a_n \in \mathbb{C}, \sum_{-\infty}^{\infty} \mid a_n \mid^p < \infty\}$ endowed with the norm $\| \{a_n\} \|_p = \left[\sum_{-\infty}^{\infty} \mid a_n \mid^p \right]^{1/p}$, $1 \le p < \infty$. Therefore $(l^p(\mathbb{Z}), \| . \|_p)$ is a normed space.

(7) In the last example, the norm $\| \{a_n\} \|_0 = \sup_n \mid a_n \mid$ induces a norm over the space of sequences $l^\infty(\mathbb{Z}) = \{\{a_n\}_{n\in\mathbb{N}} \mid a_n \in \mathbb{C}, \| \{a_n\} \|_\infty < \infty\}$.

(8) Consider $\rho_{k,m} = \sup_{x\in\mathbb{R}} \mid x^k \frac{d^n f}{dx} \mid$. The Schwartz space is the set of functions $\mathcal{S}(\mathbb{R}) = \{f \in C^\infty(\mathbb{R}) \mid \rho_{k,m} < \infty, \forall k, m \in \mathbb{N}\}$. The metric

$$d(f, g) = \sum_{k,m=-\infty}^{\infty} \frac{1}{2^{k+m}} \frac{\rho_{k,m}(f, g)}{1 + \rho_{k,m}(f, g)}$$

induces a metric over $\mathcal{S}(\mathbb{R})$.

(9) Let $K \subset \mathbb{R}^n$ be a compact subset and $V \subset \mathbb{R}^m$ an open subset.

(a) Let $C^0(K; \mathbb{R}^m) = \{f : K \to V \mid f \text{ continuous}\}$. The space $E = \left(C^0(K; \mathbb{R}^m), \| . \|_0 \right)$ endowed with the norm $\| f \|_0 = \sup_{x\in U} \mid f(x) \mid$ is a Banach space.

(b) Let $C^r(K; \mathbb{R}^m) = \{f : K \to V \mid d^i f \text{ continuous, for all } 0 \le i \le r\}$ $(f : U \to \mathbb{R}^m$ is a C^r-map$)$. The space $E = \left(C^r(K; \mathbb{R}^m), \| . \|_r \right)$ is a normed space with norm

$$\| f \|_{C^r} = \sum_{i=0}^{k} \| d^i f \|_0 .\qquad (10)$$

It follows from the Ascoli-Arzelà theorem that $E_r = \left(C^r(K; \mathbb{R}^m), \| . \|_r \right)$ is a Banach space. Moreover, the inclusion $F_{i,i-1} = \left((C^i(K; \mathbb{R}^m), \| . \|_{i-1} \right)$ $\hookrightarrow E_{i-1}$ is compact, that is, any sequence $\{f_n\}_{n\in\mathbb{N}} \subset F_{i,i-1}$ admits a subsequence $\{f_{n_k}\}$ converging in E_{i-1}.

The spaces $C^r(U; V), r \ge 0$ are the most important ones for the purposes of this book. The norm is defined according to the context.

All metric spaces admit a completion [28]. More precisely, if (X, d) is a metric space, then there is a metric space $(\widetilde{X}, \widetilde{d})$ and an inclusion $\iota : X \to \widetilde{X}$ such that:

(i) $\widetilde{d}(\iota(x), \iota(y)) = d(x, y)$ for every pair $x, y \in X$,
(ii) $\iota(X)$ is dense in $\widetilde{X}$,
(iii) $(\widetilde{X}, \widetilde{d})$ is complete,
(iv) $(\widetilde{X}, \widetilde{d})$ is unique up to the isomorphism of metric spaces.

Exercises

(1) Let V be a real Banach space. Verify that a norm defined by an inner product satisfies the parallelogram identity

$$|| x + y ||^2 + || x - y ||^2 = 2 \left(|| x ||^2 + || y ||^2 \right). \qquad (11)$$

(2) Show that a norm defines an inner product, if and only if it satisfies the identity of the parallelogram.
(3) In the items (1) and (2) above, study the case when V is a complex vector space.
(4) For each inclusion shown in the sequence of spaces (7), display an example to show that the inclusion cannot be an equality.
(5) Show that $C^1([a, b])$ is not a Banach space when fitted with the norm $|| . ||_0$. Extend the result to $C^r([a, b])$, $1 \leq k < \infty$.
(6) Let $K \subset \mathbb{R}^n$ be a compact set. Show that $E_r = \left(C^r(K; \mathbb{R}^m), || . ||_{C^r} \right)$ is complete and the inclusion $F_{i,i-1} = \left((C^i(K; \mathbb{R}^m), || . ||_{i-1} \right) \hookrightarrow E_{i-1}$ is compact.

Definition 13 Let U be a subset of a Banach space E. U is locally compact if every point has a compact neighborhood. That is, given $r > 0$, if U is open and bounded, then U admits a finite cover of balls of radius r.

It follows from the definition that if E is locally compact and $U \subset E$, then $\overline{U}$ is compact.

Definition 14 Let V be a vector space. A subset $U \subset V$ is convex if the segment $r : [0, 1] \to U, r(t) = x + t(y - x)$ is contained in U for every pair of points $x, y \in U$.

Definition 15 The set U is connected if we have a continuous curve $\gamma : [0, 1] \to U$ such that $\gamma(0) = x$, $\gamma(1) = y$ and $\gamma([0, 1]) \subset U$ for every pair of points $x, y \in U$.

4 Calculus Theorems

In this section, some statements about calculus are established to fix the notation.

4.1 One Real Variable Functions

Theorem 5 (Intermediate Value) *Let f be a continuous real-valued function on a closed and bounded interval $[a, b]$. Then f attains all values between $f(a)$ and $f(b)$.*

Theorem 6 (Mean Value) *Let $f : [a, b] \to \mathbb{R}$ be a real-valued function of class C^1. So there is a point $c \in (a, b)$ such that*

$$f(b) - f(a) = f'(c)(b - a). \tag{12}$$

The Fundamental Theorem of Calculus (FTC) is a milestone in Calculus;

Theorem 7 (FTC) *Let $f : [a, b] \to \mathbb{R}$ be a function of class C^0 and $F : [a, b] \to \mathbb{R}$ the function given by*

$$F(x) = \int_a^x f(t)dt.$$

So F is differentiable and $F'(x) = f(x)$, for all $x \in [a, b]$.

4.2 Functions of Several Real Variables

Let $U \subset \mathbb{R}^n$ be an open subset.

Definition 16 The tangent plane at a point $p \in U$ is the set

$$T_p U = \left\{ \vec{v} \in \mathbb{R}^n \mid \exists\, \gamma : (-\epsilon, \epsilon) \to U,\ \gamma(0) = p \text{ and } \gamma'(0) = \vec{v} \right\}.$$

Taking any vector $\vec{v} \in \mathbb{R}^n$ and $\gamma(t) = p + t\vec{v}$, it follows that $\vec{v} \in T_p U$, hence $T_p U = \mathbb{R}^n$. Therefore $T_p U$ is a vector space isomorphic to $\mathbb{R}^n$.

Consider the map $f = (f_1, \ldots, f_m) \in C^k(U; \mathbb{R}^m)$, $k \geq 1$, and the curve $\gamma : (a, b) \to U$ of class C^1. The composition defines a curve $h : (a, b) \to \mathbb{R}^m$, $h = f \circ \gamma = (h_1, \ldots, h_m)$, and $h_i : (a, b) \to \mathbb{R}$ is C^1. The Mean Value Theorem applied to each coordinate gives the numbers $c_1, \ldots, c_m \in (a, b)$ such that

$$h(b) - h(a) = \left(h_1'(c_1), \ldots, h_m'(c_m) \right).$$

In this case, we get the mean value inequality

$$| h(b) - h(a) | \leq \sup_i | h_i'(c_i) | \, | b - a | . \tag{13}$$

Changing the coordinate system in $\mathbb{R}^n$ changes the volume element and, consequently also changes the expression of the integral. The change of the variable in multiple integrals is performed according to the following.

Theorem 8 *Let $U \subset \mathbb{R}^n$ be an open subset, $\Phi : U \to \Phi(U)$ a C^1-diffeomorphism and let $f : \Phi(U) \to \mathbb{R}$ be a continuous function. So the function $(f \circ \Phi)$. $| \det(\Phi) |$ is integrable in U and*

$$\int_{\Phi(U)} f(y)dy = \int_U (f \circ \Phi)(x). \, | \det(d\Phi_x) | \, dx, \quad y = \Phi(x). \qquad (14)$$

5 Proper Maps

This section is rather important for studying polynomial maps.

Let X and Y be two topological spaces. Let $V \subset Y$ be an arbitrary subset and define

$$f^{-1}(V) = \{x \in X \mid f(x) \in V\}.$$

Definition 17 Let X and Y be two topological spaces. A map $f : X \to Y$ is called *proper* if the inverse image $f^{-1}(K) \subset X$ is compact for all compact subsets $K \subset Y$.

Examples:

(1) The restriction of a proper map is always proper.
(2) Every polynomial $P(x) \in \mathbb{R}[x]$ is proper. Take a compact set $K \subset \mathbb{R}$; then $f^{-1}(K)$ is closed since K is closed. We have to check if $f^{-1}(K)$ is bounded. Suppose it is not bounded; then we have $y_0 \in K$ such that $\lim_{x \to \infty} P(x) = y_0$. This cannot happen, since $\lim_{x \to \infty} | P(x) | = \infty$. Therefore $f^{-1}(K)$ is compact since it is closed and bounded.
(3) Let $P : \mathbb{R}^2 \to \mathbb{R}$ be the polynomial $P(x, y) = xy$. So P is not proper, since $f^{-1}(0)$ is the union of the two coordinate axes.
(4) The immersion $\gamma : \mathbb{R} \to T^2$ defined by Eq. (42)[2] is not proper if $r \in \mathbb{R} \backslash \mathbb{Q}$.
(5) For all $n \in \mathbb{N}$, consider the function $f(x) = \sin(nx)$. Since $f^{-1}(0) = \mathbb{Z}$, f is not proper.
(6) If X is compact and $f : X \to Y$ is continuous, then f is proper.

Given two locally compact Hausdorff spaces X and Y, the lack of being proper for a C^0-map $f : X \to Y$ means that we have a convergent sequence $\{y_n\}_{n \in \mathbb{N}} \subset f(X)$ so that the set $f^{-1}(\{y_n\})$ contains a subsequence $\{x_n\}_{n \in \mathbb{N}} \subset X$ such that $\lim | x_n | = \infty$.

Proposition 3 *If X and Y are locally compact Hausdorff spaces, then any proper map $f : X \to Y$ has closed range.*

Proof Let $\{x_n\}_{n \in \mathbb{N}} \subset X$ be a sequence such that its image defines a convergent subsequence $\{y_n = \lim x_n\}_{n \in \mathbb{N}} \subset f(X)$ and $y = \lim y_n$. Let $K_0 \subset Y$ be a compact set containing y. Now, consider the compact subsets $K_f = \overline{K_0 \cap f(X)}$ and $f^{-1}(K_f)$.

---

[2]See in Chap. 1.

Since $f^{-1}(K_f)$ is compact, the sequence $\{x_n\}_{n\in\mathbb{N}} \subset X$ admits a convergent subsequence $\{x_{n_k}\}$. Let $x = \lim x_{n_k}$. Given continuity, we have $y = f(x_{n_k}) \in f(X)$. Hence $f(X)$ is closed.

$\square$

The interesting cases will be when Y is connected, $f : X \to Y$ is proper and $f(X)$ is open, then f must be surjective.

Exercises Let X and Y be locally compact Hausdorff spaces.

(1) Show that if a map $f : X \to Y$ is a local homeomorphism and is proper, then f is surjective.
(2) Assume X and Y are non-compact. Take compact subsets $K_X \subset X$ and $K_Y \subset Y$ and let $\widehat{X} = X \cup K_X$ and $\widehat{Y} = Y \cup K_Y$ be the compactifications, respectively. Show that if a map $f : X \to Y$ admits a continuous extension $\widehat{f} : \widehat{X} \to \widehat{Y}$, then f is proper
(3) Show that the immersion $\gamma : \mathbb{R} \to \mathrm{T}^2$ given by the Example 42 in Chap. 1 is not proper if $r \in \mathbb{R}\backslash\mathbb{Q}$.

6 Equicontinuity and the Ascoli-Arzelà Theorem

Consider (X, d) a complete metric space and $F \subset X$ a subset.

Definition 18 Assume F is a Banach space. A subset $\mathcal{C} \subset C^0(X; F)$ is equicontinuous at $x_0 \in X$ if, for any $f \in \mathcal{C}$, given $\epsilon > 0$, there is $\delta > 0$ such that for all $x \in X$ if $d(x, x_0) < \delta$, then $\| f(x) - f(x_0) \|_F < \epsilon$. $\mathcal{C}$ is equicontinuous in X if it is equicontinuous for every $x \in X$.

The concept of equicontinuity is a subtle one; it must be observed in the definition that the value of δ is independent of $f \in \mathcal{C}$. Indeed, if the set $\mathcal{C}$ is equicontinuous, then all $f \in \mathcal{F}$ is uniformly continuous, since given an $\epsilon > 0$, there is a $\delta > 0$, regardless of the point that serves all functions belonging to $\mathcal{C}$. The sequence $f_n : [0, 1] \to \mathbb{R}$, $f_n(x) = x^n$ is not equicontinuous, since near $x = 1$ there is no such δ.

Let $K \subset X$ be a compact subset and let $\{f_n\}_{n\in\mathbb{N}} \subset C^0(K, \mathbb{C})$ be a uniformly bounded sequence; that is, we have $M > 0$ such that $| f_n(x) | < M$ for all $x \in K$ and $n \in \mathbb{N}$. If $\{f_n\}_{n\in\mathbb{N}}$ is a sequence converging uniformly to $f \in C^0(K, \mathbb{C})$, then $\{f_n\}_{n\in\mathbb{N}}$ is equicontinuous as we can see as follows: consider $n_0 \in \mathbb{N}$ such that $\| f_n - f_m \|_0 < \epsilon$ for all $n, m \geq n_0$. The equicontinuity then follows from the inequality

$$| f_n(y) - f_n(x) | \leq | f_n(y) - f_{n_0}(y) | + | f_{n_0}(y) - f_{n_0}(x) | + | f_{n_0}(x) - f_n(x) | .$$

According to Theorem 4, all continuous functions defined on a compact set are uniformly continuous. Taking $\delta > 0$ such that $d(x, y) < \delta$ for all $x, y \in K$, we get $| f_{n_0}(y) - f_{n_0}(x) | < \epsilon$.

Theorem 9 (Ascoli-Arzelà) *Let K be a compact subset of a metric space X and let F be a Banach space. Let C be a subset and $E = (C^0(K, F), \|\cdot\|_0)$. So $C \subset E$ is relatively compact if and only if the following conditions are verified;*

(i) C is equicontinuous.
(ii) For $x \in K$, the set $C(x) = \{f(x) \mid f \in C\}$ is relatively compact in X.

Proof The conditions (i) and (ii) are necessary and sufficient given the following arguments.
(i) necessity ($\Rightarrow$).
Since C is relatively compact in $C^0(K; F)$, given $\epsilon > 0$, there is a finite subset $\{f_1, \ldots, f_p\} \subset C$ such that $C \subset \cup_{i=1}^{p} B_i$, and $B_i = B_\epsilon(f_i) = \{f \in C^0(K; F) \mid \| f - f_i \|_F < \epsilon\}$. Once K is compact, we can take a finite set $\{x_1, \ldots, x_p\} \subset K$ and a finite cover $\cup_{i=1}^{p} V_i$, in which the V_i are open neighborhoods of x_i. Given $\epsilon > 0$, there is $\delta_1, \ldots, \delta_p$ in such way that we can choose $V_i = \{x \in K \mid d(x, x_i) < \delta_i\}$. Let $\delta = \min_{1 \le i \le p}\{\delta_i\}$. It follows that C is equicontinuous and $C(x)$ is relatively compact.
(ii) sufficient ($\Leftarrow$).
Given $\epsilon > 0$, consider the finite cover $\cup_{i=1}^{p} V_i$ in K defined in the last item. Once each set $C(x_i)$ is relatively compact, it follows that a finite union $\mathcal{U} = C(x_1) \cup \cdots \cup C(x_p)$ is relatively compact. Consider the open balls $B_\epsilon(a_1), \ldots, B_\epsilon(a_t)$ of ray ϵ and centered at the points $a_1, \ldots, a_t$, respectively, such that $\mathcal{U} \subset \cup_{i=1}^{t} B_\epsilon(a_i)$. We assume that $f(x_i) \in B_\epsilon(a_i)$, or equivalently, $\| f(x_i) - a_i \| < \epsilon$. Defining $C_i = \{f \in C; \| f(x_i) - a_i \| < \epsilon\}$, it is straightforward that C admits a finite cover $\cup_{i=1}^{p} C_i$. In order to prove that C_i is bounded for every i, it will be enough to verify that the diameter is upper bounded by 4ϵ. Now if $f, g \in C_i$ and $x \in K$, then $x \in B_\epsilon(a_i)$, and therefore

$$\| f(x) - g(x) \| \le \| f(x) - f(x_i) \| + \| f(x_i) - a_i \| + \| a_i - g(x_i) \| + \| g(x_i) - g(x) \|$$
$$\le 4\epsilon.$$

$\square$

The Ascoli-Arzelà theorem is useful to prove several results. In most cases we studied in this book either $F = \mathbb{R}^n$ or $F = \mathbb{C}^n$, so condition (ii) can be replaced by condition (ii') in which we assume that C is a bounded set. F being finite dimensional, every bounded subset of F is locally compact. An important application is to prove the compactness of the embedding $F_{i,i-1} \hookrightarrow E_{i-1}$ mentioned earlier.

Theorem 10 *Let $K \subset \mathbb{R}$ be a compact subset and $\{f_n\}_{n\in\mathbb{N}} \subset C^0(K, \mathbb{C})$. If $\{f_n\}_{n\in\mathbb{N}}$ converges uniformly to f, then*

$$\int_a^b f(x)dx = \lim_{n\to\infty} \int_a^b f_n(x)dx.$$

Proof See Ref. [38]. $\square$

Theorem 11 *Let K be a compact set and let $\{f_n\}_{n\in\mathbb{N}} \subset C^1(K, \mathbb{C})$ be a sequence such that $\{f_n(x_0)\}$ converges to $x_0 \in K$. If $\{f_n'\}_{n\in\mathbb{N}}$ converges uniformly in K, then $\{f_n\}$*

converges uniformly to a function $f \in C^1(K, \mathbb{C})$. *Moreover,* $f'(x) = \lim_{n\to\infty} f'_n(x)$
for all $x \in K$.

Proof Essentially, the argument follows from the fact that the sequence $\{f'_n\}$ is
continuous, so the claim follows from the Fundamental Theorem of Calculus. See
Ref. [38].   □

7   Functional Analysis Theorems

Functional analysis starts off when we observe that certain topological properties of
finite dimensional spaces and also some properties of the linear operators between
them are no longer true when we consider vector spaces of infinite dimension. For
example, convergence becomes a very relevant concept, and so many topological
properties are sensitive to the norm defined on the space. Topological properties are
very different in infinite dimension, e.g., the open ball is no longer a locally closed
subset and the closed ball, as well as the spheres, are not compact. Not every vector
subspace is closed, so a complementary subspace may not exist and the decomposition as direct sums may fail. In general, the Range and Nullity Theorem is false. In
the finite dimension all norms are equivalent, which is false in the infinite dimension.

Next, we outline some results used throughout the text.

7.1   Riesz and Hahn-Banach Theorems

For the purposes of our applications in this book, the following versions of Riesz's
Representation Theorem (RRT) are enunciated; they can be found in several analysis
textbooks. A more general version of the theorem is in [28].

Theorem 12 *Let E be a Hilbert space and $f : E \to \mathbb{R}$ a bounded linear functional.
So we have some $v_f \in H$ such that for all $g \in H$,*

$$f(g) = \; <v_f, g> \; .$$

Moreover, $\mid f \mid = \mid v_f \mid_E$.

There is no Riesz representation theorem for an arbitrary Banach space.

Theorem 13 *Let X be a Hausdorff and locally compact topological space. Con-
sider $C^0(X) = \{f : X \to \mathbb{R} \mid f \text{ continuous}\}$ and $T : C^0(X) \to \mathbb{R}$ a bounded linear functional such that $T(f) \geq 0$ whenever $f \geq 0$. So we have a Radon measure
μ contained in the σ-algebra of Borel subsets $\mathcal{B}(X)$ such that*

$$T(f) = \int_X f d\mu, \; f \in C^0(X).$$

Theorem 14 (Hahn-Banach) *Let E be a normed vector space and let $F \subset E$ be a subspace. If $\theta : F \to \mathbb{R}$ is a continuous linear functional such that $\mid \theta \mid < C$, then there is an extension $\tilde{\theta} : E \to \mathbb{R}$ of θ such that $\mid \tilde{\theta} \mid < C$.*

Proof See Ref. [28]. □

Corollary 1 *Let E be a normed vector space and $v \in E$ a non-null vector. So we have a functional $\theta : E \to \mathbb{R}$ such that $\theta(v) \neq 0$.*

Proof Let F be the subspace generated by v. Consider $\theta_v : F \to \mathbb{R}$, given by $\theta_v(v) = 1$, and extend it linearly (use the Hahn-Banach Theorem) to the functional $\theta : E \to \mathbb{R}$. □

In the finite dimension, the Hahn-Banach theorem is trivial: if $v \neq 0$, then one of its coordinates must be non-null, and then we assume $v_i \neq 0$. Then take $\theta = \pi_i$.

The following theorem gives a necessary and sufficient topological condition for the dimension of a normed space to be finite.

Theorem 15 *Let E be a normed space. E has finite dimension if, and only if, it is locally compact.*

Proof The need stems from Heine-Borel's theorem, which states that any limited and closed set of a finite-dimensional normed space is compact. We will prove sufficiency. Assume E is locally compact and consider $B = \{x \in E; \mid x \mid \leq 1\}$. In this way, there is a finite set of points $\{p_1, \ldots, p_m\} \subset B$ such that $B \subset \cup_{i=1}^{m} B_r(p_i)$. Consider F the vector subspace generated by the linear combination of $p_1, \ldots, p_m$. The claim is proved if $F = E$. By contradiction, suppose that $F \subsetneq E$. Let $x \notin F$, since F is a closed subspace of E, then there is $a \in F$ such that $\text{dist}(x, F) = \min_{y \in F} \mid x - y \mid = \mid x - a \mid$. Assume the point $\frac{x-a}{\mid x-a \mid}$ is contained in the ball $B_r(p_l)$;

$$\left| \frac{x - a}{\mid x - a \mid} - x_l \right| \leq r.$$

In this way,

$$\mid x - (a + \mid x - a \mid x_l) \mid \leq r. \mid x - a \mid.$$

The vector $a_l = a + \mid x - a \mid x_l$ belongs to F. Now taking $r = 1/2$, we get

$$\mid x - a_l \mid \leq \frac{\mid x - a \mid}{2},$$

and $\text{dist}(x, a_l) < \text{dist}(x, F)$. Hence $F = E$. □

Corollary 2 *Let E be a normed space. Let $\{x_n\}_{n \in \mathbb{N}} \subset E$ be an arbitrary sequence and let $F(n)$ be the subspace generated by $\{x_1, \ldots, x_n\}$. E has infinite dimension if, and only if, there is a sequence $\{y_n\}$ with the following properties; for all $x \in F(n-1)$,*

(i) $y_n \in F(n)$,
(ii) $\mid y_n \mid = 1$,
(iii) $\mid y_n - x \mid \geq 1/2$.

7.2 Topological Complementary Subspace

Let E and F be Banach spaces and let $V \subset E$ be a linear subspace.

Definition 19 A linear subspace $W \subset E$ is:

(i) an algebraic complement of V if $V \cap W = \varnothing$ and $E = V + W$. Indeed, $E = V \oplus W$.
(ii) a topological complement of V if W is an algebraic complement of V and is also a closed subset in E.

Proposition 4 *A linear subspace $V \subset E$ has a topological complement if and only if we have a projection $P : E \to E$ with $P(E) = V$.*

Proof (i) ($\Rightarrow$) Suppose V has a topological complement $W \subset E$. Since W is closed, every $x \in E$ can be written uniquely as $x = v_x + w_x$. Define $P : E \to E$ by $P(x) = v_x$. Therefore P is linear and $P^2 = P$, since $P^2(x) = P(P(x)) = P(v_x) = v_x$. Clearly, $P(E) = V$.
(ii) ($\Leftarrow$) Suppose there is an idempotent linear operator $P : E \to E$ with $P(E) = V$. Let $W = \text{Ker}(P)$. Then we have that $W = \text{Ker}(P) = (I - P)(E)$ is closed since $W = P^{-1}(0)$ and P is a continuous map. Furthermore,

$$E = P(E) \oplus (I - P)(E) = V \oplus W.$$

Hence W is a topological complement of V.   □

Proposition 5 *If $V \subset E$ is finite dimensional, then V has a topological complement.*

Proof Assume that $\dim(V) = n$ and let $\beta = \{e_1, \ldots, e_n\}$ be a basis for V. Every $v \in V$ can be written as a linear combination

$$v = v_1 e_1 + \cdots + v_n e_n,$$

with the coefficients $v_i \in \mathbb{C}$ (or $v_i \in \mathbb{R}$). The linear functionals $\theta_i : V \to \mathbb{C}$ given by $\theta_i(v) = v_i, 1 \leq i \leq n$ are continuous since V is finite dimensional. So each functional can be extended to a functional $\Theta_i : E \to \mathbb{C}$ such that:

(a) $\Theta_i(v) = \theta_i(v)$, for all $v \in V$;
(b) $\mid \Theta_i \mid = \mid \theta_i \mid$.

Let $W \subset E$ be defined by

$$W = \bigcap_{i=1}^{n} \text{Ker}(\Theta_i).$$

Let's check that W is a topological complement of V.

- W is closed since $\text{Ker}(\Theta_i)$ is closed.

- $V \cap W = \{0\}$. Assume $x \in V \cap W$, so $x = \sum_{i=1}^{n} a_i e_i \in V$ and $\Theta_i(x) = a_i = 0$, for all $1 \le i \le n$, since $x \in W$. Therefore $V \cap W = \{0\}$.

- $E = V + W$. From the decomposition

$$x = \underbrace{\sum_{i=1}^{n} \Theta_i(x) e_i}_{\in V} + \underbrace{\left[x - \sum_{i=1}^{n} \Theta_i(x) e_i \right]}_{\in W},$$

we have $E = V + W$. Now we can project E onto W by $P : E \to W$, $P(x) = x - \sum_{i=1}^{n} \Theta_i(x) e_i$. Therefore W is a topological complement of V. □

8   The Contraction Lemma

Let (X, d) be a complete metric space. A map $\phi : X \to X$ is a contraction if there is $\lambda \in \mathbb{R}$ such that $0 < \lambda < 1$ and

$$d\big(\phi(x'), \phi(x)\big) \le \lambda d(x', x), \quad \text{for all } x', x \in X.$$

Fix $x \in X$ and consider the sequence $\{\phi^n(x)\}_{n \in \mathbb{N}}$ given by $\phi^n(x) = \overbrace{(\phi \circ \cdots \circ \phi)}^{n}(x)$. For every pair $x', x \in X$, we have the inequalities

(i) $d\big(\phi^n(x'), \phi^n(x)\big) < \lambda^n . d(x', x),$

(ii) $d\big(\phi^{n+1}(x), \phi^n(x)\big) < d\big(\phi^n(\phi(x)), \phi^n(x)\big) < \lambda^n . d(\phi(x), x).$

Proposition 6 *If $\phi : X \to X$ is a contraction, then the sequence $\{\phi^n(x)\}_{n \in \mathbb{N}}$ converges for all $x \in X$.*

Proof It is sufficient to prove that $\{\phi^n(x)\}_{n \in \mathbb{N}}$ is a Cauchy sequence, which is achieved by proving the estimate

$$d(\phi^n(x), \phi^m(x)) \leq \frac{d(\phi(x), x)}{1 - \lambda} \lambda^m.$$

Taking $n = m + p$, we get

$$d(\phi^n(x), \phi^m(x)) \leq$$
$$\leq d(\phi^{m+p}(x), \phi^{m+p-1}(x)) + d(\phi^{m+p-1}(x), \phi^{m+p-2}(x)) + \cdots + d(\phi^{m+1}(x), \phi^m(x))$$
$$\leq d(\phi^{m+p-1}(\phi(x)), \phi^{m+p-1}(x)) + \cdots + d(\phi^m(\phi(x)), \phi^m(x)) \leq$$
$$\leq (\lambda^{m+p-1} + \lambda^{m+p-2} + \cdots + \lambda^m).d(\phi(x), x) \leq \frac{d(\phi(x), x)}{1 - \lambda} \lambda^m.$$

Define $C = \frac{d(\phi(x),x)}{1-\lambda}$. Once $0 < \lambda < 1$, given $\epsilon > 0$, there is $n_0 \in \mathbb{N}$ such that $C \mid \lambda \mid^n < \epsilon$ for all $m > n_0$. Therefore

$$d(\phi^n(x), \phi^m(x)) \leq \epsilon, \quad \text{for all } n, m > n_0.$$

Hence the sequence $\{\phi^n(x)\}$ converges for all $x \in X$.   □

Due to the pointwise convergence of the sequence $\{\phi^n(x)\}$ for all $x \in X$, define

$$\xi(x) = \lim_{n \to \infty} \phi^n(x). \tag{15}$$

Proposition 7 *The point $\xi(x)$ does not depend on x; moreover, it is the only fixed point of ϕ.*

Proof The fact that $\xi(x)$ is a fixed point of ϕ is an immediate consequence of

$$\xi(x) = \lim \phi^{n+1}(x) = \lim \phi(\phi^n(x)) = \phi(\lim \phi^n(x)) = \phi(\xi(x)).$$

Evaluating the limit $n \to \infty$ on both sides of the inequality

$$d(\phi^n(x'), \phi^n(x)) < \lambda^n.d(x', x),$$

it follows that $\xi(x') = \xi(x) = \xi$.   □

Bounding the claims proved so far, we have a very useful consequence;

Lemma 1 (Contraction Lemma) *Let (X, d) be a complete metric space and let $\phi : X \to X$ be a contraction, so there is always a unique fixed point of ϕ in X.*

Corollary 3 *If we have $n_0 \in \mathbb{N}$ such that $\phi^{n_0} : X \to X$ is a contraction, then $\phi : X \to X$ has a unique fixed point.*

Proof Let ξ be the unique fixed point of ϕ^{n_0}, so

$$\phi(\xi) = \phi^{1+n_0}(\xi) = \phi^{n_0}(\phi(\xi)).$$

Therefore $\phi(\xi)$ is a fixed point of ϕ^{n_0}, hence $\phi(\xi) = \xi$. □

Remark 1 The condition $0 < k < 1$ cannot be extended to include the case $k = 1$, as the following example shows: let $X = [1, \infty)$, $d(x, y) = |x - y|$ and $\phi(x) = x + \frac{1}{x}$.

(i) $|\phi(x) - \phi(y)| < |x - y|$, for all $x, y \in X$,
(ii) there is no $k \in (0, 1)$ such that $|\phi(x) - \phi(y)| < k. |x - y|$,
(iii) ϕ does not have a fixed point.

The contraction lemma has two relevant applications: (i) the inverse function theorem and (Ii) the existence and uniqueness theorem of solutions for an ordinary differential equation with fixed initial condition (FIC).

There are others forms of fixed point theorems also useful to prove the existence of solutions. For example, the hypothesis of the Browder Fixed Point Theorem only requires that the function be continuous.

Theorem 16 (Browder) *Let $\overline{B_r} = \{x \in \mathbb{R}^n; |x| \leq 1\}$ and consider $\phi : \overline{B_r} \to \overline{B_r}$ a continuous map. So f has at least one fixed point.*

Proof See Ref. [21]. □

Appendix B
Differentiable Manifolds, Lie Groups

In Appendix B, we will introduce elementary concepts regarding differentiable manifolds and Lie groups. We will not study this in depth; we introduce basic definitions and some elementary concepts useful to the reader. When we studied the Frobenius integrability theorem in Chap. 4, Lie groups were a source of examples.

1 Differentiable Manifolds

The concept of a differentiable submanifold in $\mathbb{R}^n$ can be extended to a more abstract concept of a differentiable manifold in which there is no need to use $\mathbb{R}^n$ in the definition; we need only a differentiable structure. A differentiable manifold structure on a set allows for the development of concepts, techniques and applications of differentiation and integration as whole.

A k-dimensional submanifold $M^k \subset \mathbb{R}^n$ is defined as a set with the property that it is locally a graph of a map $f : U \subset \mathbb{R}^k \to \mathbb{R}^{n-k}$, $f = (f_1, \ldots, f_{n-k})$, with $U \subset \mathbb{R}^k$ diffeomorphic to $\mathbb{R}^k$. This property allows us to introduce a local coordinate system (U, f) on M^k. It is a procedure analogous to the one in analytic geometry when Cartesian coordinates are introduced on the plane. Let us look at the example of the sphere S^2; the spherical coordinates define a coordinate system associating with a point $p \in S^2$ a pair $p = (\theta, \psi)$ of real numbers. θ is called the longitude and ψ is the latitude of the point p. Cartographers use spherical coordinates to describe positions on the map of the earth, for example, for location (see GPS) of geographic accidents and measuring distances. There is no global chart on the sphere since it is not diffeomorphic to the plane; we next show that the sphere can be covered with two local charts (U_1, f_1), (U_2, f_2) such that;

(i) $S^2 = U_1 \cup U_2$,
(ii) $f_2 \circ f_1^{-1} : f_1(U_1 \cap U_2) \to f_2(U_1 \cap U_2)$ is a diffeomorphism.

We use the stereographic projection to construct these two local charts on S^2 (Fig. 1).

© Springer Nature Switzerland AG 2021   341
C. M. Doria, *Differentiability in Banach Spaces, Differential Forms and Applications*,
https://doi.org/10.1007/978-3-030-77834-7

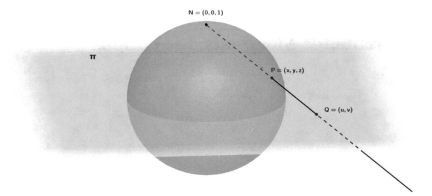

Fig. 1 Stereographic projection

Example 2 Let S^2 be the sphere with radius $R = 1, \pi = \{z = 0\} = \mathbb{R}^2 \times \{0\}$ and let $N = (0, 0, 1)$ be the focus (in this case the North Pole). The stereographic projection on the plane $\{z = 0\}$, with the focus on N, defines a map $\pi_e^N : S^2 - \{N\} \to \mathbb{R}^2$;

$$\pi_e^N(x, y, z) = Q = \left(\frac{x}{1 - z}, \frac{y}{1 - z} \right) = (u, v). \tag{1}$$

The inverse $(\pi_e^N)^{-1} : \mathbb{R}^2 \to S^2 - \{N\}$ is given by

$$(\pi_e^N)^{-1}(u, v) = \left(\frac{2u}{1 + u^2 + v^2}, \frac{2v}{1 + u^2 + v^2}, \frac{u^2 + v^2 - 1}{1 + u^2 + v^2} \right). \tag{2}$$

Therefore $\pi_e^N : S^2 - \{N\} \to \mathbb{R}^2$ is a diffeomorphism. Considering the focus at the South Pole $S = (0, 0, -1)$, we have

$$\pi_e^S(x, y, z) = \left(\frac{x}{1 + z}, \frac{y}{1 + z} \right) = (u', v'). \tag{3}$$

and

$$(\pi_e^S)^{-1}(u', v') = \left(\frac{2u'}{1 + u'^2 + v'^2}, \frac{2v'}{1 + u'^2 + v'^2}, \frac{1 - u'^2 - v'^2}{1 + u'^2 + v'^2} \right). \tag{4}$$

The transition map $\pi_e^S \circ (\pi_e^N)^{-1} : \mathbb{R}^2 - \{0\} \to \mathbb{R}^2 - \{0\}$ is the diffeomorphism

$$\pi_e^S \circ (\pi_e^N)^{-1}(u, v) = \left(\frac{u}{u^2 + v^2}, \frac{v}{u^2 + v^2} \right) \tag{5}$$

with the inverse

$$\pi_e^N \circ (\pi_e^S)^{-1}(u', v') = \left(\frac{u'}{u'^2 + v^2}, \frac{v'}{u'^2 + v'^2} \right).$$

Therefore the set $\mathcal{A}_{S^2} = \{(\pi_e^N, S^2 - \{N\}), (\pi_e^S, S^2 - \{S\})\}$ is an atlas for S^2.

Keeping the S^2 example in mind, we give the following definition;

Definition 20 A set $M^k \subset \mathbb{R}^n$ is a differentiable submanifold of $\mathbb{R}^n$ if it admits a differentiable atlas $\mathcal{A} = \{(U_\lambda, f_\lambda) \mid \lambda \in \Lambda\}$, and we have the following:

(i) $M = \cup_\lambda U_\lambda$.
(ii) We have an open subset $V_\lambda \in \mathbb{R}^n$ such that $U_\lambda = V_\lambda \cap M$ is open in M and
$f_\lambda : U_\lambda \to \mathbb{R}^k$ is a diffeomorphism for all $\lambda \in \Lambda$. The pair (U_λ, f_λ) is a local chart on M.
(iii) $f_{\lambda'} \circ f_\lambda^{-1} : f_\lambda(U_\lambda \cap U_{\lambda'}) \to f_{\lambda'}(U_\lambda \cap U_{\lambda'})$ is a diffeomorphism for all pairs $\lambda, \lambda' \in \Lambda$.
A differentiable atlas on M defines a differentiable structure.

Examples: Let's introduce some examples of differentiable manifolds.

(1) $S^1 = \{(x, y) \mid \mathbb{R}^2 \mid x^2 + y^2 = 1\}$. We use the stereographic projection to show that S^1 is a differentiable manifold. Letting $N = (0, 1)$ be the North Pole, the straight line defined by the points N and $(u, 0) \in \mathbb{R} \times \{0\}$ is

$$r(t) = (0, 1) + t(u - 0, 0 - 1) = (tu, 1 - t).$$

The intercept of $r(t)$ with S^1 occurs when $t^2u^2 + (1 - t)^2 = 1$, so we obtain the equation $t[t(u^2 + 1) - 2] = 0$ for which the only interesting solution is $t_0 = \frac{2}{u^2+1}$. Therefore the point of intersection of $r(t)$ with S^1 is

$$(x, y) = \left(\frac{2u}{u^2 + 1}, \frac{u^2 - 1}{u^2 + 1} \right).$$

It follows that $u = \frac{x}{1-y}$. So we now define the chart $(\widehat{U}_1, \phi_1), \phi_1 : \widehat{U}_1 \to \mathbb{R} \times \{0\}$ to be

$$\widehat{U}_1 = S^1 - \{N\} \quad \text{and} \quad \phi_1(x, y) = \frac{x}{1 - y}.$$

Analogously, we define the stereographic projection using the South Pole $S = (0, -1)$ to introduce the chart

$$\widehat{U}_2 = S^1 - \{S\} \quad \text{and} \quad \phi_2(x, y) = \frac{x}{1 + y}.$$

Both ϕ_1 and ϕ_2 are diffeomorphisms. Next, we check for the ϕ_2 map;

(a) ϕ_2 is surjective;

Given $u' \in \mathbb{R}$, since $y \neq -1$, we have $\phi_2(u'(1+y), y) = u'$, and $(u'(1+y), y) \in \widehat{U}_2$, i.e.,

$$(u')^2(1+y)^2 + y^2 = 1 \quad \Rightarrow \quad (u')^2 = \frac{1-y}{1+y} \quad \Rightarrow \quad y = \frac{1-(u')^2}{1+('u)^2}.$$

Consequently,

$$\phi_2^{-1}(u') = \left(\frac{2u'}{1+(u')^2}, \frac{1-(u')^2}{1+(u')^2} \right).$$

(b) ϕ_2 is injective;

Let $\frac{x}{1+y} = \frac{z}{1+w}$, $x^2 + y^2 = 1$ and $z^2 + w^2 = 1$. So

$$x^2(1+w)^2 = z^2(1+y)^2 \quad \Rightarrow \quad x^2 + 2x^2 w + x^2 w^2 = z^2 + 2z^2 y + z^2 y^2$$

$$\Rightarrow x^2 + 2x^2 w + x^2(1-z^2) = z^2 + 2z^2 y + z^2(1-x^2) \quad \Rightarrow \quad \frac{1+w}{1+y} = \frac{z^2}{x^2}.$$

Since $\frac{1+w}{1+y} = \frac{z}{x}$, we have $z = x$ and $w = y$.

(c) ϕ_2 is a homeomorphism.

The maps ϕ_2 and ϕ_2^{-1} are rational maps, so they are continuous.

So we have $S^1 = \widehat{U}_1 \cup \widehat{U}_2$ and $\widehat{U}_1 \cap \widehat{U}_2 = (S^1 - \{N\}) \cap (S^1 - \{S\})$. The coordinate change (transition map) in this case is given by

$$\phi_{21} : \mathbb{R} - \{0\} \to \mathbb{R} - \{0\}, \quad \phi_{21}(u) = \frac{1}{u}.$$

Therefore $\mathfrak{A} = \{(\widehat{U}_1, \phi_1), (\widehat{U}_2, \phi_2)\}$ defines a differentiable atlas on S^1.

(2) The 2-Torus $T^2 = S^1 \times S^1$.

Considering the product topology, we take the local charts on T^2 as the product of local charts on S^1. So T^2 is a differentiable manifold with an atlas that is

$$\mathfrak{A} = \{(\widehat{U}_1 \times \widehat{U}_1, \phi_1 \times \phi_1), (\widehat{U}_1 \times \widehat{U}_2, \phi_1 \times \phi_2), (\widehat{U}_2 \times \widehat{U}_1, \phi_2 \times \phi_1), (\widehat{U}_2 \times \widehat{U}_2, \phi_2 \times \phi_2)\}.$$

We could easily extend the argument to show the n-Torus $S^1 \times \overset{n}{\ldots} \times S^1$ is an n-manifold.

(3) The Real Projective Plane $\mathbb{R}P^n$.

Consider in $\mathbb{R}^{n+1} - \{0\}$ the equivalent relation: the two points $x, y \in \mathbb{R}^{n+1}$ are equivalent $x \sim y$ if $y = tx$ for some $t \in \mathbb{R}\setminus\{0\}$. The Projective Plane is the quotient space $\mathbb{R}P^n = \mathbb{R}^{n+1} / \sim$, and indeed, the space of lines through the origin in $\mathbb{R}^{n+1}$. We consider the quotient topology on $\mathbb{R}P^n$. For $1 \leq i \leq n+1$, consider the plane

$$\pi^0(i) = \{(x_1, \ldots, x_{i-1}, 0, x_{i+1}, \ldots, x_{n+1}) \mid x_i \in \mathbb{R}\}.$$

The set $V_i = \mathbb{R}^{n+1} \setminus \pi^0(i) = \{x = (x_1, \ldots, x_{n+1}) \in \mathbb{R}^{n+1} \mid x_i \neq 0\}$ is an open subset in $\mathbb{R}^{n+1}$; therefore $U_i = V_i / \sim$ is open in $\mathbb{R}P^{n+1}$. To construct a local chart on $\mathbb{R}P^n$, we consider the planes $\pi^1(i) = \{(x_1, \ldots, x_{i-1}, 1, x_{i+1}, \ldots, x_{n+1}) \mid x_i \in \mathbb{R}\}$. We will work on the case $i = n$; let $\pi^1(n+1) = \{(x_1, \ldots, x_n, 1)\} \subset \mathbb{R}^{n+1}$ be a hyperplane and $l \subset V_{n+1}$ a line. Every line r through the origin is given by $r(t) = t.\vec{v}$, with $\vec{v} = (v_1, \ldots, v_{n+1})$ and $\mid \vec{v} \mid = 1$. Since $l \subset V_{n+1}$, we have $v_n \neq 0$. So

$$r \cap \pi^1(n+1) = \left(\frac{v_1}{v_{n+1}}, \ldots, \frac{v_{n+1}}{v_n}, 1 \right).$$

Let $\phi_{n+1} : \mathbb{R}P^n \to \pi(n)$ be the map $\phi_{n+1}(v) = (\frac{v_1}{v_{n+1}}, \ldots, \frac{v_{n+1}}{v_n}, 1)$. The pair (U_{n+1}, ϕ_{n+1}) defines a local chart on $\mathbb{R}P^n$. An atlas on $\mathbb{R}P^n$ is obtained by taking all the local charts (U_i, ϕ_i), $1 \leq i \leq n+1$, constructed using the same method for the case $i = n$. A point $p \in \mathbb{R}P^n$ is specified by its projective coordinates

$$p = [x_1 : x_2 : \ldots, x_{n-1} : x_n].$$

Then the following items are easily verified:

(a) $\mathbb{R}P^n = \{[x_1 : x_2 : \cdots : x_{n-1} : 0]\} \cup \{[x_1 : x_2 : \cdots : x_{n-1} : 1]\}$,
(b) $\mathbb{R}P^{n-1} = \{[x_1 : x_2 : \cdots : x_{n-1} : 0]\}$ and $\mathbb{R}^n = \{[x_1 : x_2 : \cdots : x_{n-1} : 1]\}$.

Therefore

$$\mathbb{R}P^n = \mathbb{R}P^{n-1} \cup \mathbb{R}^n. \tag{6}$$

The same procedures can be applied to show that the projective plane $\mathbb{C}P^n$ is a manifold. In the complex case, the projective plane corresponds to the space of planes through the origin, and $\dim(\mathbb{C}P^n) = 2n$. Similar to the real case, we have

$$\mathbb{C}P^n = \mathbb{C}P^{n-1} \cup \mathbb{C}^n. \tag{7}$$

The existence of local charts is equivalent to the introduction of local coordinates; differentiability assures us of the existence of the tangent plane at each point. If M is a tetrahedron in $\mathbb{R}^3$, then clearly the conditions in items (ii) and (iii) are not satisfied, since there is no tangent plane at the vertices of the tetrahedron. When M is the Boy's surface, defined in (39), the condition (ii) is not satisfied due to the existence of self-intersections. Therefore the tetrahedron and the Boy's surface are not differentiable submanifolds of $\mathbb{R}^3$. To understand this apparent confusion, it is necessary to discern the intrinsic topological nature of the surface from the way it is embedded in $\mathbb{R}^n$. The submanifold concept extends to a differentiable manifold, making it unnecessary to be a subset of $\mathbb{R}^n$. The concepts and techniques of differential and integral calculus can be developed using only the fact that changes in coordinates (transition maps) are differentiable.

Definition 21 A Hausdorff topological space M is an n-dimensional manifold if it admits an atlas $\mathcal{A}_M = \{(U_\lambda, \phi_\lambda) \mid \lambda \in \Lambda\}$ such that

(i) $M = \cup_\lambda U_\lambda$;
(ii) $U_\lambda \subset M$ is an open subset;
(iii) $\phi_\lambda : U_\lambda \to \mathbb{R}^n$ is a homeomorphism.

Definition 22 A Hausdorff topological space M is an n-dimensional differentiable manifold if M is an n-dimensional manifold and if the atlas $\mathcal{A} = \{(U_\lambda, \phi_\lambda) \mid \lambda \in \Lambda\}$ satisfies the following condition: for every pair $\lambda, \lambda' \in \Lambda$,

$$\phi_{\lambda'} \circ \phi_\lambda^{-1} : \phi_\lambda(U_\lambda \cap U_{\lambda'}) \to f_{\lambda'}(U_\lambda \cap U_{\lambda'})$$

is a diffeomorphism. In this case, $\mathcal{A}_M$ is a differentiable atlas, and equivalent to a differentiable structure on M.

The difference between the concept of a manifold and a differentiable manifold is subtle, so are the consequences. For example, the tetrahedron is not a differentiable submanifold of $\mathbb{R}^3$, but it is a differentiable manifold. Due to Whitney's Embedding Theorem [21], any n-dimensional differentiable manifold can be embedded into $\mathbb{R}^{2n}$ as a differentiable submanifold. To show a non-trivial example of an n-dimensional manifold not admitting a differentiable structure goes much further than the scope of this text.

Remark Let $[M]$ be the equivalence class of n-dimensional manifolds homeomorphic to M. An n-dimensional manifold is differentiable if $M' \in [M]$ admits a differentiable structure. There are examples of manifolds admitting an infinite number of non-equivalent differentiable structures.

Exercises

(1) Let $M \subset \mathbb{R}^n$ be a submanifold. Show that the Definition 20 is equivalent to the definition given in (12) in Chap. 1.
(2) Consider n $\mathbb{R}^n \backslash \{0\}$ the following equivalent relation: $u \sim v$ if $\exists t \in \mathbb{R}$ such that $v = tu$. Show that the Projective Plane $\mathbb{R}P^n = (\mathbb{R}^n \backslash \{0\})/\sim$ is an n-dimensional differentiable manifold.
(3) An orthogonal k-frame in $\mathbb{R}^n$ is an orthonormal basis $\{u_1, \dots, u_k\}$ of $\mathbb{R}^n$. The Stiefel manifold $V_k(\mathbb{R}^n)$ is the set of orthonormal k-frames in $\mathbb{R}^n$. Show the following items;
   ((i) $V_1(\mathbb{R}^n) = S^{n-1}$,
   (ii) $V_k(\mathbb{R}^n)$ is a $\left(nk - \frac{k(k+1)}{2}\right)$-dimensional differentiable manifold.
(4) The Grassmannian manifold $G_k(\mathbb{R}^n)$ is the set of k-dimensional subspaces of $\mathbb{R}^n$. Show that the following items are true;
   (i) $G_1(\mathbb{R}^n)$ is diffeomorphic to $\mathbb{R}P^{n-1}$.
   (ii) $G_k(\mathbb{R}^n)$ is an $[k(n-k)]$-dimensional differentiable manifold.

2   Bundles: Tangent and Cotangent

In $\mathbb{R}^n$, the tangent plane at $p \in \mathbb{R}^n$ is the subspace

$$T_p\mathbb{R}^n = \{v \in \mathbb{R}^n \mid v = \gamma'(0),\, p = \gamma(0),\, \gamma : (-\epsilon, \epsilon) \to \mathbb{R}^n\},$$

and the curve $\gamma : (-\epsilon, \epsilon) \to \mathbb{R}^n$ is a C^∞ curve. For all $p \in \mathbb{R}^n$, the tangent plane $T_p\mathbb{R}^n$ is a vector space isomorphic to $\mathbb{R}^n$. If $U \subset \mathbb{R}^n$ is an open subset, then $T_pU \simeq T_p\mathbb{R}^n$ for all $p \in U$.

The concept of a tangent space is fundamental in Differential Topology, since many of the techniques are based on using the differential operator. To define the tangent plane of a manifold M at a given point $p \in M$, we have to fix a local chart containing p. Let M be a differentiable manifold, $\mathfrak{A}$ a differentiable atlas for M and $x \in M$. Fix a local chart $(\widehat{U}_\alpha, \phi_\alpha) \in \mathfrak{A}$ such that $x \in \widehat{U}_\alpha$. Setting $p = \phi_\alpha(x)$, we define

$$T_xM = T_pU_\alpha.$$

Considering that x may be contained in several charts, we must verify that the definition does not depend on the chart. Let $(\widehat{U}_\beta, \phi_\beta) \in \mathfrak{A}$ be another chart such that $x \in \widehat{U}_\beta$. Let $q = \phi_\beta(x)$. According to the definition, we have that $T_x = T_qU_\beta$. Since $\phi_{\beta\alpha} : \phi_\alpha(\widehat{U}_\alpha \cap \widehat{U}_\beta) \to \phi_\beta(\widehat{U}_\alpha \cap \widetilde{U}_\beta)$ is a diffeomorphism and $q = \phi_{\beta\alpha}(p)$, the linear map

$$d\phi_{\beta\alpha} : T_pU_\alpha \to T_qU_\beta$$

is an isomorphism of vector spaces. Therefore the definition of the tangent plane T_xM is consistent up to a diffeomorphism. In Chap. 4, the tangent plane was identified with the spaces of the linear differential operators acting in the set $C^\infty(M; \mathbb{R})$ of functions.

Once we have the tangent plane T_xM at $x \in M$, we consider the cotangent plane T_x^*M (dual vector space).

Definition 23   The Tangent Bundle of a differentiable manifold M is the space

$$TM = \bigcup_{x \in X} T_xM = \{(x, v) \mid x \in M,\, v \in T_xM\}$$

endowed with the topology such that the projection $\pi : TM \to M$, $\pi(x, v) = x$ is continuous.[3] For $x \in M$, the fiber on x is the vector space defined by the inverse image $\pi^{-1}(x) = T_xM$.

---

[3]If $U \subset M$ is open, then $\pi^{-1}(U) \subset TM$ is open.

To describe locally the tangent bundle of a manifold M, we must fix a local chart. Let $\mathfrak{A} = \{(\widehat{U}_\alpha, \phi_\alpha) \mid \alpha \in \Lambda\}$ be an atlas for M. For any $\alpha \in \Lambda$ and $x \in \widehat{U}_\alpha$, the fiber on $x \in \widehat{U}_\alpha$ is

$$T_x M = \{v \in \mathbb{R}^n \mid v \in T_{\phi_\alpha}(x) U_\alpha\} \simeq \mathbb{R}^n.$$

So for any $\alpha \in \Lambda$, there is an isomorphism $v_\alpha : \pi^{-1}(x) \to \mathbb{R}^n$. The following theorem is an elementary result in the theory of vector bundles (see in [34]);

Theorem 17 *Let M be an n-dimensional differentiable manifold. Consider the cover $M = \cup_{\lambda \in \Lambda} U_\lambda$ given by open subsets such that U_λ is contractible for all $\lambda \in \Lambda$. So TM admits a cover $TM = \cup_{\lambda \in \Lambda} \pi^{-1}(U_\lambda)$ such that $\pi^{-1}(U_\lambda$ is diffeomorphic to $U_\lambda \times \mathbb{R}^n$).*

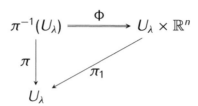

Every n-dimensional differentiable manifold admits a cover by contractible open subsets, so the space TM is locally isomorphic by $U_\lambda \times \mathbb{R}^n$.

Analogously, on the cotangent bundle T^*M, we define the projection $\pi^* : T^*M \to M$, with fibers that are $(\pi^*)^{-1}(x) = T_x^* M$. Locally, T^*M is diffeomorphic to $U_\lambda \times (R^n)^*$.

Let $(M^m, \mathfrak{A}_M)$ and $(N^n, \mathfrak{A}_N)$ be differentiable manifolds.

Definition 24 A map $f : M \to N$ is differentiable if for any two local charts $(\widehat{U}, \phi) \in \mathfrak{A}_M$ and $(\widehat{V}, \psi) \in \mathfrak{A}_N$, the map $f_{VU} = \psi \circ f \circ \phi^{-1} : U \to V$ is differentiable, with $U \subset \mathbb{R}^m$ and $V \subset \mathbb{R}^n$ being open subsets.

$$
\begin{array}{ccc}
\widehat{U} & \xrightarrow{\ f\ } & \widehat{V} \\
\downarrow{\scriptstyle \phi} & & \downarrow{\scriptstyle \psi} \\
U & \xrightarrow{\ f_{VU}\ } & V.
\end{array}
$$

The map $f_{VU} : U \to V$ is a differentiable map between Euclidean spaces and therefore can be analyzed with the usual techniques. The differential of f_{VU} defines a linear transformation $df_{VU}(p) : \mathbb{R}^m \to \mathbb{R}^n$ at the point $p \in U$. If we fix the canonical bases of $\mathbb{R}^m$ and $\mathbb{R}^n$, then we obtain an application $df_{VU} : U \to M(m, n)$.

Let $M \subset \mathbb{R}^n$ be an open subset. A differentiable section of the fibered space TM is a differentiable map $s : M \to \Gamma(U)$ such that $\pi \circ s = \pi$. So we have $s(x) \in \pi^{-1}(x)$ for all $x \in M$, i.e., $s(x) \in T_x M$.

Definition 25 Let M be a differentiable manifold and let $\Gamma(M)$ be the space of differentiable sections on TM ;

(i) a vector field on M is a section $s \in \Gamma(M)$.
(ii) a differentiable 1-form is a section $s \in \Gamma(T^*M)$.

3 Lie Groups

A group G is a topological group if G is a manifold and the group multiplication $G \times G \to G, (g, g') \to g^{-1}g'$ is continuous.

Definition 26 A topological group G is a Lie group if it admits a differentiable atlas $\mathcal{A}_G$ and the map $G \times G \to G, (g, g') \to g^{-1}g'$, defined by the group multiplication is differentiable.

Lie groups appear in several areas of mathematics. Its topological, geometric and representation theory permeate several topics. We will define some very elementary properties. What makes G a distinct differentiable manifold is the group operation. The 2-dimensional sphere S^2 does not carry a group structure, while the 1-dimensional sphere S^1 and the 3-dimensional sphere S^3 both carry a group structure, and so they are Lie groups. The following maps simplify the work on Lie groups;

Definition 27 Let G be a Lie group and $g \in G$;

(i) A left translation by g is the diffeomorphism $L_g : G \to G, L_g(x) = g.x$.
(ii) A right translation by g is the diffeomorphism $R_g : G \to G, R_g(x) = x.g$.
(iii) For any $g \in G$, the adjoint diffeomorphism $\mathrm{Ad}_g : G \to G$ is given by $\mathrm{Ad}_g(x) = g.x.g^{-1}$.

For any $g \in G$, the differentials of the diffeomorphism defined above induce the following isomorphisms;
(1) $(L_g)_* = dL_g : T_x G \to T_{gx} G$,
(2) $(R_g)_* = dR_g : T_x G \to T_{xg} G$,
(3) $(\mathrm{Ad}_g)_* = d\mathrm{Ad}_g : T_x G \to T_{gxg^{-1}} G$

Classical examples of Lie groups are the linear groups (matrix groups) $GL_n(\mathbb{R})$, $SL_n(\mathbb{R})$, O_n, $O(p, q)$, $SO(p, q)$, $Sp_{2n}(\mathbb{R})$, $GL_n(\mathbb{C})$, U_n and SU_n. A finite group is a 0-dimensional Lie group, indeed, a discrete group.

Appendix C
Tensor Algebra

The basic algebraic concepts for studying the Exterior Algebra in Chap. 6 are discussed in this appendix.

1 Tensor Product

The tensorial product of vector spaces arises naturally in several branches of mathematics, especially in Linear Algebra and its ramifications in Differential Geometry, and in the Theory of Cohomology and Representation Theory, among many other areas.

Consider V, W finite-dimensional real vector spaces[4] and let $V \times W$ be the Cartesian product of V with W. The free abelian group generated by the elements of $V \times W$ is

$$F = \left\{ \sum_{n \in \mathbb{N}} r_n(x_n, y_n) \mid r_n \in \mathbb{R}, (x_n, y_n) \in V \times W \right\}.$$

Let G be the subgroup of F generated by the following elements: let $x, x' \in V$, $y, y' \in W$ and $a \in \mathbb{R}$;

(i) $(x + x', y) - (x, y) - (x', y)$,
(ii) $(x, y + y') - (x, y) - (x, y')$,
(iii) $(ax, y) - a(x, y)$,
(iv) $(x, ay) - a(x, y)$.

The tensor product $V \otimes_\mathbb{R} W$ of V and W is the quotient F/G. The homomorphism $j : F \to F/G$ is induced by the linear extension of the projection $j(x, y) = x \otimes y$. Since F is generated by $V \times W$, the quotient F/G is generated by the elements

---

[4] They could be complex vector spaces.

© Springer Nature Switzerland AG 2021
C. M. Doria, *Differentiability in Banach Spaces, Differential Forms and Applications*,
https://doi.org/10.1007/978-3-030-77834-7

$x \otimes y = j(x, y)$. The following properties follow from the definition: for any $x, y \in V \times W$ and $k \in \mathbb{R}$;

(i) $(x + x') \otimes y = [(x + x', y] = [(x, y)] + [(x', y)] = x \otimes y + x' \otimes y$,
(ii) $x \otimes (y + y') = x \otimes y + x \otimes y'$,
(iii) $k(x \otimes y) = (kx) \otimes y = x \otimes (ky)$.

The space $V \otimes_{\mathbb{R}} W$ is generated by the elements $x \otimes y$, $x \in V$ and $y \in W$. To describe $V \otimes W$ in terms of a basis, we consider $\beta_V = \{e_i \mid 1 \leq i \leq n\}$ and $\beta_W = \{f_j \mid 1 \leq j \leq m\}$ the basis of V and W, respectively. For any $x \in V$ and $y \in W$ given as $x = \sum_{i=1}^{n} x^i e_i$ and $y = \sum_{j=1}^{m} y^j f_j$, we have

$$x \otimes y = \sum_{i=1}^{n} \sum_{j=1}^{m} x^i y^j e_i \otimes f_j.$$

The set $\{e_i \otimes f_j \mid e_i \in \beta V, f_j \in \beta_W\}$ is a basis of $V \otimes_{\mathbb{R}} W$. Given any abelian group P, a bilinear function $f : V \times W \to P$ can be extended to a linear map $f_* : V \otimes_{\mathbb{R}} W \to P$.

Proposition 8 (Universal Property of the Tensor Product) *Let F be the abelian group generated by $V \times W$, and P a free abelian group such that $f : F \to P$ is a homomorphism satisfying the following conditions:*

(1) $f(x + x', y) = f(x, y) + f(x', y), \forall x, x' \in V, y \in W$;
(2) $f(x, y + y') = f(x, y) + f(x, y')), \forall x \in v, y, y' \in W$;
(3) $f(ax, y) = f(x, ay) = af(x, y), \forall a \in \mathbb{R}, x \in V, y \in W$.

Then we have a unique homomorphism $p_f : F/G \to P$ such that the following diagram commutes:

$$
\begin{array}{ccc}
F & \xrightarrow{f} & P \\
{\scriptstyle j}\downarrow & \nearrow & \\
F/G. & &
\end{array}
$$

Proof Let $p_f : F/G \to P$ be the map $p_f(x \otimes y) = f(x, y)$. It is well-defined since

(1) $f(x + x', y) - f(x, y) - f(x', y) = 0 \Rightarrow p_f((x + x') \otimes y) = p_f(x \otimes y) + p_f(x' \otimes y)$;
(2) $f(x, y + y') - f(x, y) - f(x, y') = 0 \Rightarrow p_f(x \otimes (y + y')) = p_f(x \otimes y) + p_f(x \otimes y')$;
(3) $af(x, y) = f(ax, y) = f(x, ay) \Rightarrow f^*((ax) \otimes y) = f^*(x \otimes (ay)) = af^*(x \otimes y)$,

For all $g \in G$, we have $f(g) = 0$, hence $g \in \mathrm{Nucl}(f)$. $\qquad\qquad\square$

The construction of the tensor product extends to the tensor product of k vector spaces $V_1 \times \ldots V_k$. In this case, we get the multilinear property: given any $x_i \in V_i$, $i = 1, \ldots, k$, and $a, b \in \mathbb{R}$,

$$x_1 \otimes \cdots \otimes (a.x_i + b.x_i') \otimes \cdots \otimes x_k = a. (x_1 \otimes \cdots \otimes x_i \otimes \cdots \otimes x_k) + b. (x_1 \otimes \cdots \otimes x_i' \otimes \cdots \otimes x_k).$$

Proposition 9 *Let V, W, S be real vector spaces. So we have the isomorphisms*

$$(1)\ \mathbb{R} \otimes V \to V, \qquad (2)\ V \otimes W \to W \otimes V,$$
$$k \otimes x \to kx. \qquad\qquad x \otimes y \to y \otimes x.$$

$$(3)\ (V \oplus W) \otimes S \to (V \oplus S) \otimes (W \oplus S), \qquad (4)\ (V \otimes W) \otimes S \to V \otimes (W \otimes S),$$
$$(x + y) \otimes z \to x \otimes z + y \otimes z. \qquad\qquad (x \otimes y) \otimes z \to x \otimes (y \otimes z).$$

Proof The proof is straightforward from the definition.

$\square$

Exercises

1. Let $\{e_i \mid 1 \leq i \leq n\}$ and $\{f_j \mid 1 \leq j \leq m\}$ be the basis of V and W, respectively. Show that $\{e_i \otimes f_j \mid 1 \leq i \leq n,\ 1 \leq j \leq m\}$ is a basis of $V \otimes_{\mathbb{R}} W$.
2. Let $\mathcal{A}$ and $\mathcal{B}$ be two algebras over a field $\mathbb{K}$. Define the tensor product $\mathcal{A} \otimes_{\mathbb{K}} \mathcal{B}$ and show that $\mathcal{A} \otimes_{\mathbb{K}} \mathcal{B}$ is also a $\mathbb{K}$-algebra.
3. Consider the $\mathbb{K}$-algebra $M_{nm}(\mathbb{K})$ of $n \times m$ matrices over a field $\mathbb{K}$, and $M_n(\mathbb{K}) = M_{nn}(\mathbb{K})$. Show that there are the following isomorphisms;

   (a) $M_{nm}(\mathbb{R}) \simeq \mathbb{R}^n \otimes \mathbb{R}^m$.
   (b) $M_n(\mathbb{R}) \otimes_{\mathbb{R}} \mathbb{K} \simeq M_n(\mathbb{K})$, $\mathbb{K} = \mathbb{C}, \mathbb{H}$.
   (c) $\mathbb{C} \otimes_{\mathbb{R}} \mathbb{C} \simeq \mathbb{C} \oplus \mathbb{C}$.
   (d) $M_2(\mathbb{C}) \simeq \mathbb{C} \otimes_{\mathbb{R}} \mathbb{H}$.
   (e) $M_4(\mathbb{R}) \simeq \mathbb{H} \otimes_{\mathbb{R}} \mathbb{H}$.

2   Tensor Algebra

The tensor product of $\mathbb{K}$-vector spaces, $\mathbb{K} = \mathbb{R}, \mathbb{C}$, introduces naturally the definition of Tensorial Algebra associated with a $\mathbb{K}$-vector space V. Let

$$V^{\otimes n} = \overbrace{V \otimes \cdots \otimes V}^{n}, \quad n = 1, 2, \ldots$$

and $V^0 = \mathbb{K}$. Armed with isomorphisms

$$\pi : \mathbb{K} \otimes V^{\otimes n} \to V^{\otimes n} \quad , \quad \pi' : V^{\otimes n} \otimes K \to V^{\otimes n}$$
$$k \otimes x \to kx \qquad\qquad\qquad x \otimes k \to xk$$

and the tensor product $\phi : V^{\otimes m} \otimes V^{\otimes n} \to V^{\otimes m+n}$,

$$(x_1 \otimes \cdots \otimes x_m) \otimes (x_{m+1} \otimes \cdots \otimes x_{(m+n)}) \to x_1 \otimes \cdots \otimes x_m \otimes x_{m+1} \otimes \cdots \otimes x_{(m+n)},$$

the vector space

$$T(V) = \bigoplus_{n=0}^{\infty} V^{\otimes n} = K \oplus V \oplus V^{\otimes 2} \oplus \ldots V^{\otimes n} \oplus \ldots$$

defines the algebra $(T(V), +, \otimes)$ over $\mathbb{K}$. Each element $v \in T(V)$ can be decomposed into

$$v = \sum_{n \in \mathbb{N}} v^{(n)}, \quad v^{(n)} \in V^{\otimes n}.$$

Definition 28 $(T(V), +, \otimes)$ is the Tensor Algebra generated by the vector space V.

The Tensor Algebra $T(V)$ is a graduated algebra since $V^{\otimes m} \otimes V^{\otimes n} \subset V^{\otimes m+n}$.

Proposition 10 (Universal Property of the Tensor Algebras) *Let $\mathcal{A}$ be an algebra over $\mathbb{K}$ and $f : V \to \mathcal{A}$ a homomorphism. So we have a unique homomorphism $f_* : T(V) \to \mathcal{A}$ such that the diagram below commutes:*

$$\begin{array}{ccc} V & \xrightarrow{f} & A \\ \downarrow & \nearrow & \\ T(V). & & \end{array}$$

Proof Consider the homomorphism $f_{(n)} : V^{(n)} \to \mathcal{A}, n \in \mathbb{N}$, given by

$$f^{(0)}(k) = k, \quad f^{(n)}(x_1 \otimes x_2 \otimes \cdots \otimes x_n) = f(x_1).f(x_2) \ldots f(x_n).$$

For $n \in \mathbb{N}$, we define the homomorphism $f_* : T(V) \to \mathcal{A}$ by $f_* |_{V^{(n)}} = f^{(n)}$. It follows that f_* is a homomorphism of algebras. Since V generates $T(V)$, f_* is the unique homomorphism of $T(V)$ over $\mathcal{A}$ that matches f on V. $\qquad\square$

Let V be endowed with an inner product and let V^* be the dual space with the induced inner product. We can extend the inner product to the vector space $(\otimes_{i=1}^{n} V) \otimes (\otimes_{j=1}^{m} V^*)$ as follows; let $u_i, v_i \in V$, $1 \le i \le n$ and $f_j, h_j \in V^*$, $1 \le j \le m$,

$$< u_1 \otimes \cdots \otimes u_n \otimes f_1 \otimes \cdots \otimes f_m, v_1 \otimes \cdots \otimes v_n \otimes h_1 \otimes \cdots \otimes h_m >=$$
$$=< u_1, v_1 > \cdot \ldots < u_n, v_n > \cdot < f_1, h_1 > \cdot \ldots < f_m, h_m > . \tag{1}$$

Fix a basis $\beta = \{e_1, \ldots, e_n\}$ of V, and consider $G = (g_{ij})$ the matrix representing the inner product with respect to the basis β. Then we have

$$< e_{i_1} \otimes \cdots \otimes e_{i_n} \otimes e^*_{j_1} \otimes \cdots \otimes e^*_{j_m}, e_{l_1} \otimes \cdots \otimes e_{l_n} \otimes e^*_{k_1} \otimes \cdots \otimes e^*_{k_m} >=$$
$$= g_{i_1 l_1} \ \ldots \ g_{i_n l_n} \cdot g^{j_1 k_1} \ \ldots \ g^{j_m k_m} . \tag{2}$$

If β is an orthonormal basis, then

$$\{e_{i_1} \otimes \cdots \otimes e_{i_n} \otimes e^*_{j_1} \otimes \cdots \otimes e^*_{j_m} \mid 1 \leq i_1, \ldots, k_m \leq n\}$$

is also an orthonormal basis of the tensor algebra $T(V \otimes V^*)$.

References

1. Sheldon Axler, Paul Bourdon and Wade Ramsey, *Harmonic Function Theory*, GTM 137, Springer-Verlag, 2001.
2. H. Bass, E. Connell and D. Wright, *The Jacobian Conjecture: Reduction of Degree and Formal Expansion of the Inverse*, Bull. AMS **7**, 287–330 (1982).
3. A. Bialynicki-Birula and M. Rosenlicht, *Injective Morphism of Real Algebraic Varieties*, Proc. AMS 13 (1962), 200–203.
4. Raoul Bott and Loring W. Tu, *Differential Forms in Algebraic Topology*, GTM 82, Springer-Verlag, 1982.
5. F. Braun and B. Oréfice-Okamoto, *On polynomial submersions of degree 4 and the real Jacobian conjecture in $\mathbb{R}^2$*, J. Math. Anal. Appl. **433** (2016), 75–87.
6. R. Creighton Buck, *Advanced Calculus*, International Series in Pure and Appl. Math., McGraw-Hill, 1978.
7. Manfredo P. do Carmo, *Geometria Riemanniana*, IMPA, Proj. Euclides, Rio de Janeiro, 2005.
8. Lokenath Debnath and Piotr Mikusiński, *Introduction to Hilbert Spaces with Applications*, Elsevier Academic Press, 2005.
9. S. K. Donaldson and Peter Kronheimer, *The Geometry of Four-Manifolds*, Oxford Math. Monographies, 1990.
10. S. K. Donaldson, *Riemann Surfaces*, Oxford Graduate Texts in Mathematics **22**, Oxford, 2011.
11. L. M. Drużkowski, *An Effective approach to Keller's Jacobian Conjecture*, Math. Ann. **264**, 303–313 (1983).
12. J. Eells and Luc Lemaire, *Selected Topics in Harmonic Maps*, CBMS 50, AMS, 1983.
13. J. Eells, *Elliptic Operators on Manifolds*, College of Differential Geometry, ICTP, 1989, Italy.
14. John Erdos, *Operators on Hilbert Spaces*, Lectures Notes, http://www.mth.kcl.ac.uk/~jerdos/OpTh/FP.htm.
15. Per Enflo, *A counterexample to the approximation problem in Banach spaces*, Acta Mathematica Vol. 130 (1), 1973.
16. A. Van Den Essen, *Polynomial Automorphisms*, Progress in Math., Birkhäuser, 2000.
17. D. A. Fleisch, *A Student's Guide to Maxwell's Equations*, Cambridge University Press, 2008.
18. Gerald B. Folland, *Introduction to Partial Differential Equations*, Princeton University, 1995.
19. D. Gilbarg and N. S. Trudinger, *Elliptic Partial Differential Equations of Second Order*, Springer-Verlag, NY, 1983.
20. Seymour Goldberg, *Unbounded Linear Operators*, McGraw-Hill, 1966.
21. Victor Guillemin and Alan Pollack, *Differential Topology*, Prentice Hall, Inc, Englewood Cliffs, New Jersey, 1974.

© Springer Nature Switzerland AG 2021

C. M. Doria, *Differentiability in Banach Spaces, Differential Forms and Applications*,
https://doi.org/10.1007/978-3-030-77834-7

22. Allen Hatcher, *Algebraic Topology*, Cambridge Univ. Press, 2001.
23. Morris Hirsch and Stephen Smale, *Differential Equations, Dynamical Systems, and Linear Algebra*, Series of Monog. and Texts in Pure and Applied Mathematics, Academic Press, New York, 1970.
24. C. S. Hönig, *Análise Funcional e o Problema de Sturm-Liouville*, Ed Edgard Blücher LTDA, USP-SP, 1978.
25. E. M. G. M. Hubbers, *The Jacobian Conjecture: Cubic Homogeneous Maps in Dimension Four*, Master's Thesis, University of Nijmegen, Toernooiveld, 6525 ED Nijmegen, The Netherlands, 1994.
26. O. H. Keller, *Ganze Cremona-Transformationem Monats*, Math. Physik **47**, (1939), 299–306.
27. Erwin Kreyszig, *Introductory Functional Analysis with Applications*, Wiley Classics Library, 1978.
28. Serge Lang, *Analysis I*, Addison-Wesley Publ. Co., 1969.
29. Seymour Lipschutz, *General Topology*, Schaum's Outline Series, McGraw-Hill, 1971.
30. Elon L. Lima, *Curso de Análise*, Projeto Euclides, Vol. 2, IMPA, Rio de Janeiro, 1978.
31. Artur Lopes and Claus I. Doering, *Equações Diferenciais Ordinárias*, CMU, IMPA, Rio de Janeiro, 2012.
32. Jerrold E. Marsden, Ralph Abraham and Tudor Ratiu, *Manifolds, Tensor Analysis, and Applications*, Springer, 1988.
33. W. S. Massey, *Algebraic Topology: An Introduction*, Springer-Verlag, GTM 56, New York, 1977.
34. John W. Milnor and James D. Stasheff, *Characteristic Classes*, Annals of Mathematics Studies 76, Princeton Univ. Press, 1974.
35. Susumu Oda and K. Yoshida, *A short proof of the Jacobian conjecture in the case of degree ≤ 2*, C.R.Math. Rep., Acad. Sci. Canada, Vol. V (1983), 159-162.
36. Sergey Pinchuk, *A counterxample to the strong real Jacobian Conjecture*, Math. Zeitschrift 217, 1-4 (1994).
37. A. V. Pukhlikov, *A "Real" Proof of the Fundamental Theorem of Algebra* (Russian), Matematicheskoe Prosvesschenie **1** (1997), 85–89.
38. Walter Rudin, *Principles of Mathematical Analysis*, Int. Student Ed., McGraw-Hill, 1964.
39. Konrad Schmüdgen, *Unbounded Self-Adjoint Operators*, GTM 265, Springer, 2012.
40. A. Scorpan, *The Wild World of 4-Manifolds*, AMS, 2005.
41. P. A. Smith, *Transformation of Finite Period*, Annals of Math. (2) **39** (1938), 127–164.
42. Michael Spivak, *Calculus on Manifolds*, Westview Press, 1965.
43. A. J. Tromba, *Some Theorems on Fredholm Maps*, Proceed. AMS, Vol. 34, 2, 1972.
44. S. S. Wang, *A Jacobian criterion for separability*, Journ. of Algebra **65**, 453–494 (1980).
45. Frank Warner, *Foundations of Differentiable Manifolds and Lie Groups*, GTM 94, Springer-Verlag, 1983.
46. Katrin Wehrheim, *Holomorphic Curves Course Notes*, MRSI, 2009.
47. Credits - in the elaboration of the text, the following free computer programs were used: (i) Latex - free Software by LaTeX Project Public License (LPPL) (ii) Geogebra 5 - GeoGebra - Dynamic Mathematics for Everyone http://www.geogebra.org/, free Software. (iii) PPLANE 2005.10 Copyright 2005, John C. Polking, Rice University Author: John Polking, Professor, Dept of Mathematics, Rice University. Supporting Programmer: Joel Castellanos, Rice University. http://math.rice.edu/~dfield/dfpp.html (iv) Jos Leys, Mathematical Imaginary, URL http://www.josleys.com/.

Index

A
Analytic function, 16

B
Bilinear form, 322
    Hermitian, 322
    non-degenerate, 322
    sesquilinear, 322
Boy's Surface, 48
Bundle section, 349

C
Clifford algebra, 237
Cocycle condition, 247, 250
Cohomology
    De Rham, 250, 280
Commutator, 197
Continuous
    function, 326
    uniformly , 326
Contraction map, 337
Convergence
    strong, 87
    weak, 87
Coordinate
    cylindrical, 46
    polar, 42
    spherical, 43
Coordinate system, 200
Curve
    coordinate, 200
    integral, 180

D
Derivative
    directional, 2
    Fréchet, 135
    Gâteaux, 135
    partial, 3
    total, 11
Diffeomorphism, 42
Differentiable
    function, 4, 5
    map, 41
Differential, 10
Differential complex, 250
Differential form, 243, 349
    closed, 250
    exact, 250
Dirac delta, 314
Dirichlet problem, 290
Distribution
    integrable, 201
    involutive, 201
    maximal, 201
    Maxwell-Boltzmann, 36
Dual
    basis, 324
    space, 323
Dynamical system, 196

E
Equation
    Cauchy-Riemann, 43
    Euler-Lagrange, 154
Equicontinuous, 332
Exterior
    algebra, 236
    product, 234

© Springer Nature Switzerland AG 2021
C. M. Doria, *Differentiability in Banach Spaces, Differential Forms and Applications,*
https://doi.org/10.1007/978-3-030-77834-7

Printed in the United States
by Baker & Taylor Publisher Services